Oxford Lecture Series in
Mathematics and its Applications 6

*Series editors*

John Ball   Dominic Welsh

# OXFORD LECTURE SERIES
# IN MATHEMATICS AND ITS APPLICATIONS

# Combinatorial Designs and Tournaments

Ian Anderson

*Department of Mathematics*
*University of Glasgow*

CLARENDON PRESS · OXFORD

1997

Oxford University Press, Great Clarendon Street, Oxford OX2 6DP
Oxford New York
Athens Auckland Bangkok Bogata Bombay Buenos Aires
Calcutta Cape Town Dar es Salaam Delhi Florence Hong Kong
Istanbul Karachi Kuala Lumpur Madras Madrid Melbourne
Mexico City Nairobi Paris Singapore Taipei Tokyo Toronto
and associated companies in
Berlin Ibadan

Oxford is a trade mark of Oxford University Press

Published in the United States
by Oxford University Press, Inc., New York

This book incorporates and expands on Combinatorial designs:
Construction methods published by Ellis Horwood in 1990.

A catalogue record for this book is available from the British Library

Library of Congress Cataloging in Publication Data
Data available

ISBN 0 19 850029 7

Typeset by Technical Typesetting Ireland, Belfast, N. Ireland
Printed in Great Britain by
Bookcraft (Bath) Ltd., Midsomer Norton, Avon

*To*
*Christine Helen Grace*

# Preface

This is a considerably revised version of my book *Combinatorial designs: construction methods* which was published by Ellis Horwood in 1990, but which has for some time been out of print. While I have retained much of the basic material on blocks designs and Latin squares, I have put more emphasis on the construction of a variety of tournament designs, showing how block designs and Latin squares can often be used in the construction of these tournaments. This has involved the writing of a new chapter on league schedules, and the addition of extra material to the chapters on bridge and whist tournaments. To compensate, it has been necessary to omit some of the material on designs and Latin squares; for example, the proofs of the multiplier theorem for difference sets and the Bruck–Chowla–Ryser theorem, which can readily be found elsewhere, and Wilson's work on transversal designs, have been removed.

The resulting book has two aims: first, to present some of the basic material on block designs and orthogonal Latin squares, emphasizing in particular methods of constructing cyclic examples by means of difference systems; and, secondly, to give an account of the construction of league schedules, tournaments with various balance requirements, bridge tournaments (via balanced Room squares), and whist tournaments. Much of this material is not readily accessible in book form, and is of particular interest in the way that it uses combinatorial designs and orthogonal Latin squares in the constructions.

This book is not aimed at design theory researchers. It is, rather, aimed at students, both undergraduate and postgraduate, and professional mathematicians who wish to learn about the subject and then pass on that knowledge to their students. References are given throughout which can be followed up as necessary, and it is my hope that the frequent references to the nineteenth-century mathematicians Kirkman, Anstice, and Moore will help the reader to view the subject as one with a long and fascinating history. If the reader wishes to study designs in greater depth, then the classic *Combinatorial theory* by Marshall Hall, second edition 1986, will be invaluable, as will be the recent encyclopaedic volume, *The C.R.C. handbook of combinatorial designs*, edited by Charles Colbourn and Jeff Dinitz.

I should like to thank Oxford University Press for agreeing to add this book to their list; indeed, it was a great mistake not to have asked them to publish the original edition. I am grateful to Yn Sheng Liaw and Jim Lewis for drawing my attention to various errors in the original edition. Finally, I wish to express my sincere thanks to Norman Finizio for the help, advice, and encouragement that he has given me in the preparation of this new edition, as well as for all the pleasure I have experienced while working with him in the fascinating area of whist tournaments.

*Glasgow*                                                                                    I.A.
August 1996

# Contents

# List of abbreviations

| | |
|---|---|
| BCR | Bruck–Chowla–Ryser theorem |
| BIBD | Balanced incomplete block design |
| BRS | Balanced Room square |
| BTD | Balanced tournament design |
| DWh | Directedwhist tournament design |
| FBTD | Factored balanced tournament design |
| FPP | Finite projective plane |
| GDD | Group divisible design |
| GF($q$) | Galois (finite) field of order $q$ |
| H($s, 2n$) | Howell design |
| $I_n$ | Unit $n \times n$ matrix, or $\{1, \ldots, n\}$ |
| KTS | Kirkman triple system |
| MESRS | Maximum empty subarray Room square |
| MOLS | Mutually orthogonal Latin squares |
| N($n$) | Largest $m$ for which $m$ MOLS of order $n$ exist |
| OA | Orthogonal array |
| PBD | Pairwise balanced design |
| PBTD | Partitioned balanced tournament design |
| SAMDRR | Spouse-avoiding mixed doubles round robin tournament |
| SOLS | Self-orthogonal Latin square(s) |
| S($t, k, v$) | Steiner system |
| STS | Steiner triple system |
| SQS | Steiner quadruple system |
| TD | Transversal design |
| $t - (v, k, \lambda)$ | $t$-design |
| TWh | Triplewhist tournament design |
| ($v, k, \lambda$) BIBD | BIBD with parameters $v$, $k$, $\lambda$ |
| Wh($n$) | Whist tournament design for $n$ players |
| $Z_n$ | The integers modulo $n$ |
| $\binom{n}{r}$ | Binomial coefficient |

# 1

# Introduction to basics

## 1.1 Block designs

The modern study of block designs is often said to have begun with the publication in 1936 of a paper by the statistician F. Yates. In that paper he considered collections of subsets of a set with certain balance properties; one of his examples was the following.

**Example 1.1.1** $\{a, b, c\}$, $\{a, b, d\}$, $\{a, c, e\}$, $\{a, d, f\}$, $\{a, e, f\}$, $\{b, c, f\}$, $\{b, d, e\}$, $\{b, e, f\}$, $\{c, d, e\}$, $\{c, d, f\}$. Here there are 10 subsets (blocks) of the 6-element set $\{a, b, c, d, e, f\}$ such that:

(i)  each block contains the same number of elements, namely three, and
(ii) each pair of elements occurs in a block the same number of times, namely twice.

The second property, that of *balance*, is desirable in many experiments where for fairness all comparisons between pairs of elements should occur equally often. Designs with this property of balance became known as balanced incomplete block designs; some were listed in the statistical tables of Fisher and Yates in 1938. We now make our formal definition, using *k*-subset as an abbreviation for *k*-element subset.

**Definition 1.1.1** A *balanced incomplete block design* (BIBD) is a collection of *k*-subsets (called *blocks*) of a *v*-set $S, k < v$, such that each pair of elements of $S$ occur together in exactly $\lambda$ of the blocks.

The use of $v$ for the number of elements reflects the statistical applications where the elements are varieties. 'Incomplete' refers to the condition $k < v$, ensuring that elements are missing from each block. The number of blocks is denoted by $b$, and the design is often described as a $(v, k, \lambda)$ *design* or a $(v, k, \lambda)$ BIBD or a $(v, k, \lambda)$ *configuration*. For instance, the example of Yates above is a $(6, 3, 2)$ design in which $b = 10$. Note that the design has the further property that every element occurs in the same number of blocks; this is a common property of all BIBDs.

**Theorem 1.1.1** *In a $(v, k, \lambda)$ design with $b$ blocks each element occurs in $r$ blocks where*

(i) $$\lambda(v - 1) = r(k - 1), \tag{1.1}$$

(ii) $$bk = vr. \tag{1.2}$$

**Proof** Consider any element $x$, and suppose that it occurs in $r$ blocks. In each of these blocks $x$ makes a pair with $k - 1$ other elements; so altogether there are $r(k - 1)$ pairs involving $x$. But $x$ is paired with each of the $v - 1$ other elements exactly $\lambda$ times, so $r(k - 1) = \lambda(v - 1)$. This shows that $r$ is independent of the choice of $x$, being uniquely determined by $v$, $k$, and $\lambda$. Next note that each block has $k$ elements, so there are $bk$ appearances of elements in blocks. But each of the $v$ elements appears in $r$ blocks, so $bk = vr$.

Sometimes a $(v, k, \lambda)$ design is described as $(v, b, r, k, \lambda)$ design, but it is important to realize that, once $v$, $k$ and $\lambda$ are known, the other two parameters are uniquely determined. The example of Yates is a $(6, 10, 5, 3, 2)$ design.

**Example 1.1.2** No $(11, 6, 2)$ design can exist. For if one did, eqn (1.1) would give $r = 4$, and then eqn (1.2) would give the impossibility $6b = 44$.

Many infinite families of block designs were constructed by R. C. Bose in his remarkable 1939 paper to which much reference will be made. But, long before this, BIBDs had been studied by other mathematicians. Away back in 1844, a prize problem in the *Lady's and Gentleman's Diary* had asked for a system of $k$-subsets of a $v$-set such that no $t$-subset occurs more than once in any of the $k$-subsets. The special case of $t = 2$, with each 2-subset, or pair, of elements occurring in exactly one $k$-subset, is just a $(v, k, 1)$ design. In 1847 the Rev. T. Kirkman showed that a $(v, 3, 1)$ design could exist only when $v \equiv 1$ or $3 \pmod 6$, and constructed such designs in all possible cases. These designs, now known as *Steiner triple systems*, will be studied in detail in Chapter 6.

**Example 1.1.3** (The seven-point plane)  The blocks $\{1, 2, 4\}, \{2, 3, 5\}, \{3, 4, 6\},$

$\{4, 5, 7\}$, $\{5, 6, 1\}$, $\{6, 7, 2\}$, $\{7, 1, 3\}$ form a $(7, 3, 1)$ design which can be thought of as consisting of the seven 'lines' in Fig. 1.1.

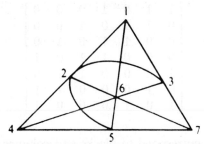

**Fig. 1.1** The seven lines of the seven-point plane.

These $(v, 3, 1)$ designs should really be called Kirkman triple systems, but they have come to be known as *Steiner triple systems* due to the fact that Kirkman's paper was overlooked for many years and Steiner posed the problem of the existence of such designs in a geometrical context six years later in 1853. Since Steiner did not solve the problem himself—a solution was published by Reiss in 1859—it is all the more unfair that Kirkman's name has not become attached to the designs.

Kirkman has in fact good claim to the title of 'father of design theory'. References to his work will abound. In 1850 he proved, among other things, that BIBDs with $\lambda = 1$, $k = p + 1$, $v = p^2 + p + 1$ exist for all prime numbers $p$. Such designs are called finite projective planes.

**Definition 1.1.2** A *finite projective plane of order n* is an $(n^2 + n + 1, n + 1, 1)$ design, $n \geqslant 2$.

The design in Example 1.1.3 is a finite projective plane of order 2. Kirkman's existence result was later generalized; using the theory of finite fields, Veblen and Bussey established, in 1906, the existence of a finite projective plane of order $q$ whenever $q$ is a prime power. These designs turn out to be closely related to mutually orthogonal Latin squares, the study of which was initiated by Euler in the late eighteenth century: this story will be told later.

The study of block designs is greatly helped by the use of related matrices.

**Definition 1.1.3** The *incidence matrix* of a $(v, b, r, k, \lambda)$ design is a $b \times v$ matrix $A = (a_{ij})$ defined by

$$a_{ij} = \begin{cases} 1 & \text{if the } i\text{th block contains the } j\text{th element;} \\ 0 & \text{otherwise.} \end{cases}$$

*Introduction to basics*

For example, the incidence matrix of the seven-point plane of Fig. 1.1 is

$$
\begin{bmatrix}
1 & 1 & 0 & 1 & 0 & 0 & 0 \\
0 & 1 & 1 & 0 & 1 & 0 & 0 \\
0 & 0 & 1 & 1 & 0 & 1 & 0 \\
0 & 0 & 0 & 1 & 1 & 0 & 1 \\
1 & 0 & 0 & 0 & 1 & 1 & 0 \\
0 & 1 & 0 & 0 & 0 & 1 & 1 \\
1 & 0 & 1 & 0 & 0 & 0 & 1
\end{bmatrix}.
$$

Observe that rows correspond to blocks and columns to elements. Note too that $A$ depends on the orderings of the blocks and the elements; however, the important properties of $A$ will be independent of this.

**Theorem 1.1.2** *If $A$ is the incidence matrix of a $(v, k, \lambda)$ design then*

$$
A'A = (r - \lambda)I + \lambda J, \tag{1.3}
$$

*where $A'$ denotes the transpose of $A$, $I$ is the $v \times v$ identity matrix, and $J$ is the $v \times v$ matrix every entry of which is 1.*

**Proof** Since $A'$ is a $v \times b$ matrix, $A'A$ is a $v \times v$ matrix. The $(i, j)$ entry of $B = A'A$ is the scalar product of the $i$th row of $A'$ and the $j$th column of $A$, i.e. of the $i$th and $j$th columns of $A$. If $i = j$, $b_{ij}$ is therefore just the number of 1s in the $i$th column, i.e. the number of blocks containing the $i$th element; thus $b_{ii} = r$. If $i \neq j$, the scalar product is a sum of 1s corresponding to rows in which both columns $i$ and $j$ have a 1; thus $b_{ij}$ is the number of blocks containing the $i$th and $j$th elements, i.e. $b_{ij} = \lambda$.

One important consequence of eqn (1.3) is that a simple proof of a fundamental result of the statistician R.A. Fisher can be obtained from it. This result asserts that a BIBD must have at least as many blocks as elements. Fisher originally proved this by purely combinatorial arguments (see Exercises 1, number 9), but the following matrix proof is elegant.

**Theorem 1.1.3** (Fisher's inequality, 1940) *In any $(v, k, \lambda)$ design,*

$$
b \geq v. \tag{1.4}
$$

**Proof** Let $A$ be the incidence matrix. We first show that $A'A$ is non-singular by showing that its determinant is non-zero. Now

$$|A'A| = \begin{vmatrix} r & \lambda & \lambda & \cdots & \lambda \\ \lambda & r & \lambda & \cdots & \lambda \\ \lambda & \lambda & r & \cdots & \lambda \\ \vdots & & & & \\ \lambda & \lambda & \lambda & \cdots & r \end{vmatrix}$$

$$= \begin{vmatrix} r & \lambda & \lambda & \cdots & \lambda \\ \lambda - r & r - \lambda & 0 & & 0 \\ \lambda - r & 0 & r - \lambda & & 0 \\ \vdots & & & & \\ \lambda - r & 0 & 0 & \cdots & r - \lambda \end{vmatrix},$$

on subtracting the first row from each of the other rows. Now add to the first column the sum of all the other columns to obtain

$$|A'A| = \begin{vmatrix} r + (v-1)\lambda & \lambda & \lambda & \cdots & \lambda \\ 0 & r - \lambda & 0 & \cdots & 0 \\ 0 & 0 & r - \lambda & \cdots & 0 \\ \vdots & & & & \\ 0 & 0 & 0 & \cdots & r - \lambda \end{vmatrix}$$

$$= \{r + (v-1)\lambda\}(r-\lambda)^{v-1} = rk(r-\lambda)^{v-1}$$

on using eqn (1.1). But $k < v$, so, by (1.1), $r > \lambda$; so $|A'A| \neq 0$. But $A'A$ is a $v \times v$ matrix, so the rank $\rho$ of $A'A$ is $\rho(A'A) = v$. Finally, since $\rho(A'A) \leqslant \rho(A)$, and since $\rho(A) \leqslant b$ (for $A$ has just $b$ rows), $v \leqslant \rho(A) \leqslant b$.

**Example 1.1.4** No $(16,6,1)$ design can exist. For $v = 16$, $k = 6$, $\lambda = 1$ leads to $r = 3$ and hence to $b = 8 < v$.

Conditions (1.1), (1.2), and (1.4) are *necessary* for the existence of a $(v,b,r,k,\lambda)$ design, but they are not *sufficient*. For example no $(43,7,1)$ design exists.

If $b = v$, the extreme case of Fisher's inequality, then $r = k$ and the design is said to be *symmetric*. This is not because its incidence matrix is symmetric—usually it is not—but because of a symmetry between blocks and elements in some of its properties, as described in the next theorem. This theorem was first proved by Bose in his 1939 paper, again by a purely combinatorial argument (see Exercises 1, number 7), but we present here an elegant proof involving incidence matrices. It is an adaptation of the

*Introduction to basics*

matrix approach of Ryser, due to Murphy in 1975. Note that for a symmetric design eqn (1.3) becomes

$$A'A = (k - \lambda)I + \lambda J. \tag{1.5}$$

It turns out that we then also have

$$AA' = (k - \lambda)I + \lambda J. \tag{1.6}$$

so that, if $i \neq j$, the scalar product of the $i$th and $j$th rows of $A$ is $\lambda$, i.e. the $i$th and $j$th blocks have $\lambda$ elements in common. It is because of the symmetry between blocks and elements in the statements

> every block contains $k$ elements,
>
> every element occurs in $k$ blocks

and

> every pair of elements occur in $\lambda$ blocks,
>
> every pair of blocks intersect in $\lambda$ elements

that the name 'symmetric' design is given.

**Theorem 1.1.4** *If $A$ is the incidence matrix of a symmetric design then $AA' = A'A$. Thus every pair of blocks intersect in $\lambda$ elements.*

**Proof** Note first that $AJ = JA = kJ$, so that $A'J = (JA)' = (kJ)' = kJ$, and similarly $JA' = kJ$. Also, $J^2 = vJ$. Use will be made of the fact that a matrix commutes with its inverse. By eqn (1.5),

$$\left(A' - \sqrt{\left(\frac{\lambda}{v}\right)}J\right)\left(A + \sqrt{\left(\frac{\lambda}{v}\right)}J\right) = A'A + \sqrt{\left(\frac{\lambda}{v}\right)}(A'J - JA) - \frac{\lambda}{v}J^2$$

$$= A'A - \lambda J = (k - \lambda)I.$$

Thus $1/(k - \lambda)(A + \sqrt{(\lambda/v)}J)$ is the inverse of $A' - \sqrt{(\lambda/v)}J$ and hence commutes with it. Thus

$$(k - \lambda)I = \left(A + \sqrt{\left(\frac{\lambda}{v}\right)}J\right)\left(A' - \sqrt{\left(\frac{\lambda}{v}\right)}J\right)$$

$$= AA' + \sqrt{\left(\frac{\lambda}{v}\right)}(JA' - AJ) - \frac{\lambda}{v}J^2 = AA' - \lambda J,$$

whence $AA' = (k - \lambda)I + \lambda J = A'A$.

This argument in fact yields a more general result. By a $(0, 1)$ matrix will be meant a matrix all of whose entries are 0 or 1.

**Theorem 1.1.5** *If $A$ is a $v \times v$ $(0,1)$ matrix satisfying eqn (1.5), then $A$ satisfies eqn (1.6).*

**Proof** By eqn (1.5), each column of $A$ contains $k$ 1s; so $JA = kJ$. Thus $A'J = (JA)' = (kJ)' = kJ$. Proceeding as in the above proof we therefore again reach

$$AA' - (k - \lambda)I - \lambda J = \sqrt{\left(\frac{\lambda}{v}\right)} (AJ - JA') \qquad (1.7)$$

but we cannot immediately assert that $AJ - JA' = 0$ as before, for we do not know that $AJ = kJ$. Denote the matrix on the left-hand side of (1.7) by $H$, and the matrix on the right by $K$. Then $H' = H$ so that $H$ is symmetric, whereas $K' = -K$, so that $K$ is skew-symmetric. Since $H = K$ it follows that $H = H' = K' = -K = -H$, so that $H = 0$. Thus $AA' - (k - \lambda)I - \lambda J = 0$, i.e. (1.6) holds.

We now turn to a few specific examples of symmetric designs.

**Example 1.1.5** Construction of a $(16, 6, 2)$ design. For each $i \leqslant 16$, let $A_i$ consist of all the numbers other than $i$ which occupy a square of the array shown in the same row or column as $i$. Then $|A_i| = 6$ for each $i$; for example, $A_3 = \{1, 2, 4, 7, 11, 15\}$. Further, any two numbers occur together in precisely two sets $A_i$; for example, 2 and 14 occur together in $A_6$ and $A_{10}$, whereas 6 and 16 occur in $A_8$ and $A_{14}$.

| 1 | 2 | 3 | 4 |
|---|---|---|---|
| 5 | 6 | 7 | 8 |
| 9 | 10 | 11 | 12 |
| 13 | 14 | 15 | 16 |

**Example 1.1.6** Construction of a $(13, 4, 1)$ design, i.e. a finite projective plane of order 3. The following blocks form such a design.

$$\{1, 2, 4, 10\}, \{2, 3, 5, 11\}, \{3, 4, 6, 12\}, \{4, 5, 7, 13\}, \{5, 6, 8, 1\},$$
$$\{6, 7, 9, 2\}, \{7, 8, 10, 3\}, \{8, 9, 11, 4\}, \{9, 10, 12, 5\},$$
$$\{10, 11, 13, 6\}, \{11, 12, 1, 7\}, \{12, 13, 2, 8\}, \{13, 1, 3, 9\}.$$

The blocks can be thought of as 'lines' of a geometry as shown in Fig. 1.2.

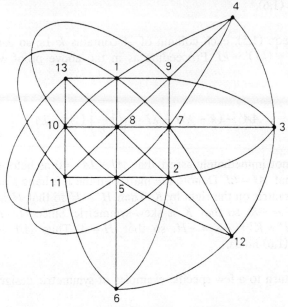

*Fig. 1.2* The 13 lines of a finite projective plane of order 3.

Note that each block is obtained from the previous one by adding 1 (mod 13) to each element. Consequently, as in the case of Example 1.1.3, the incidence matrix is a *circulant* matrix, i.e. one in which each row is obtained by shifting the previous row one position to the right. Such an elegant representation of the design clearly depends on special properties possessed by the first block which generates all the rest; it is for example instructive to consider why $\{1, 3, 5\}$ could not be taken as the first block in Example 1.1.3 in place of $\{1, 2, 4\}$. Suitable choices of a first block will be discussed further in the next chapter.

We end this introductory section by considering how new designs can be obtained from old. One method is to replace each block by its complement.

**Definition 1.1.4** Let $D$ be a $(v, b, r, k, \lambda)$ design on a set $S$ of $v$ elements. Then the *complementary design* $\overline{D}$ has as its blocks the complements $S - B$ of the blocks $B$ of $D$.

**Example 1.1.7** If $D$ is the seven-point plane of Example 1.1.3 then $\overline{D}$ consists of the blocks $\{3, 5, 6, 7\}, \{4, 6, 7, 1\}, \ldots, \{2, 4, 5, 6\}$.

It turns out that $\overline{D}$ is itself a block design.

**Theorem 1.1.6** *Suppose that $D$ is a $(v, b, r, k, \lambda)$ design. Then $\overline{D}$ is a $(v, b, b - r, v - k, b - 2r + \lambda)$ design provided that $b - 2r + \lambda > 0$.*

**Proof** Clearly there are $b$ blocks of size $v - k$ in $\overline{D}$. An element of $S$ occurs in a block of $\overline{D}$ precisely when it does not occur in the corresponding block of $D$; so it occurs in $b - r$ blocks of $\overline{D}$. Finally, the number of blocks of $\overline{D}$ containing $x$ and $y$ is equal to the number of blocks of $D$ containing neither $x$ nor $y$, and this number is

$b - $ (number of blocks of $D$ containing at least one of $x, y$)

$\quad = b - $ (number of blocks $D$ containing $x +$ number containing $y$

$\qquad -$ number containing both)

$\quad = b - (r + r - \lambda) = b - 2r + \lambda.$

**Corollary 1.1.7** *If $D$ is a symmetric $(v, k, \lambda)$ design with $v - 2k + \lambda > 0$ then $\overline{D}$ is a symmetric $(v, v - k, v - 2k + \lambda)$ design.*

For example, $\overline{D}$ in Example 1.1.7 is a symmetric $(7, 4, 2)$ design. It is often convenient to denote $k - \lambda$ in a symmetric design by $n$:

$$n = k - \lambda. \tag{1.8}$$

This fits in with the notation $(n^2 + n + 1, n + 1, 1)$ for a finite projective plane of order $n$. Note that, if $\bar{n}$ refers to the complementary design $\overline{D}$, then

$$\bar{n} = (v - k) - (v - 2k + \lambda) = k - \lambda = n;$$

so $n$ is unaltered under complementation. Thus if we are studying symmetric designs with a given value of $n$ we can, by taking complements if necessary, assume that $k \leqslant \frac{1}{2}v$. Then $v - 2k + \lambda$ is definitely positive.

Another method of obtaining new designs from old is the following. Suppose a symmetric $(v, k, \lambda)$ design $D$ is given, and all the elements of one block $B$ are thrown away. The number of blocks is then reduced by 1, and these remaining blocks are all of size $k - \lambda$ since, by Theorem 1.1.4, any block of $D$ intersects $B$ in $\lambda$ elements. So a $(v - k, v - 1, r, k - \lambda, \lambda)$ design is obtained.

**Definition 1.1.5** The $(v - k, v - 1, r, k - \lambda, \lambda)$ design obtained from a symmetric $(v, k, \lambda)$ design by deleting all the elements of one block is called a *residual design*.

**Example 1.1.8** Start with a finite projective plane of order $n$, i.e. an $(n^2 + n + 1, n + 1, 1)$ design. Then any residual design is an $(n^2, n^2 + n, n + 1, n, 1)$ design.

**Definition 1.1.6** An $(n^2, n, 1)$ design is called an *affine plane* of order $n$.

**Example 1.1.9** Take the plane of Example 1.1.6 and remove the elements of the block $\{3, 4, 6, 12\}$. A $(9, 12, 4, 3, 1)$ design is obtained with blocks $\{1, 2, 10\}$, $\{2, 5, 11\}$, $\{5, 7, 13\}$, $\{5, 8, 1\}$, $\{7, 9, 2\}$, $\{7, 8, 10\}$, $\{8, 9, 11\}$, $\{9, 10, 5\}$, $\{10, 11, 13\}$, $\{11, 1, 7\}$, $\{13, 2, 8\}$, $\{1, 13, 9\}$. This is an affine plane of order 3; it is also a Steiner triple system. Relabelling the elements 10, 11, 13 by 3, 4, 6, respectively, we can rewrite this design as a collection of 3-subsets of $\{1, \ldots, 9\}$: $\{1, 2, 3\}$, $\{2, 4, 5\}$, $\{5, 6, 7\}$, $\{1, 5, 8\}$, $\{2, 7, 9\}$, $\{3, 7, 8\}$, $\{4, 8, 9\}$, $\{3, 5, 9\}$, $\{3, 4, 6\}$, $\{1, 4, 7\}$, $\{2, 6, 8\}$, $\{1, 6, 9\}$.

In this opening section we have been introduced to the parameters $v$, $b$, $r$, $k$, $\lambda$ associated with a block design, and we have seen that they have to satisfy certain relationships. We have also constructed designs corresponding to certain sets of parameters; for instance in Example 1.1.9 we constructed a $(9, 12, 4, 3, 1)$ design. A natural question to ask is the following: given parameters $v$, $b$, $r$, $k$, $\lambda$, can there exist *different* designs with these parameters? Here we must clarify what we mean by different. If we simply relabel the elements or list the blocks in a different order, we still have essentially the same design. So we shall say that two $(v, b, r, k, \lambda)$ designs $D_1$, $D_2$ are *isomorphic* (or the same) if there exists a one–one mapping from the set of elements of $D_1$ to the set of elements of $D_2$ such that the blocks of $D_1$ are mapped onto the blocks of $D_2$. If no such mapping exists, then $D_1$ and $D_2$ are said to be *non-isomorphic* (or different) designs.

**Example 1.1.10** The following $(7, 3, 1)$ design is isomorphic to the seven-point plane of Example 1.1.3:

$$\{1, 2, 3\}, \{1, 4, 5\}, \{1, 6, 7\}, \{2, 4, 6\}, \{2, 5, 7\}, \{3, 4, 7\}, \{3, 5, 6\}.$$

For the mapping $1 \to 2$, $2 \to 1$, $3 \to 4$, $4 \to 7$, $5 \to 6$, $6 \to 3$, $7 \to 5$ transforms the blocks into

$$\{2, 1, 4\}, \{2, 7, 6\}, \{2, 3, 5\}, \{1, 7, 3\}, \{1, 6, 5\}, \{4, 7, 5\}, \{4, 6, 3\}.$$

There is, in fact, essentially only one $(7, 3, 1)$ design (see Exercises 1, number 6).

**Example 1.1.11** A $(7, 3, 2)$ design $D_1$ could be constructed by taking each block of a $(7, 3, 1)$ design twice. However, the following 14 triples also form a $(7, 3, 2)$ design:

$$\{1, 2, 4\}, \{2, 3, 5\}, \{3, 4, 6\}, \{4, 5, 7\}, \{5, 6, 1\}, \{6, 7, 2\}, \{7, 1, 3\},$$
$$\{1, 3, 4\}, \{2, 4, 5\}, \{3, 5, 6\}, \{4, 6, 7\}, \{5, 7, 1\}, \{6, 1, 2\}, \{7, 2, 3\}.$$

Since this second design $D_2$ has no repeated blocks, it cannot be isomorphic to $D_1$. So there exist two different $(7, 3, 2)$ designs.

It is important to note that the parameters do not uniquely determine a design; there may be two or more non-isomorphic designs with the same parameters.

## 1.2 Resolvability

The blocks of the affine plane of order 3 in Example 1.1.9 can be grouped into four groups, each of three blocks, so that the blocks in each group together contain each element exactly once:

$$\{1,2,3\} \qquad \{1,5,8\} \qquad \{1,4,7\} \qquad \{1,6,9\}$$
$$\{4,8,9\} \qquad \{3,4,6\} \qquad \{2,6,8\} \qquad \{2,4,5\}$$
$$\{5,6,7\} \qquad \{2,7,9\} \qquad \{3,5,9\} \qquad \{3,7,8\}.$$

This is an example of a resolvable design.

**Definition 1.2.1** A BIBD is *resolvable* if the blocks can be arranged into $r$ groups so that the $(b/r) = (v/k)$ blocks of each group are disjoint and contain in their union each element exactly once. The groups are called the *resolution classes* or the *parallel classes*.

Interpreting geometrically, we can think of the blocks of the above example as lines, and think of each group of disjoint blocks covering the whole set as a collection of parallel lines. The lines fall into four parallel classes. Unlike the projective planes, where every pair of lines intersect, this example reflects the existence of parallel lines in Euclidean geometry. We shall see shortly that all affine planes have parallel classes of lines; indeed, because of this property, affine planes are sometimes known as Euclidean planes.

One of the first examples of resolvable designs in the combinatorial literature arose in connection with Kirkman's schoolgirls problem. Posed in 1850, it was stated as follows.

Fifteen young ladies in a school walk out three abreast for seven days in succession; it is required to arrange them daily so that no two shall walk abreast twice.

Kirkman had observed that a $(15, 3, 1)$ design, i.e. a Steiner triple system on 15 elements, contains 35 triples; so he asked if these 35 triples could be grouped into seven resolution classes of five triples each—in other words, does a resolvable $(15, 3, 1)$ design exist? The example at the start of this section would provide a solution for the corresponding problem for nine schoolgirls (in a four-day week). Kirkman was himself able to produce a solution (we shall present it in section 7.1), and much more recently it has been shown that the corresponding problem for $6n + 3$ schoolgirls always has a solution (this will be the subject matter of Chapter 7). Here meanwhile is an elegant solution for the 15 schoolgirls problem: label the girls A, $B_1, \ldots, B_7$, $C_1, \ldots, C_7$ and note how each day's arrangement is obtained from the previous day's by permuting the $B_i$ and $C_i$ cyclically, leaving A unchanged.

**Example 1.2.1** A resolvable $(15, 3, 1)$ design, solving Kirkman's schoolgirls problem.

| Day 1: | $AB_1C_1$ | $B_2B_4C_3$ | $B_3B_7C_5$ | $B_5B_6C_2$ | $C_4C_6C_7$ |
|--------|-----------|-------------|-------------|-------------|-------------|
| Day 2: | $AB_2C_2$ | $B_3B_5C_4$ | $B_4B_1C_6$ | $B_6B_7C_3$ | $C_5C_7C_1$ |
| Day 3: | $AB_3C_3$ | $B_4B_6C_5$ | $B_5B_2C_7$ | $B_7B_1C_4$ | $C_6C_1C_2$ |
| Day 4: | $AB_4C_4$ | $B_5B_7C_6$ | $B_6B_3C_1$ | $B_1B_2C_5$ | $C_7C_2C_3$ |
| Day 5: | $AB_5C_5$ | $B_6B_1C_7$ | $B_7B_4C_2$ | $B_2B_3C_6$ | $C_1C_3C_4$ |
| Day 6: | $AB_6C_6$ | $B_7B_2C_1$ | $B_1B_5C_3$ | $B_3B_4C_7$ | $C_2C_4C_5$ |
| Day 7: | $AB_7C_7$ | $B_1B_3C_2$ | $B_2B_6C_4$ | $B_4B_5C_1$ | $C_3C_5C_6$ |

Resolvable designs are of use in experimental design. Suppose for example that 15 types of washing machine are to be compared by use of the $(15, 3, 1)$ design above. Each block gives three machines to be compared directly by one person, so each pair of machines will be compared directly precisely once during the experiment. If there is only one model of each type of machine available, the whole experiment can be carried out in seven sessions by using the blocks of day $i$ simultaneously in the $i$th session.

A very familiar example of a resolvable design arises in the fixture list for a football league. Suppose that $2n$ teams enter a league and that each team is to play each other team exactly once. Further, the fixtures are to be arranged for $2n - 1$ days, so that each team is involved in one game on each day. Now, a fixture list giving the games for each day is just a resolvable $(2n, 2, 1)$ design in which the $i$th resolution class consists of the pairs of teams playing each other on the $i$th day.

**Example 1.2.2** League schedule for four teams A, B, C, D.

| Day 1: | A v B | C v D |
|--------|-------|-------|
| Day 2: | A v C | B v D |
| Day 3: | A v D | B v C. |

This corresponds to the resolvable $(4, 2, 1)$ design on $\{A, B, C, D\}$ with blocks $\{A, B\}, \{C, D\}, \{A, C\}, \{B, D\}, \{A, D\}, \{B, C\}$.

**Theorem 1.2.1** *For each positive integer n there exists a resolvable* $(2n, 2, 1)$ *design, i.e. there exists a fixture list for a league of* $2n$ *teams with games scheduled for* $2n - 1$ *days.*

**Proof** Label the teams $\infty, 1, 2, \ldots, 2n - 1$, and on day $i$ play $i$ v $\infty$, $(i - 1)$ v $(i + 1)$, $(i - 2)$ v $(i + 2), \ldots, (i - (n - 1))$ v $(i + (n - 1))$, each integer being reduced (mod $2n - 1$) to lie in the interval $[1, 2n - 1]$. Then team $h$ will play against team $k$ on day $i$ where $h \equiv i - j$, $k \equiv i + j \pmod{2n - 1}$ for some $j$. But $h \equiv i - j$ and $k \equiv i + j$ if and only if $2i \equiv h + k$ and $2j \equiv k - h$ (mod $2n - 1$); these congruences have unique solutions for $i$ and $j$ since $2n - 1$ is odd.

Diagrammatically, we can represent these fixtures by chords of a circle. Place $2n - 1$ points equally spaced round a circle, and label them $1, \ldots, 2n - 1$; also label the centre $\infty$. The games of day $i$ are then represented by a line joining $i$ and $\infty$, and by parallel chords joining the other points in pairs. The diagrams of Fig. 1.3 illustrate the case of eight teams. Ignore for the moment the arrows: they are relevant in Chapter 8 when we study venue sequences.

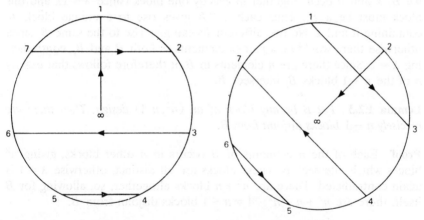

**Fig. 1.3** Diagrams representing games played on days 1 and 2 in a league of 8 teams.

**Example 1.2.3** Fixture list for a league of eight teams.

| | | | |
|---|---|---|---|
| Day 1: | $\infty$ v 1 | 2 v 7 | 3 v 6 | 4 v 5 |
| Day 2: | $\infty$ v 2 | 3 v 1 | 4 v 7 | 5 v 6 |
| Day 3: | $\infty$ v 3 | 4 v 2 | 5 v 1 | 6 v 7 |
| Day 4: | $\infty$ v 4 | 5 v 3 | 6 v 2 | 7 v 1 |
| Day 5: | $\infty$ v 5 | 6 v 4 | 7 v 3 | 1 v 2 |
| Day 6: | $\infty$ v 6 | 7 v 5 | 1 v 4 | 2 v 3 |
| Day 7: | $\infty$ v 7 | 1 v 6 | 2 v 5 | 3 v 4. |

Note the *cyclic* nature of the fixture list: add 1 (mod 7) to each entry to get the next day's fixtures, leaving $\infty$ unchanged. The cyclic structure should recall to mind Examples 1.1.3, 1.1.6 and 1.2.1; such cyclic constructions will be used a great deal in this book in the study of designs, tournaments, and orthogonal Latin squares, and Chapter 2 will provide the necessary background.

We end this introduction to resolvability by returning to affine planes and showing that these designs are in fact always resolvable. To begin with, a lemma is proved which should be reminiscent of the Euclidean property that, given a point $P$ not on a line $l$, there is a unique line through $P$ parallel (disjoint) to $l$.

**Lemma 1.2.2** *Let $B$ be any block in an $(n^2, n, 1)$ design, and let $x$ be any element not in $B$. Then there is a unique block containing $x$ and disjoint from $B$.*

**Proof** The $(n^2, n, 1)$ design has $r = n + 1$, so there are $n + 1$ blocks, say $B_1, \ldots, B_{n+1}$, containing $x$. How many of these intersect $B$? For each $b \in B$, $x$ and $b$ occur together in exactly one block (since $\lambda = 1$), and this block must be a $B_i$. Thus each $b \in B$ gives rise to a unique block $B_i$ containing it and $x$. No two different $b$s can give rise to the same $B_i$ since otherwise there would be a pair of elements in both $B$ and $B_i$, contradicting $\lambda = 1$. Since there are $n$ elements in $B$, it therefore follows that exactly $n$ of the $n + 1$ blocks $B_i$ intersect $B$.

**Lemma 1.2.3** *Let $B$ be any block of an $(n^2, n, 1)$ design. Then there are precisely $n - 1$ blocks disjoint from $B$.*

**Proof** Each of the $n$ elements of $B$ occurs in $n$ other blocks, giving $n^2$ blocks which intersect $B$. These blocks are all distinct, otherwise $\lambda = 1$ is again contradicted. There are $n^2 + n$ blocks altogether, so, allowing for $B$ itself, there are $n^2 + n - n^2 - 1 = n - 1$ blocks disjoint from $B$.

**Theorem 1.2.4** *Every affine plane is resolvable.*

**Proof** Take any block $B_1$ and any $x \notin B_1$. By Lemma 1.2.2 there is a unique block $B_2$ containing $x$ and disjoint from $B_1$. Now take any $y \notin B_1 \cup B_2$. There is a unique block $B_3$ containing $y$ and disjoint from $B_1$. But $B_3$ is also disjoint from $B_2$, for if $z \in B_2 \cap B_3$ then $B_2$ and $B_3$ would be two blocks containing $z$ and disjoint from $B_1$, contradicting Lemma 1.2.2. Proceed in this way until $n$ disjoint blocks are obtained which partition the whole set of $n^2$ elements. This gives one resolution class.

Next take any block $C_1$ not yet used, and in the same way construct another resolution class. None of the blocks in this class can have already occurred in the first class, because of Lemma 1.2.3. So now two disjoint classes have been obtained. Proceed in this way until no blocks are left unused. This completes the proof.

Finally, recall that affine planes were obtained from finite projective planes as residual designs. This is in fact the only way in which affine planes arise, for the resolvability of affine planes enables us to reverse the process, in effect adding a 'line at infinity'.

**Theorem 1.2.5** *An affine plane of order n exists if and only if a finite projective plane of order n exists.*

**Proof** Suppose an $(n^2, n^2 + n, n + 1, n, 1)$ design exists. It is resolvable, and it has $n + 1$ classes. Adjoin a new element $\infty_i$ to each block of the $i$th resolution class, $i = 1, \ldots, n + 1$, and introduce a new block $\{\infty_1, \ldots, \infty_{n+1}\}$. The resulting design consists of blocks all of size $n + 1$. Further, it is balanced with $\lambda = 1$: for any given $i$, $\infty_i$ will occur once in a block with each element $x$ of the affine plane since $x$ occurs in precisely one of the blocks of the $i$th class; and each pair $\infty_i$, $\infty_j$ occurs once in the new block.

**Example 1.2.4** Take the resolvable $(4, 2, 1)$ design of Example 1.2.2 to obtain the $(7, 3, 1)$ design with blocks

$$\{A, B, \infty_1\}, \{C, D, \infty_1\}, \{A, C, \infty_2\}, \{B, D, \infty_2\}$$
$$\{A, D, \infty_3\}, \{B, C, \infty_3\}, \{\infty_1, \infty_2, \infty_3\}.$$

## 1.3 Latin squares

Closely related to block designs are Latin squares.

**Definition 1.3.1** A *Latin square* on $n$ symbols is an $n \times n$ array such that each of the $n$ symbols occurs once in each row and in each column. The number $n$ is called the *order* of the square.

**Example 1.3.1**

$$\begin{bmatrix} A & B & C & D \\ D & A & B & C \\ C & D & A & B \\ B & C & D & A \end{bmatrix}$$

is a Latin square of order 4.

Latin squares can be used to present league schedules. Suppose that a league schedule for $2n$ teams has been arranged in $2n - 1$ rounds. Then define a $2n \times 2n$ array $A = (a_{ij})$ by

$$a_{ii} = 2n, \qquad a_{ij} = k, \quad (i \neq j)$$

where the $i$th and $j$th teams play each other in round $k$. Then, since each team plays precisely one game in each round, each row (and column) of $A$ contains each of $1,\ldots,2n$ precisely once, so that $A$ is a Latin square. For example, the fixtures of the league schedule of Example 1.2.3, with $\infty$ replaced by 8, can be presented by

$$
\begin{bmatrix}
8 & 5 & 2 & 6 & 3 & 7 & 4 & 1 \\
5 & 8 & 6 & 3 & 7 & 4 & 1 & 2 \\
2 & 6 & 8 & 7 & 4 & 1 & 5 & 3 \\
6 & 3 & 7 & 8 & 1 & 5 & 2 & 4 \\
3 & 7 & 4 & 1 & 8 & 2 & 6 & 5 \\
7 & 4 & 1 & 5 & 2 & 8 & 3 & 6 \\
4 & 1 & 5 & 2 & 6 & 3 & 8 & 7 \\
1 & 2 & 3 & 4 & 5 & 6 & 7 & 8
\end{bmatrix}.
$$

Note that this Latin square is *symmetric* about the main diagonal in the sense that $a_{ij} = a_{ji}$ *for each pair* $i, j$. Conversely, any symmetric Latin square of order $2n$ with constant main diagonal can be interpreted as giving a resolvable league schedule for $2n$ teams.

   Latin squares are used in the design of statistical experiments. One of the earliest examples of this is due to de Palluel who, in 1788, used the square of Example 1.3.1 in an experiment involving 16 sheep, four each of four different breeds, with one of each breed on each of four different diets. If rows correspond to breeds and columns to diets, and each letter corresponds to a different time, then the square shows how to select four sheep for killing on each of four different days, so that the four sheep killed on any one day include one of each breed and one on each diet.

   More recent applications make use of the concept of *orthogonality* of Latin squares, an idea which goes back to Euler. In 1779 he published a paper concerning the problem of 36 officers. These officers were chosen so that there were six from each of six different regiments: further, among the six officers from each regiment there was one officer from each of six given ranks. Euler's problem was to arrange the 36 officers in a $6 \times 6$ array so that each row and column contained one officer of each regiment and one of each rank. He conjectured (correctly) that such an arrangement is impossible.

   If 6 is replaced by 4 and 36 by 16, and a corresponding problem is posed involving 16 officers, four from each of four regiments $\alpha$, $\beta$, $\gamma$, $\delta$, and one of each rank $a$, $b$, $c$, $d$ from each regiment, then this corresponding problem can be solved. Here is a solution.

**Example 1.3.2**

$$\begin{bmatrix} \alpha a & \beta b & \gamma c & \delta d \\ \gamma d & \delta c & \alpha b & \beta a \\ \delta b & \gamma a & \beta d & \alpha c \\ \beta c & \alpha d & \delta a & \gamma b \end{bmatrix}.$$

Since there is to be one officer of each regiment in each row and column, the letters $\alpha$, $\beta$, $\gamma$, $\delta$ must form a Latin square; similarly for $a$, $b$, $c$, $d$. Further, since there is just one officer of each rank from each regiment, the pairs (Greek letter, Latin letter) in the array must all be distinct; for example $\beta c$ occurs just once. This leads to the following definitions.

**Definition 1.3.2** If $A = (a_{ij})$ and $B = (b_{ij})$ are any two $n \times n$ arrays, the *join* $(A, B)$ of $A$ and $B$ is the $n \times n$ array whose $(i, j)$th entry is the pair $(a_{ij}, b_{ij})$.

**Example 1.3.3** The join of

$$\begin{bmatrix} 1 & 2 & 3 \\ 2 & 1 & 2 \\ 4 & 1 & 3 \end{bmatrix} \text{ and } \begin{bmatrix} 1 & 1 & 1 \\ 2 & 2 & 2 \\ 3 & 3 & 3 \end{bmatrix} \text{ is } \begin{bmatrix} (1,1) & (2,1) & (3,1) \\ (2,2) & (1,2) & (2,2) \\ (4,3) & (1,3) & (3,3) \end{bmatrix}.$$

**Definition 1.3.3** The Latin squares $A, B$ of order $n$ are *orthogonal* if all the entries in the join of $A$ and $B$ are distinct. If $A, B$ are orthogonal, $B$ is called an *orthogonal mate* of $A$.

Note that all the entries of $(A, B)$ being different is equivalent to all possible pairs occurring exactly once; for there are only $n^2$ possible pairs. Note too that the condition for orthogonality can be expressed as:

$$\text{if } a_{ij} = a_{IJ} \text{ and } b_{ij} = b_{IJ} \text{ then } i = I \text{ and } j = J. \tag{1.9}$$

More generally, Latin squares $A_1, \ldots, A_r$ are *mutually orthogonal* if they are orthogonal in pairs. The abbreviation MOLS will be used for mutually orthogonal Latin squares.

**Example 1.3.4** Three MOLS of order 4.

$$\begin{bmatrix} 1 & 2 & 3 & 4 \\ 2 & 1 & 4 & 3 \\ 3 & 4 & 1 & 2 \\ 4 & 3 & 2 & 1 \end{bmatrix}, \begin{bmatrix} 1 & 2 & 3 & 4 \\ 4 & 3 & 2 & 1 \\ 2 & 1 & 4 & 3 \\ 3 & 4 & 1 & 2 \end{bmatrix}, \begin{bmatrix} 1 & 2 & 3 & 4 \\ 3 & 4 & 1 & 2 \\ 4 & 3 & 2 & 1 \\ 2 & 1 & 4 & 3 \end{bmatrix}.$$

Euler called the join of two MOLS a *Graeco-Latin* square because of his use of Greek and Latin letters, as in Example 1.3.2, and the name Latin square for a single array was apparently introduced by Fisher in 1926. Euler conjectured in his 36-officer problem that no two MOLS of order 6 exist; more generally, he conjectured in 1782 that there is no $n \equiv 2 \pmod 4$ for which two MOLS of order $n$ exist. Although he was correct about 6 he was wrong about all larger $n \equiv 2 \pmod 4$: the work of Bose, Shrikhande, and Parker which shows this will be described in Chapter 4.

**Definition 1.3.4** Denote by $N(n)$ the largest value of $r$ for which there exist $r$ MOLS of order $n$.

**Theorem 1.3.1** $N(n) \geqslant 2$ *for all integers* $n \geqslant 3$ *other than 6;* $N(6) = 1$.

In the proof of this theorem it is those $n \equiv 2 \pmod 4$ which cause all the problems. The easiest values of $n$ to deal with are the odd ones.

**Theorem 1.3.2** $N(n) \geqslant 2$ *whenever n is odd.*

**Proof** Let $A, B$ be the $n \times n$ arrays with entries $1, \ldots, n$, defined by

$$a_{ij} \equiv j - i + 1 \pmod n, \qquad b_{ij} \equiv i + j - 1 \pmod n.$$

Then $A$ and $B$ are clearly Latin squares. For example,

$$a_{ij} = a_{ik} \Rightarrow j - i + 1 \equiv k - i + 1 \pmod n \Rightarrow j \equiv k \pmod n \Rightarrow j = k,$$

so that the entries in the $i$th row of $A$ are all distinct. To show orthogonality, condition (1.9) must be verified; so suppose that $a_{ij} = a_{IJ}$ and $b_{ij} = b_{IJ}$. Then $j - i + 1 \equiv J - I + 1$ and $i + j - 1 \equiv I + J - 1 \pmod n$, and adding these congruences gives $2j \equiv 2J \pmod n$ while subtracting one from the other gives $2i \equiv 2I \pmod n$. Since $n$ is odd these imply $j \equiv J$ and $i \equiv I \pmod n$, whence $i = I$ and $j = J$.

**Example 1.3.5** Take $n = 5$ in the above construction to obtain the following two MOLS of order 5.

$$\begin{bmatrix} 1 & 2 & 3 & 4 & 5 \\ 5 & 1 & 2 & 3 & 4 \\ 4 & 5 & 1 & 2 & 3 \\ 3 & 4 & 5 & 1 & 2 \\ 2 & 3 & 4 & 5 & 1 \end{bmatrix} \text{ and } \begin{bmatrix} 1 & 2 & 3 & 4 & 5 \\ 2 & 3 & 4 & 5 & 1 \\ 3 & 4 & 5 & 1 & 2 \\ 4 & 5 & 1 & 2 & 3 \\ 5 & 1 & 2 & 3 & 4 \end{bmatrix}.$$

These can be represented by the following Graeco-Latin square

$$\begin{bmatrix} 11 & 22 & 33 & 44 & 55 \\ 52 & 13 & 24 & 35 & 41 \\ 43 & 54 & 15 & 21 & 32 \\ 34 & 45 & 51 & 12 & 23 \\ 25 & 31 & 42 & 53 & 14 \end{bmatrix}.$$

Observe in passing that the Graeco-Latin square obtained from the construction of Theorem 1.3.2 always has the property that the pairs $(x, y)$ and $(y, x)$ occur in the same column (Exercises 1, number 25).

Another way of looking at the above construction is to consider differences. If we can construct two first columns so that the differences $a_{i1} - b_{i1}$ are precisely all the integers (mod $n$) exactly once, then the squares $A, B$ developed cyclically from them, with $a_{ij} \equiv a_{i1} + (j - 1)$, $b_{ij} \equiv b_{i1} + (j - 1)$, will be orthogonal; the pair $(h, k)$ will occur in the join of $A$ and $B$ in the $i$th row, where $a_{i1} - b_{i1} \equiv h - k$. If $n$ is odd, this is easily achieved by taking $a_{i1} = 2 - i$ and $b_{i1} = i$; the differences are $2 - 2i$, $i = 1, \ldots, n$, i.e. $0, n - 2, n - 4, \ldots, 3, 1, n - 1, \ldots, 2$. The squares obtained are precisely those given by the construction in the theorem. Note that we could have taken $a_{i1} = -i$, but $2 - i$ was preferred because it gives the 'natural' first row $1 \ldots n$.

Fundamental to the idea of orthogonality is that of a transversal.

**Definition 1.3.5** A *transversal* of a Latin square of order $n$ is a set of $n$ positions, no two in the same row or column, containing each of the $n$ symbols exactly once.

For example, the main diagonal positions form a transversal of the second square of Example 1.3.5, but not of the first. The diagonal positions from top right to bottom left form a transversal of the first square. Clearly, if $A$ has an orthogonal mate $B$ then the positions in which one particular symbol occurs in $B$ must be a transversal of $A$. The following result is therefore clear.

**Theorem 1.3.3** *A Latin square of order n has an orthogonal mate if and only if it has n disjoint transversals.*

**Example 1.3.6** The Latin square

$$\begin{bmatrix} 2 & 1 & 3 \\ 3 & 2 & 1 \\ 1 & 3 & 2 \end{bmatrix}$$

has three disjoint transversals: (i) positions $(1,3),(2,2),(3,1)$; (ii) positions $(1,2),(2,1),(3,3)$; (iii) positions $(1,1),(2,3),(3,2)$. Assigning $1,2,3$ respectively to these transversals yields the orthogonal mate

$$\begin{bmatrix} 3 & 2 & 1 \\ 2 & 1 & 3 \\ 1 & 3 & 2 \end{bmatrix}.$$

Not every Latin square has a transversal. For example, $\begin{bmatrix} 1 & 2 \\ 2 & 1 \end{bmatrix}$ has none.

More generally, Euler showed that the $n \times n$ array $A = (a_{ij})$ defined on the set $\{1,\ldots,n\}$ by

$$a_{ij} \equiv i+j-1 \pmod{n}$$

has no transversal, and hence no orthogonal mate, if $n$ is even (see Exercises 1, number 24). However, if a Latin square of order $n$ does have a transversal, it can be used to construct a Latin square of order $n+1$. For if $A$ is an $n \times n$ Latin square on $\{1,\ldots,n\}$, replace every entry on the transversal by $n+1$, move each replaced entry to the right and to the bottom, thereby forming an extra row and column, putting $n+1$ in the new bottom right corner position. For example, from the square of Example 1.3.6 with the top right to bottom left diagonal as transversal, the following Latin square is obtained:

$$\begin{bmatrix} 2 & 1 & 4 & 3 \\ 3 & 4 & 1 & 2 \\ 4 & 3 & 2 & 1 \\ 1 & 2 & 3 & 4 \end{bmatrix}.$$

Mutually orthogonal Latin squares will be discussed more fully in Chapters 4 and 5. In particular, Theorem 1.3.1 will be proved, and it will be shown that, for every $n \geqslant 1$ apart from 2, 3 and 6, there in fact exists a Latin square of order $n$ which is orthogonal to its transpose! Then in subsequent chapters important connections between MOLS and other designs such as Kirkman triple systems and whist tournaments will be discussed. But perhaps the most important interrelation is that between MOLS and finite projective planes; this is the final topic of this introductory section.

First we observe the following easy upper bound on $N(n)$.

**Lemma 1.3.4** *For each* $n \geqslant 2$, $N(n) \leqslant n-1$.

**Proof** Suppose that $A_1,\ldots,A_r$ are $r$ MOLS of order $n$. We can relabel the entries at will, so can assume that each has first row $1\ldots n$. Consider the $r$ entries in the $(2,1)$ position. None of these can be 1 since 1 has already occurred in the first column of each square. Further, no two can be equal, for the join of any two of the squares already has each repeated pair in the first row. So $r \leqslant n - 1$.

**Definition 1.3.6** A set of $n - 1$ MOLS of order $n$ is called a *complete* set.

In the light of Lemma 1.3.4 it is natural to ask for what values of $n$ a complete set exists. To answer this question we first look at a way of obtaining designs from MOLS.

Orthogonal Latin squares are used in statistical experiments where the design to be used is not necessarily balanced, but has the property that each pair of elements occur together in at most one block. Such a design is called a $(0, 1)$ design since each concurrency $\lambda_{xy}$ is 0 or 1, where $\lambda_{xy}$ is the number of blocks containing both $x$ and $y$.

Suppose that $A_1, \ldots, A_t$ are $t$ MOLS of order $n$ on $\{1, \ldots, n\}$. Take an $n \times n$ square whose $n^2$ positions are labelled $1, \ldots, n^2$ as follows:

$$S = \begin{bmatrix} 1 & 2 & 3 & \cdots & n \\ n+1 & n+2 & n+3 & \cdots & 2n \\ \vdots & & & & \\ n^2-n+1 & n^2-n+2 & n^2-n+3 & \cdots & n^2 \end{bmatrix}.$$

Then consider the collection of blocks $R_i$, $C_i$, $B_{r,m}$ defined by

$$R_i = \{(i-1)n + 1, \ldots, in\} = \text{set of labels in } i\text{th row of } S,$$

$$C_i = \{i, n+i, \ldots, n(n-1) + i\} = \text{set of labels in } i\text{th column of } S,$$

$$B_{r,m} = \{x : x \text{ is the label in } S \text{ of a position in which } A_r \text{ has entry } m\}$$

$$(1 \leqslant i \leqslant n, 1 \leqslant r \leqslant t, 1 \leqslant m \leqslant n).$$

Altogether we have here $(t + 2)n$ blocks each of size $n$. It follows from the orthogonality of the $A_r$ that no pair of elements can occur in more than one block. Suppose for example that $x$ and $y$ both occur in $B_{r_1,m_1}$ and $B_{r_2,m_2}$. Then $A_{r_1}$ has the same entry $m_1$ in positions $x$ and $y$, while $A_{r_2}$ has entry $m_2$ in these positions; so the pair $(m_1, m_2)$ occurs twice in the join of $A_{r_1}$ and $A_{r_2}$, contradicting orthogonality.

Thus the blocks form a $(0, 1)$ design. Note that the number of pairs of elements in blocks is

$$(t + 2)n \binom{n}{2} = \tfrac{1}{2}n^2(n-1)(t+2).$$

This number must be no more than $\binom{n^2}{2}$, so

$$\tfrac{1}{2}n^2(n-1)(t+2) \leqslant \tfrac{1}{2}n^2(n^2-1),$$

giving another proof that $t \leqslant n - 1$. However, note further that if $t = n - 1$ then *every* pair of elements occurs in a block, so the design must be an $(n^2, n, 1)$ design, i.e. an affine plane of order $n$. The blocks $B_{r,m}$ for fixed $r$ will form a resolution class, as will the $R_i$ and the $C_i$.

**Example 1.3.7**   Take the three MOLS of order 4 of Example 1.3.4.

$$R_1 = \{1,2,3,4\}, \qquad R_2 = \{5,6,7,8\}, \qquad R_3 = \{9,10,11,12\},$$

$$R_4 = \{13,14,15,16\}, \qquad C_1 = \{1,5,9,13\}, \qquad C_2 = \{2,6,10,14\},$$

$$C_3 = \{3,7,11,15\}, \qquad C_4 = \{4,8,12,16\}.$$

The blocks obtained from the first square $A_1$ are

$$B_{11} = \{1,6,11,16\}, \qquad B_{12} = \{2,5,12,15\},$$

$$B_{13} = \{3,8,9,14\}, \qquad B_{14} = \{4,7,10,13\}.$$

Similarly the other blocks are

$$B_{21} = \{1,8,10,15\}, \qquad B_{22} = \{2,7,9,16\}, \qquad B_{23} = \{3,6,12,13\},$$

$$B_{24} = \{4,5,11,14\}, \qquad B_{31} = \{1,7,12,14\}, \qquad B_{32} = \{2,8,11,13\},$$

$$B_{33} = \{3,5,10,16\}, \qquad B_{34} = \{4,6,9,15\}.$$

Altogether 20 blocks are obtained, forming a $(16, 4, 1)$ affine design from which, by Theorem 1.2.5, a finite projective plane of order 4 can be obtained.

We now establish that the above argument is reversible.

**Theorem 1.3.5**   *A complete set of $n - 1$ MOLS of order $n$ exists if and only if an affine plane of order $n$ exists (i.e. if and only if a finite projective plane of order $n$ exists).*

**Proof**   All that remains to be done is to show how to obtain a complete set of MOLS from an affine plane. So suppose an $(n^2, n, 1)$ design exists; by Theorem 1.2.4 it is resolvable. Label the elements so that $R_1 = \{1,\ldots,n\}$, $R_2 = \{n + 1,\ldots,2n\},\ldots, R_n = \{n^2 - n + 1,\ldots,n^2\}$ form one resolution class and $C_1 = \{1, n + 1,\ldots,n^2 - n + 1\},\ldots, C_n = \{n, 2n,\ldots,n^2\}$ form another. There are $n - 1$ further classes. For the $h$th of these further classes, define $A_h = (a_{ij}^{(h)})$ by $a_{ij}^{(h)} = l$ where $(i - 1)n + j$ lies in the $l$th block of the $h$th class. We claim that $A_1,\ldots, A_{n-1}$ are MOLS.

First, each $A_h$ is a Latin square. For suppose that $a_{ij}^{(h)} = a_{iJ}^{(h)} = l$. Then the $l$th block of the $h$th class contains both $(i - 1)n + j$ and $(i - 1)n + J$. But this is impossible since these two elements occur together just once in the design, and they have already occurred together in $R_i$. A similar argument holds for columns; so $A_h$ is indeed a Latin square.

To establish orthogonality, suppose now that $a_{ij}^{(h)} = a_{IJ}^{(h)} = l$ and $a_{ij}^{(k)} = a_{IJ}^{(k)} = m$, $h \neq k$. Then $(i-1)n + j$ and $(I-1)n + J$ occur together in the $l$th block of the $h$th class and in the $m$th block of the $k$th class, contradicting $\lambda = 1$.

Thus complete sets of MOLS are equivalent to finite projective planes. The existence of a complete set implies the existence of a plane, and vice versa. For which values of $n$ these exist is not fully known, but it will be shown that both exist whenever $n$ is a prime power.

## 1.4 Pairwise balanced designs

The eventual disproof of the Euler conjecture concerning MOLS of order $n \equiv 2 \pmod 4$ made use of balanced designs in which not all blocks have the same size.

**Definition 1.4.1** A *pairwise balanced design* PBD($v, K, \lambda$) is a collection of subsets (blocks) of a $v$-set $S$ such that

(i) the size of each block is in $K$ and is less than $v$;
(ii) each pair of elements of $S$ occur together in exactly $\lambda$ of the blocks.

Clearly, if $K = \{k\}$, the concept reduces to that of a BIBD.

**Example 1.4.1** $\{1,2,4\}, \{2,3,5\}, \{3,4,6\}, \{5,6,1\}, \{4,5\}, \{2,6\}, \{1,3\}$ form a PBD $(6, \{2,3\}, 1)$.

Pairwise balanced designs were in fact discussed long before 1960. Away back in 1850 Kirkman proved that if $n = p_1^{\alpha_1} p_2^{\alpha_2} \dots p_r^{\alpha_r}$ then a PBD($n, \{p_1, \dots, p_r\}, 1$) exists. As a special case he took $n = p^2$, thereby showing the existence of an affine ($p^2, p, 1$) design and hence of a finite projective plane of order $p$ for each prime $p$.

**Lemma 1.4.1** *Suppose that a PBD($v, K, 1$) exists and that the smallest and largest block sizes are $h$ and $k$ respectively. Then $v \geqslant k(h-1) + 1$.*

**Proof** Let $B = \{x_1, \dots, x_k\}$ be a block of size $k$. Since $k < v$ we can choose an element $x_0 \notin B$. For each $i = 1, \dots, k$ let $B_i$ denote the unique block containing $x_0$ and $x_i$, and let $A_i = B_i - \{x_0\}$. Then the sets $A_i$ are disjoint, and each has at least $h - 1$ elements. Thus $v \geqslant 1 + k(h-1)$, the 1 corresponding to $x_0$.

Just as eqns (1.1) and (1.2) have to hold for a BIBD, there are similar restrictions on the possible parameters of a PBD.

**Lemma 1.4.2** *If a PBD($v, K, \lambda$) exists with $b_i$ blocks of size $k_i$ for each $k_i \in K$, then*

$$\lambda v(v - 1) = \sum b_i k_i (k_i - 1). \qquad (1.10)$$

**Proof** There are $\binom{v}{2} = \frac{1}{2}v(v - 1)$ pairs of elements, each occurring $\lambda$ times, giving $\frac{1}{2}\lambda v(v - 1)$ pairs in the blocks altogether. But each block of size $k_i$ contains $\frac{1}{2}k_i(k_i - 1)$ pairs, so the total number of pairs in the blocks is therefore also $\sum_i \frac{1}{2}b_i k_i(k_i - 1)$.

We now turn to possible methods of constructing PBDs.

### Method of removing elements

Take a BIBD and remove some of its elements. This was the way in which Example 1.4.1 was obtained from the seven-point plane.

**Example 1.4.2** Take the $(13, 4, 1)$ design of Example 1.1.6 and delete one element from all the blocks containing it to obtain a PBD($12, \{3, 4\}, 1$).

### Method of adjoining elements

Take a *resolvable* ($v, b, r, k, 1$) design and any $h \leqslant r$; for each $i \leqslant h$ adjoin a new element $\infty_i$ to each block of the $i$th resolution class. Also, if $h > 1$, introduce a new block $\{\infty_1, \ldots, \infty_h\}$. This gives a PBD($v + h, \{k, k + 1, h\}, 1$).

**Example 1.4.3** Start with resolvable $(15, 35, 7, 3, 1)$ design as in Example 1.2.1, and take $h = r = 7$ to obtain a PBD($22, \{4, 7\}, 1$). Note that, because $h = r$, no blocks of size 3 are left.

**Example 1.4.4** Start with resolvable $(16, 4, 1)$ design as in Example 1.3.7. Here $r = 5$. Take $h = 4$ to obtain a PBD($20, \{4, 5\}, 1$).

Note that this method of adding elements is precisely how finite projective planes are obtained from affine planes. Indeed, in the last example above the choice $h = 5$ would have led to a finite projective plane. A third method of constructing PBDs will become available once difference systems have been introduced in Chapter 2.

There is one type of PBD which will be of particular importance in later chapters, namely *group divisible designs*. To introduce a simple example, recall how the seven-point plane can be obtained by starting with $\{1, 2, 4\}$ and then adding $i$ (mod 7), $i = 1, \ldots, 6$, to each element to obtain the other six blocks. Suppose we now carry out the same procedure, this time mod 8; then the following blocks are obtained:

$$\{1, 2, 4\}, \{2, 3, 5\}, \{3, 4, 6\}, \{4, 5, 7\}, \{5, 6, 8\}, \{6, 7, 1\}, \{7, 8, 2\}, \{8, 1, 3\}.$$

These blocks contain every pair of elements once, except for the pairs $\{1,5\}$, $\{2,6\}$, $\{3,7\}$, $\{4,8\}$. If these are taken as four further blocks, a PBD$(8,\{2,3\},1)$ is obtained which we can now consider from a slightly different point of view. Note that the 2-element blocks form a *partition* of the whole set; they are disjoint and their union is the whole set. If we call these sets *groups* and the 3-element sets *blocks* then we have an example of a group divisible design.

**Definition 1.4.2** A *group divisible design* GDD$(v, k, m; \lambda_1, \lambda_2)$ consists of a collection of $m$-subsets, called *groups*, and a collection of $k$-subsets, called *blocks*, of a $v$-set $S$, such that:

(i) the groups form a partition of $S$;
(ii) each pair of elements from the same group occur together in exactly $\lambda_1$ blocks;
(iii) each pair of elements from different groups occur together in exactly $\lambda_2$ blocks.

For example, the discussion preceding the definition gives a GDD$(8, 3, 2; 0, 1)$. Clearly, a GDD$(v, k, m; 0, 1)$ always yields a PBD$(v, \{k, m\}, 1)$ on interpreting groups as further blocks. Clearly, too, $m$ must divide $v$.

In this book we shall be considering only GDDs with $\lambda_1 = 0$ and $\lambda_2 = 1$; we shall therefore use the simpler notation GDD$(v, k, m)$.

In general, PBDs turn out to be extremely useful aids in the construction of various types of combinatorial designs. For example it will be shown that MOLS of order $v$ can be constructed from MOLS of orders $k_1, \ldots, k_m$ if a PBD$(v, \{k_1, \ldots, k_m\}, 1)$ exists. Again, they are needed in the solution to the general Kirkman schoolgirls problem and in the construction of whist tournaments.

As an illustration of the usefulness of PBDs, we give an example of how they can be used in the construction of doubles schedules. Suppose that $v$ players wish to take part in a series of doubles tennis matches, and that they wish a number of games to be arranged so that each player partners each other player in exactly one game and opposes each other player in exactly two games. It is easy to construct suitable games for four or five players:

| Schedule for four players | Schedule for five players |
|---|---|
| 4, 1 v 2, 3 | 1, 2 v 3, 5 |
| 4, 2 v 3, 1 | 2, 3 v 4, 1 |
| 4, 3 v 1, 2 | 3, 4 v 5, 2 |
| | 4, 5 v 1, 3 |
| | 5, 1 v 2, 4 |

Note in passing that the schedule given for five players has the cyclic property that each game is obtained from the previous one by adding 1 (mod 5), and the schedule for four players is similarly constructed (mod 3), with 4 behaving like $\infty$.

For the general problem with $v$ players, suppose that there are $b$ games altogether. Each player is to play $r = v - 1$ games, with a different partner each time, where $4b$ = total number of appearances of players in games = $vr = v(v - 1)$; so $v(v - 1)$ must be a multiple of 4. This happens only when $v \equiv 0$ or 1 (mod 4), and it will be shown later by the construction of whist tournaments, which are just *resolvable* doubles tournaments, that a solution always exists for such $v$. But meanwhile consider the case $v = 20$. It has been shown in Example 1.4.4 that a PBD(20, {4, 5}, 1) exists; use it as follows. Replace each block of the PBD by a doubles schedule for its four or five players as illustrated above. The union of all these schedules will provide a solution to the problem of constructing a schedule for 20 players: given any two players, they occur together in precisely one block, and hence will play together in the games of the schedule derived from that block and no other.

This problem will be studied further in section 2.3 and more fully in Chapter 11. See also Exercises 7, numbers 7 and 8.

## 1.5 Systems of distinct representatives

One method of constructing a Latin square of order $n$ on $\{1, \ldots, n\}$ might be to write down the first row at random, then choose a second row avoiding repetitions in any columns, and so on. For example, in the case $n = 5$, a possible start might be

$$
\begin{array}{ccccc}
1 & 2 & 3 & 4 & 5 \\
3 & 5 & 1 & 2 & 4.
\end{array}
$$

It is then easy to select a possible third row, e.g.

$$
\begin{array}{ccccc}
2 & 4 & 5 & 1 & 3,
\end{array}
$$

but the following question naturally arises: if this row by row method of construction is used, do we ever get stuck, or is it always possible to find a further row until a square is obtained?

In finding a third row above, the problem was essentially the following. Let $A_i$ denote the set of numbers which are eligible for inclusion in the third row and the $i$th column. Then

$$
A_1 = \{2, 4, 5\}, \quad A_2 = \{1, 3, 4\}, \quad A_3 = \{2, 4, 5\}, \quad A_4 = \{1, 3, 5\}, \quad A_5 = \{1, 2, 3\}.
$$

Choosing a third row is equivalent to selecting distinct elements from the sets $A_1, \ldots, A_5$; in the choice of third row above, the choice was $2 \in A_1$, $4 \in A_2$, $5 \in A_3$, $1 \in A_4$, $3 \in A_5$.

**Definition 1.5.1** A *system of distinct representatives* (SDR) for a collection of sets $A_1, \ldots, A_m$ is a collection of distinct elements $x_1, \ldots, x_m$ such that $x_i \in A_i$ for each $i$. Each $x_i$ is said to *represent* $A_i$.

The obvious question is: when does a given collection of sets possess an SDR? There is one obvious necessary condition. The sets

$$A = \{1,2,3\}, B = \{1,2\}, C = \{3,4,5\}, D = \{2,3\}, E = \{1,3\}$$

do *not* possess an SDR since the four sets $A, B, D, E$ contain only three elements in their union and hence do not have enough elements to provide *distinct* representatives. So the following condition is clearly necessary:

for all $k$, any $k$ sets contain at least $k$ elements in their union. (1.11)

The remarkable fact that this condition is also sufficient to guarantee the existence of an SDR was proved by Philip Hall in 1935.

**Theorem 1.5.1** *The sets $A_1, \ldots, A_m$ possess an SDR if and only if, for each $k \leqslant m$, any $k$ of the sets contain at least $k$ elements in their union.*

**Proof** (Halmos and Vaughan, 1950) The necessity of the condition is obvious, so suppose now that condition (1.11) holds. The proof proceeds by induction on $m$. The case $m = 1$ is trivial, so suppose that the theorem is true for every $m < h$ and consider the case $m = h$. There are two cases to consider.

(i) First suppose that for each $k < h$ the union of any $k$ sets in fact contains at least $k + 1$ elements. Select any of the sets and choose any of its elements $x_0$ as its representative, removing $x_0$ from all the other sets containing it. Then any $k$ of the remaining $h - 1$ sets must have at least $k$ elements in their union, and it follows from the induction hypothesis that these $h - 1$ sets possess an SDR which, together with $x_0$, form an SDR for the given sets.

(ii) Next suppose that, for some $l < h$, there are $l$ sets containing exactly $l$ elements in their union. By the induction hypothesis these $l$ sets possess an SDR. Consider the remaining $h - l$ sets, with these $l$ elements removed. They will possess an SDR by the induction hypothesis unless, for some $t \leqslant h - l$, there are $t$ of them with fewer than $t$ elements in their union. But then these $t$ sets and the $l$ sets already represented would constitute a collection of $t + l$ sets with fewer than $t + l$ elements in their union, contradicting (1.11). So the remaining $h - l$ sets do indeed possess an SDR which, with the SDR for the $l$ sets already dealt with, will be an SDR for the $h$ given sets.

This theorem has been popularized under the name of the *marriage problem*. Consider the set $A_i$ to be the set of ladies which the $i$th gentleman is willing to marry. Then the theorem asserts that, if each man makes a list of the ladies he is willing to marry, then (assuming the ladies have no say in the matter!) each man can marry a lady on his list (so that no two men marry the same lady) if and only if, for all $k$, the union of any $k$ lists always contains at least $k$ names.

Hall's theorem can be used to confirm that the row by row method of constructing Latin squares always works. It is convenient first to derive the following corollary to Hall's theorem.

**Theorem 1.5.2**  *Let $A_1, \ldots, A_n$ be n subsets of an n-set S such that $|A_i| = m$ for each i and such that each element of S occurs in exactly m of the sets $A_i$. Then $A_1, \ldots, A_n$ possess an SDR.*

**Proof**  Consider any $k$ of the sets $A_i$. Including repetitions they contain $km$ elements in their union. No element can occur more than $m$ times, so the number of distinct elements in their union is at least $(1/m)km = k$. Thus condition (1.11) holds.

**Definition 1.5.2**  If $r \leqslant n$, an $r \times n$ *Latin rectangle* on $\{1, \ldots, n\}$ is an $r \times n$ array such that no element occurs more than once in any row or column.

The following establishes the row by row construction of Latin squares.

**Theorem 1.5.3**  *Any $r \times n$ Latin rectangle with $r < n$ can be extended to an $(r + 1) \times n$ Latin rectangle.*

**Proof**  For each $i \leqslant n$ let $A_i$ denote the set of elements which do not occur in the $i$th column of the $r \times n$ rectangle. Then $|A_i| = n - r$ for each $i$. Since each element occurs in each of the $r$ rows it occurs in $r$ columns and hence occurs in $n - r$ of the sets $A_i$. Choose $m = n - r$ in Theorem 1.5.2 to deduce that the sets $A_i$ possess an SDR.

Theorem 1.5.2 has an interesting application to matrices. Suppose that $A$ is an $n \times n$ $(0, 1)$ matrix with exactly $m$ 1s in each row and column. Then $A$ can be thought of as the incidence matrix of a collection of $n$ $m$-subsets of an $n$-set, where each element occurs in exactly $m$ of the sets, by interpreting $a_{ij} = 1$ to mean that the $i$th set contains the $j$th element. Theorem 1.5.2 tells us that these sets possess an SDR, i.e. there exist $n$ 1s, no two in the same row or column. If we denote the matrix containing these 1s, with 0s elsewhere, by $P_1$, then $A = P_1 + B$ where $B$ is an $n \times n$ $(0, 1)$ matrix with exactly $m - 1$ 1s in each row and column. Applying the same argument to $B$, and repeating, we eventually obtain $A = P_1 + \cdots + P_m$ where each $P_i$ contains precisely one 1 in each row and column.

**Example 1.5.1**

$$A = \begin{bmatrix} 1 & 1 & 0 & 1 & 0 & 0 & 0 \\ 0 & 0 & 1 & 1 & 1 & 0 & 0 \\ 0 & 1 & 1 & 0 & 0 & 0 & 1 \\ 0 & 1 & 0 & 0 & 1 & 1 & 0 \\ 1 & 0 & 0 & 0 & 1 & 0 & 1 \\ 0 & 0 & 0 & 1 & 0 & 1 & 1 \\ 1 & 0 & 1 & 0 & 0 & 1 & 0 \end{bmatrix} = \begin{bmatrix} 0 & 1 & 0 & 0 & 0 & 0 & 0 \\ 0 & 0 & 0 & 1 & 0 & 0 & 0 \\ 0 & 0 & 0 & 0 & 0 & 0 & 1 \\ 0 & 0 & 0 & 0 & 1 & 0 & 0 \\ 1 & 0 & 0 & 0 & 0 & 0 & 0 \\ 0 & 0 & 0 & 0 & 0 & 1 & 0 \\ 0 & 0 & 1 & 0 & 0 & 0 & 0 \end{bmatrix}$$

$$+ \begin{bmatrix} 0 & 0 & 0 & 1 & 0 & 0 & 0 \\ 0 & 0 & 1 & 0 & 0 & 0 & 0 \\ 0 & 1 & 0 & 0 & 0 & 0 & 0 \\ 0 & 0 & 0 & 0 & 0 & 1 & 0 \\ 0 & 0 & 0 & 0 & 1 & 0 & 0 \\ 0 & 0 & 0 & 0 & 0 & 0 & 1 \\ 1 & 0 & 0 & 0 & 0 & 0 & 0 \end{bmatrix} + \begin{bmatrix} 1 & 0 & 0 & 0 & 0 & 0 & 0 \\ 0 & 0 & 0 & 0 & 1 & 0 & 0 \\ 0 & 0 & 1 & 0 & 0 & 0 & 0 \\ 0 & 1 & 0 & 0 & 0 & 0 & 0 \\ 0 & 0 & 0 & 0 & 0 & 0 & 1 \\ 0 & 0 & 0 & 1 & 0 & 0 & 0 \\ 0 & 0 & 0 & 0 & 0 & 1 & 0 \end{bmatrix}.$$

**Definition 1.5.3** A square $(0,1)$ matrix $P$ with exactly one 1 in each row and each column is called a *permutation matrix*.

If $P = (p_{ij})$ is a permutation matrix it can be thought of as representing the permutation which sends $i$ to $j$ whenever $p_{ij} = 1$.

**Example 1.5.2** Let $P$ be the first matrix on the right-hand side in the above example. It represents the permutation $1 \to 2$, $2 \to 4$, $3 \to 7$, $4 \to 5$, $5 \to 1$, $6 \to 6$, $7 \to 3$, i.e. the permutation $(1245)(37)$.

The following result has now been proved.

**Theorem 1.5.4** *If $A$ is an $n \times n$ $(0,1)$ matrix with exactly $m$ 1s in each row and column, then there exist permutation matrices $P_1, \ldots, P_m$ such that $A = P_1 + \cdots + P_m$.*

Such a matrix $A$ arises as the incidence matrix of a symmetric $(v, k, \lambda)$ design. Each $P_i$ then represents an SDR of the blocks. If we write the blocks as columns and in the $i$th row write the elements of the SDR defined by $P_i$ then we obtain a $k \times v$ array in which each row contains each element once, and in which each column is a block. Thus we obtain a special type of Latin rectangle.

**Definition 1.5.4** A *Youden design* is a Latin rectangle in which the columns are the blocks of a symmetric BIBD.

**Example 1.5.3**  The seven sets $\{1,2,4\}$, $\{3,4,5\}$, $\{2,3,7\}$, $\{2,5,6\}$, $\{1,5,7\}$, $\{4,6,7\}$, $\{1,3,6\}$ form a symmetric design. Its incidence matrix is just the matrix $A$ in Example 1.5.1. The three SDRs corresponding to $P_1, P_2, P_3$ are $\{2,4,7,5,1,6,3\}$, $\{4,3,2,6,5,7,1\}$, $\{1,5,3,2,7,4,6\}$ respectively, and these give the Youden design

$$\begin{bmatrix} 2 & 4 & 7 & 5 & 1 & 6 & 3 \\ 4 & 3 & 2 & 6 & 5 & 7 & 1 \\ 1 & 5 & 3 & 2 & 7 & 4 & 6 \end{bmatrix}.$$

Each column is a block; each row is an SDR.

## 1.6  Finite fields

Many of the early constructions of designs made use of the properties of prime numbers. For example, when Kirkman proved that a $(p^2, p, 1)$ design exists for every prime value of $p$, he used in effect the fact that the set $Z_p$ of integers mod $p$ forms a *finite field*: the elements form a group under addition mod $p$ with 0 as the zero element; the non-zero elements form a group under multiplication with 1 as the identity element; and the commutative and distributive laws all hold.

**Example 1.6.1**  A field of five elements. Take $Z_5 = \{0,1,2,3,4\}$, with addition and multiplication mod 5:

| + | 0 | 1 | 2 | 3 | 4 | | × | 0 | 1 | 2 | 3 | 4 |
|---|---|---|---|---|---|---|---|---|---|---|---|---|
| 0 | 0 | 1 | 2 | 3 | 4 | | 0 | 0 | 0 | 0 | 0 | 0 |
| 1 | 1 | 2 | 3 | 4 | 0 | | 1 | 0 | 1 | 2 | 3 | 4 |
| 2 | 2 | 3 | 4 | 0 | 1 | | 2 | 0 | 2 | 4 | 1 | 3 |
| 3 | 3 | 4 | 0 | 1 | 2 | | 3 | 0 | 3 | 1 | 4 | 2 |
| 4 | 4 | 0 | 1 | 2 | 3 | | 4 | 0 | 4 | 3 | 2 | 1 |

Every non-zero element has a (multiplicative) inverse; e.g. the inverse of 2 is 3, since $2 \times 3 = 1$. Further, no two non-zero elements can multiply together to give 0. These properties fail if a non-prime is chosen. For example, in $Z_6$ there is no inverse of 3 since the congruence $3x \equiv 1 \pmod 6$ has no solution; further, $2 \times 3 = 0$, so 0 can be factorized as the product of two non-zero elements.

During the nineteenth century it was established, following the work of Galois, that there exists a finite field with $q$ elements not only when $q$ is prime but whenever $q$ is a prime power. This field, denoted by $GF(q)$, and called the *Galois field* of order $q$, is *not* the set of residues mod $q$ unless $q$ is prime. For example, the integers mod 9 do not form a field since, for example, 3 has no inverse. The importance of the fields GF($q$) to the combinatorialist is that their existence enables many constructions to be extended from primes to prime powers.

Roughly speaking, GF($p^m$) can be constructed as follows. Find a polynomial $f(x)$ of degree $m$, with coefficients in $Z_p$, which is *irreducible* over $Z_p$, i.e. which cannot be written as a product of two polynomials over $Z_p$ of smaller degree, and take as the elements of the required field all polynomials over $Z_p$ of degree less than $m$. Any such polynomial will be of the form $a_{m-1}x^{m-1} + a_{m-2}x^{m-2} + \cdots + a_1x + a_0$ where there are $p$ choices for each $a_i$; hence there are $p^m$ such polynomials. Addition is performed by adding corresponding coefficients mod $p$, and multiplication is performed mod $f(x)$, i.e. by multiplying normally over $Z_p$ and then taking the remainder when the product is divided by $f(x)$.

**Example 1.6.2** Construction of a field of order 4, i.e. with $4 = 2^2$ elements. Take $f(x) = x^2 + x + 1$. (Note that $x^2 + 1$ would not be a possible choice since $x^2 + 1$ is factorizable over $Z_2$: $x^2 + 1 = (x + 1)(x + 1)$ over $Z_2$!) The elements of the field will be 0, 1, $x$ and $x + 1$, and addition and multiplication are carried out as follows.

| + | 0 | 1 | $x$ | $x+1$ | | × | 0 | 1 | $x$ | $x+1$ |
|---|---|---|-----|-------|---|---|---|---|-----|-------|
| 0 | 0 | 1 | $x$ | $x+1$ | | 0 | 0 | 0 | 0 | 0 |
| 1 | 1 | 0 | $x+1$ | $x$ | | 1 | 0 | 1 | $x$ | $x+1$· |
| $x$ | $x$ | $x+1$ | 0 | 1 | | $x$ | 0 | $x$ | $x+1$ | 1 |
| $x+1$ | $x+1$ | $x$ | 1 | 0 | | $x+1$ | 0 | $x+1$ | 1 | $x$ |

For example,

$$x + (x + 1) = 2x + 1 = 1 \text{ since } 2 \equiv 0 \pmod 2;$$
$$x(x + 1) = x^2 + x = 1(x^2 + x + 1) + 1 = 1 \pmod{x^2 + x + 1};$$
$$(x + 1)^2 = x^2 + 2x + 1 = x^2 + 1 = 1(x^2 + x + 1) + x = x.$$

**Example 1.6.3** Construction of a field of order $9 = 3^2$. Take $f(x) = x^2 + 2x + 2$ over $Z_3$: the reader should check this is irreducible. The elements of GF(9) will be 0, 1, 2, $x$, $x + 1$, $x + 2$, $2x$, $2x + 1$, $2x + 2$. Examples of multiplication are

$$x \cdot 2x = 2x^2 = 2(x^2 + 2x + 2) + 2x + 2 = 2x + 2;$$
$$(x + 1)(2x + 1) = 2x^2 + 3x + 1 = 2x^2 + 1 = 2(x^2 + 2x + 2) + 2x = 2x.$$

The multiplication table of non-zero elements is now given.

| $\times$ | 1 | 2 | $x$ | $x+1$ | $x+2$ | $2x$ | $2x+1$ | $2x+2$ |
|---|---|---|---|---|---|---|---|---|
| 1 | 1 | 2 | $x$ | $x+1$ | $x+2$ | $2x$ | $2x+1$ | $2x+2$ |
| 2 | 2 | 1 | $2x$ | $2x+2$ | $2x+1$ | $x$ | $x+2$ | $x+1$ |
| $x$ | $x$ | $2x$ | $x+1$ | $2x+1$ | 1 | $2x+2$ | 2 | $x+2$ |
| $x+1$ | $x+1$ | $2x+2$ | $2x+1$ | 2 | $x$ | $x+2$ | $2x$ | 1 |
| $x+2$ | $x+2$ | $2x+1$ | 1 | $x$ | $2x+2$ | 2 | $x+1$ | $2x$ |
| $2x$ | $2x$ | $x$ | $2x+2$ | $x+2$ | 2 | $x+1$ | 1 | $2x+1$ |
| $2x+1$ | $2x+1$ | $x+2$ | 2 | $2x$ | $x+1$ | 1 | $2x+2$ | $x$ |
| $2x+2$ | $2x+2$ | $x+1$ | $x+2$ | 1 | $2x$ | $2x+1$ | $x$ | 2 |

It can be shown that, for every prime $p$ and every $m \geqslant 2$, there exists a polynomial $f(x)$ of degree $m$ which is irreducible over $Z_p$, so that a finite field of order $p^m$ always exists. Up to isomorphism, the field is unique. For further details see, for example, Stewart (1973).

The multiplication table in Example 1.6.3 looks extremely complicated. However, hidden in it is a very important simplification, for it turns out that every non-zero element of $GF(q)$ can be expressed as a power of one particular element.

**Definition 1.6.1** A non-zero element $\theta$ of $GF(q)$ is called a *primitive element* if $\theta, \theta^2, \theta^3, \ldots, \theta^{q-1} = 1$ are precisely all the non-zero elements of $GF(q)$; i.e. if the (multiplicative) order of $\theta$ is $q-1$. If $q=p$, so that $GF(q)$ is just $Z_p$, such a $\theta$ is called a *primitive root* of $p$.

**Example 1.6.4** 3 is a primitive root of 7 since $3^1 = 3$, $3^2 = 2$, $3^3 = 6$, $3^4 = 3 \cdot 6 = 4$, $3^5 = 3 \cdot 4 = 5$, $3^6 = 3 \cdot 5 = 1$. However, 2 is not a primitive root since $2^3 = 2^6 = 1$.

It is well known (see, for example, Hardy and Wright, 1979) that every prime $p$ has a primitive root. For future use, a list of some of these primitive roots is now given.

**Example 1.6.5** Table of primitive roots.

| $p$ | 3 | 5 | 7 | 11 | 13 | 17 | 19 | 23 | 29 | 31 | 37 | 41 |
|---|---|---|---|---|---|---|---|---|---|---|---|---|
| Primitive root | 2 | 2 | 3 | 2 | 2 | 3 | 2 | 5 | 2 | 3 | 2 | 6 |

**Example 1.6.6** Consider the field $GF(4)$ constructed in Example 1.6.2. It is easy to check that $x$ is a primitive element: for

$$x^1 = x, \quad x^2 = x+1, \quad x^3 = x(x+1) = 1.$$

**Example 1.6.7** Consider the field $GF(9)$ constructed in Example 1.6.3. It turns out that $x$ is again a primitive element; using the multiplication table we have

$$x^1 = x, \quad x^2 = x + 1, \quad x^3 = x(x+1) = 2x + 1, \quad x^4 = x(2x+1) = 2,$$
$$x^5 = x \cdot x^4 = 2x, \quad x^6 = x \cdot 2x = 2x + 2, \quad x^7 = x(2x+2) = x + 2,$$
$$x^8 = x(x+2) = 1.$$

The existence of a primitive element can be used to simplify multiplication. In Example 1.6.7, $x + 1 = x^2$ and $2x + 1 = x^3$, so $(x+1)(2x+1) = x^2 \cdot x^3 = x^5 = 2x$.

It can be shown that every GF($q$) contains a primitive element, i.e. the multiplicative group of non-zero elements of GF($q$) is cyclic (see Exercises 1, number 45 for an outline of proof). However it should be pointed out that if GF($q$) is constructed from an irreducible polynomial $f(x)$ then there is no guarantee that $x$ itself can be taken as a primitive element (as happened in the two examples above). On the other hand it can also be shown that there always exists a *particular* irreducible $f$ for which $x$ is a primitive element. Such an $f$ is called a *primitive polynomial*. For future use, here are some examples:

$$
\begin{array}{ll}
p = 2: \; x^2 + x + 1 & \qquad p = 3: \; x^2 + 2x + 2 \\
\quad\quad\;\; x^3 + x + 1 & \qquad\quad\quad\;\; x^3 + 2x + 1 \\
\quad\quad\;\; x^4 + x + 1 & \\
\quad\quad\;\; x^5 + x + 1 & \qquad p = 5: \; x^2 + 2x + 3 \\
\quad\quad\;\; x^6 + x + 1 & \qquad\quad\quad\;\; x^3 + 3x + 2.
\end{array}
$$

The following theorem summarizes the properties of finite fields which will be needed later.

**Theorem 1.6.1** *Let $q = p^m$ where $m \geq 1$ and $p$ is prime. Then*

(i) *there exists a finite field* GF($q$) *of order $q$;*

(ii) *the multiplicative group of non-zero elements of* GF($q$) *is cyclic, i.e. there exists $\theta \in$ GF($q$) such that $\theta, \theta^2, \ldots, \theta^{q-1} = 1$ are the non-zero elements of* GF($q$);

(iii) GF($p^m$) *has a (unique) subfield* GF($p^l$) *if and only if $l$ divides $m$; the non-zero elements of the subfield are $\theta^{nt}$, where $1 \leq n \leq p^l - 1$, $t = (p^m - 1)/(p^l - 1)$.*

We close this brief look at finite fields by considering the squares, which turn out to be of particular importance in applications.

If $q$ is odd and $\theta$ is a primitive element of GF($q$) then $\theta^{q-1} = 1$ so that $(\theta^{\frac{1}{2}(q-1)} - 1)(\theta^{\frac{1}{2}(q-1)} + 1) = 0$, i.e. $\theta^{\frac{1}{2}(q-1)} = 1$ or $-1$. The value 1 can be ruled out since $\theta$ has order $q - 1$; so $\theta^{\frac{1}{2}(q-1)} = -1$.

If $q = 4m + 1$ then $\frac{1}{2}(q - 1) = 2m$, so $\theta^{2m} = -1$. The non-zero squares in $\mathrm{GF}(q)$ are the squares of $\theta, \theta^2, \ldots, \theta^{2m}, \theta^{2m+1}, \ldots, \theta^{4m}$, i.e. are $\theta^2, \theta^4, \ldots, \theta^{4m} = 1$, $\theta^{4m+2} = \theta^2, \ldots, \theta^{8m} = 1$, i.e. the even powers of $\theta$ (each arising twice). Thus the odd powers of $\theta$ are the non-zero non-squares. Since $\theta^{2m} = -1$, $-1$ is itself a square and so $x$ is a square if and only if $-x$ is a square.

If $q = 4m + 3$ then $\frac{1}{2}(q - 1) = 2m + 1$ so that $\theta^{2m+1} = -1$. Again the squares are the even powers of $\theta$, so this time $-1$ is not a square, and $x$ is a square if and only if $-x$ is not a square.

**Theorem 1.6.2** *Let $\theta$ be a primitive element of $\mathrm{GF}(q)$ where $q$ is odd. Then*

(i) *the non-zero squares in $\mathrm{GF}(q)$ are the even powers of $\theta$;*
(ii) *$-1$ is a square if $q \equiv 1 \pmod 4$ but is not a square if $q \equiv 3 \pmod 4$;*
(iii) *if $q \equiv 1 \pmod 4$ then $x$ is a square if and only if $-x$ is a square;*
(iv) *if $q \equiv 3 \pmod 4$ then $x$ is a square if and only if $-x$ is not a square.*

**Example 1.6.8** In $Z_7$ the squares are $1, 4$ and $2$; their negatives are $6, 3$ and $5$, the non-squares.

**Example 1.6.9** In $Z_{13}$ the squares are $1, 4, 9, 3, 12, 10$; their negatives are $12, 9, 4, 10, 1, 3$, i.e. the squares.

**Example 1.6.10** In $\mathrm{GF}(9)$ constructed in Example 1.6.3, the squares are $1$, $x^2 = x + 1$, $x^4 = 2$, $x^6 = 2x + 2$; their negatives are $-1 = 2$, $-(x + 1) = 2x + 2$, $-2 = 1$, $-(2x + 2) = x + 1$, i.e. the squares.

## 1.7   Exercises 1

1. Starting with the $(13, 4, 1)$ design of Example 1.1.6, construct a $(13, 9, 6)$ design.
2. Construct a $(9, 12, 8, 6, 5)$ design.
3. Show that no $(8, 12, 3, 2, 1)$ design can exist.
4. Let $n = k - \lambda$. Show that the following conditions on the parameters of a design are all equivalent: (i) $\lambda(v - 1) = k(k - 1)$; (ii) $k^2 - \lambda v = n$; (iii) $(v - k)\lambda = (k - 1)(k - \lambda)$; (iv) $\lambda\lambda' = n(n - 1)$, where $\lambda' = v - 2k + \lambda$.
5. Show that Fisher's inequality rules out the possible existence of the following $(v, k, \lambda)$ designs: $(16, 6, 1)$, $(21, 6, 1)$, $(25, 10, 3)$, $(34, 12, 2)$, $(36, 15, 2)$, $(36, 15, 4)$.
6. Show that there is essentially only one $(7, 3, 1)$ design, as follows. Take elements $1, \ldots, 7$ and note that without loss of generality it can be assumed that $\{1, 2, 4\}$, $\{2, 3, 5\}$, $\{1, 5, 6\}$ are blocks: deduce that the remaining blocks are uniquely determined.

7. Check *Bose*'s proof that, in a symmetric $(v, k, \lambda)$ design, each pair of blocks intersect in $\lambda$ elements. Let $B$ be any block, and suppose it has $x_1, \ldots, x_{v-1}$ elements in common with the other $v - 1$ blocks. Then

$$\sum_i x_i = \lambda(v - 1) \quad \text{and} \quad \sum_i \tfrac{1}{2}x_i(x_i - 1) = \tfrac{1}{2}(v - 1)\lambda(\lambda - 1),$$

so that $\sum_i (x_i - \lambda)^2 = 0$; thus $x_i = \lambda$ for each $i$.

8. Let $B$ be any block of a BIBD, and let $y_i$ denote the number of other blocks intersecting $B$ in $i$ elements $(i \leqslant k)$. Show that

$$\sum_i iy_i = k(r - 1) \quad \text{and} \quad \sum_i \binom{i}{2}y_i = \binom{k}{2}(\lambda - 1).$$

9. Check *Fisher*'s original proof that $b \geqslant v$. Use the notation of the previous question and let $\bar{i}$ be the average value of $i$, so that $\sum_i iy_i = (b - 1)\bar{i}$. Then $\sum_i (i - \bar{i})^2 y_i \geqslant 0$ yields

$$(b - 1)k(r - 1) + (b - 1)k(k - 1)(\lambda - 1) \geqslant k^2(r - 1)^2.$$

Now use eqns (1.1) and (1.2) to rewrite this as $(r - k)(r - \lambda)(v - k) \geqslant 0$. Since $v > k$ and $r > \lambda$, $r \geqslant k$ so that $b \geqslant v$. (Fisher, 1940)

10. By observing that $|A|^2 = k^2(k - \lambda)^{v-1}$ for a symmetric $(v, k, \lambda)$ design, show that if $v$ is even then $k - \lambda$ must be a perfect square. Deduce that no $(22, 7, 2)$ design can exist.

11. Given an affine $(n^2, n, 1)$ design, construct an $(n^2 + 1, n^3 + n, n^2, n, n - 1)$ design as follows. Take all the blocks of the first resolution class once, and all the other blocks $n - 1$ times; also, for each block of the first class obtain $n$ further blocks by replacing one element by $\infty$. Hence obtain a $(10, 30, 9, 3, 2)$ design. (Rasch and Herrendörfer, 1977)

12. A *Steiner triple system* (STS) of order $v$ is a $(v, 3, 1)$ design. Use eqns (1.1) and (1.2) to show that a STS of order $v$ can exist only if $v \equiv 1$ or $3 \pmod 6$. Deduce that a resolvable STS can exist only if $v \equiv 3 \pmod 6$. (It will be shown that these designs always exist: see Chapters 6 and 7.)

13. Let $D$ be a symmetric $(v, k, \lambda)$ design on a set $S$. Let $B$ be a $k$-subset of $S$ such that $B$ intersects each block of $D$ in at least $\lambda$ elements. Show that $B$ must be a block as follows.
    (i) For each block $B_i$ let $|B \cap B_i| = t_i$. Then $\sum_i t_i = k^2$, $\sum_i t_i(t_i - 1) = \lambda k(k - 1)$.
    (ii) Hence $\sum_i (t_i - \lambda)^2 = (k - \lambda)^2 = n^2$.
    (iii) Put $x_i = t_i - \lambda$; then $\sum_i x_i^2 = n^2$ and $\sum_i x_i = n$; thus one $x_i$ is $n$ and all the others are 0. (Lander, 1980)

14. If $D$ is a symmetric $(v, k, \lambda)$ design, let $\gamma D$ and $\rho D$ denote, respectively, the complementary and residual designs obtained from it. Let

$\delta D$ denote the *derived* design, i.e. the one obtained by choosing a block $B_0$ of $D$ and taking as the blocks of $\delta D$ the sets $B \cap B_0$, $B \neq B_0$, $B$ a block of $D$.

   (i) Show that $\delta D$ is a $(k, v-1, k-1, \lambda, \lambda-1)$ design.

   (ii) Show that $\gamma \rho D$ and $\delta \gamma D$ have the same parameters.

   (iii) Show further that if the chosen blocks in the residual, derived constructions are complementary, then $\gamma \rho D = \delta \gamma D$, i.e. $\gamma \rho \gamma = \delta$.

15. Improve Fisher's inequality to $b \geqslant v + r - 1$ for a *resolvable* design as follows. Note that in the incidence matrix the rows fall into $r$ groups corresponding to the $r$ classes, so that the sums of the rows of each group are all equal. This leads to $r-1$ dependence relations among the rows and hence reduces the upper bound for $\rho(A)$ from $b$ to $b - (r-1)$.

16. Four housewives are to take part in a comparison of 16 different washing machines. There is just one model of each type of machine available. If each pair of machines is to be compared exactly once, construct a schedule enabling the comparisons to be performed in just five sessions.

17. Construct a football league schedule for 10 teams. Explain how to obtain from it a schedule for nine teams, with just one team resting in each round. (See also Chapter 8.)

18. Suppose that a $(v, 3, 1)$ design exists. Define $A = (a_{ij})$ by $a_{ii} = i$ and, if $i \neq j$, $a_{ij} = k$ where $\{i, j, k\}$ is the unique block containing both $i$ and $j$. Show that $A$ is a symmetric Latin square.

19. Take a $6 \times 6$ Latin square $L$ on $\{A, B, C, D, E, F\}$ and also a $6 \times 6$ array $S$ with positions labelled 1 to 36 in natural order. Call two numbers *associates* if they occur in the same row or column of $S$ or correspond to the same symbol in $L$. Let $B_i = \{j \leqslant 36: i \text{ and } j \text{ are associates}\}$. Show that the $B_i$ are the blocks of a $(36, 15, 6)$ symmetric design.

20. A Latin square $A = (a_{ij})$ of order $n$ is called *idempotent* if $a_{ii} = i$ for each $i$. (Interpreted as the multiplication table of an algebra, $A$ yields $i^2 = i$ for each $i$, i.e. each element is an *idempotent*.)

   (i) Show that if $n$ is odd then $A_n = (a_{ij})$, $a_{ij} \equiv 2i - j \pmod{n}$, is idempotent.

   (ii) Show that if $n > 2$ is even, $A_{n-1}$ yields an idempotent square of order $n$ if the $(1, 2), (2, 3), \ldots, (n-1, 1)$ entries of $A_{n-1}$ are replaced by $n$ and an extra row $(n-2, n-1, 1, \ldots, n-3, n)$ and an extra column $(n-1, 1, 2, \ldots, n-2, n)$ are appended. Thus idempotent Latin squares of order $n$ exist for all $n \neq 2$.

21. Start with two MOLS of order 3 and construct a finite projective plane of order 3.

22. Start with a finite projective plane of order 3 and construct two MOLS of order 3.

23. Show that

$$\begin{bmatrix} 1 & 2 & 3 & 4 \\ 2 & 4 & 1 & 3 \\ 3 & 1 & 4 & 2 \\ 4 & 3 & 2 & 1 \end{bmatrix}$$

has no orthogonal mate.

24. Let $A_n$ denote the following cyclic Latin square:

$$\begin{bmatrix} 1 & 2 & 3 & \cdots & n-1 & n \\ 2 & 3 & 4 & \cdots & n & 1 \\ \vdots & & & & & \\ n & 1 & 2 & \cdots & n-2 & n-1 \end{bmatrix}$$

(i) Show that if $n$ is odd then $A_n$ has an orthogonal mate.
(ii) Show that if $n$ is even then $A_n$ does not have even one transversal (and hence has no orthogonal mate). (Hint: a transversal would consist of positions $(i, j(i))$ where the $j(i)$ are all distinct and where the numbers $i + j(i)$ are all distinct (mod $n$)).

25. Show that in the Graeco-Latin square constructed in Theorem 1.3.2 the pairs $(x, y)$ and $(y, x)$ occur in the same column.

26. Take a Latin square on four elements a, b, c, d, and a $4 \times 4$ array $S$ with entries $1, \ldots, 16$ in natural order. For each $i \leqslant 16$, let $B_i$ denote the set of $j \leqslant 16$ not in the same row or column as $j$ and not corresponding to the same letter. Show that the $B_i$ are the blocks of a $(16, 6, 2)$ design.

27. (i) Show that a PBD$(v, \{3, 5\}, 1)$ can exist only if $v$ is odd.
(ii) Construct one with $v = 11$ by taking a resolvable $(6, 2, 1)$ design and adding elements. Can one with $v = 23$ be similarly constructed?

28. Show that a PBD$(v, \{4, 5\}, 1)$ can exist only if $v \equiv 0$ or $1$ (mod 4). Construct one with $v = 17$.

29. Construct a PBD$(v, \{3, 4\}, 1)$ for $v = 18$ and $v = 19$.

30. How many ways are there of extending

$$\begin{bmatrix} 1 & 2 & 3 & 4 & 5 & 6 \\ 2 & 3 & 1 & 5 & 6 & 4 \\ 3 & 1 & 2 & 6 & 4 & 5 \end{bmatrix}$$

to a $4 \times 6$ Latin rectangle?

31. Construct a $5 \times 5$ Latin square row by row, in the light of Theorem 1.5.3.

32. Prove that if one row is removed from a Latin square the resulting array is a Youden design.

33. Write the blocks of Example 1.1.6 as a Youden design.
34. Suppose that $S$ has $mn$ elements and that

$$S = A_1 \cup \cdots \cup A_m = B_1 \cup \cdots \cup B_m,$$

where $|A_i| = |B_i| = n$ for each $i$. Show that the $B_i$ can be relabelled so that $A_i \cap B_i \neq \emptyset$ for each $i$.

35. Suppose that a league schedule for $2n$ teams is given, but venues have not yet been fixed. Use the previous question to show that venues can be chosen so that in each round precisely one of the two teams of each game in the first round is at home. (Wallis, 1983.)

36. Show that premultiplication (postmultiplication) of a matrix $A$ by a permutation matrix results in a permutation of the rows (columns) of $A$. Deduce that two designs $D_1, D_2$ with incidence matrices $A_1, A_2$ are isomorphic if and only if there exist permutation matrices $P, Q$ such that $A_2 = PA_1Q$.

37. Suppose that $A_1, \ldots, A_r$ possess an SDR. Show that there exists a $j \leqslant r$ such that, for any $a \in A_j$, there is an SDR in which $A_j$ is represented by $a$. (Hint: if there are $h$ sets with precisely $h$ elements in their union, take $h$ minimal and consider any one of these $h$ sets).

38. Deduce from the previous example that if $A_1, \ldots, A_r$ possess an SDR and $|A_i| \geqslant k$ for each $i$, then there are at least $k!$ different SDRs if $r \geqslant k$, and at least $k!/(k-r)!$ SDRs if $r < k$. (M. Hall, 1948)

39. What lower bound on the number of different Latin squares with first row $1, 2, \ldots, n$ does the previous example give?

40. Verify that 3 is a primitive root of each of $7, 17, 31$ but not of 11.

41. Construct GF(8) by considering $f(x) = x^3 + x^2 + 1$ over $Z_2$. Verify that $x$ is a primitive element.

42. Construct GF(16) by considering $f(x) = x^4 + x + 1$ over $Z_2$.

43. Use $f(x) = x^2 + x + 2$ over $Z_3$ to construct a field of nine elements, and verify that $1 + x$ is a primitive element. How does this relate to the construction of GF(9) given in the text? Repeat with $f(x) = x^2 + 1$.

44. Find the squares in GF(23) and verify that they obey Theorem 1.6.2.

45. Show that GF($q$) has a primitive element as follows.
    (i) If the largest order of an element in an abelian group $G$ is $k$, then every element of $G$ has order dividing $k$.
    (ii) So if $k$ is the largest order then $x^k = 1$ for all $x \in G$. But in a field any polynomial of degree $k$ can have at most $k$ zeros: so $k \geqslant q - 1$.
    (iii) Thus $k = q - 1$, i.e. a primitive element exists.

# 2

# Difference methods

## 2.1 Difference sets

Several of the designs and tournaments constructed in the first chapter had a cyclic nature. For example the seven-point plane can be obtained by starting with the block $\{1, 2, 4\}$ and adding $1, 2, \ldots, 6 \pmod 7$ to each of $1, 2, 4$; and the schedule for five players in section 1.4 can similarly be obtained from the first game $1, 2 \text{ v } 3, 5$ by adding $1, 2, 3, 4 \pmod 5$. Clearly a clever choice of the first block or game has to be made for this method to work; for example if a schedule is constructed cyclically from the first game $1, 3 \text{ v } 2, 5$ then the next game would be $2, 4 \text{ v } 3, 1$, again involving 1 and 3 as partners. In this chapter we study cyclic constructions of designs, beginning with symmetric designs.

Why is it that the design obtained by developing the block $\{1, 2, 4\}$ cyclically $\pmod 7$ is balanced? Consider, for example, the pair 3 and 6. To find a block containing them both we have to be able to choose a value of $i$ such that $\{1 + i, 2 + i, 4 + i\}$ contains 3 and 6. In fact $i = 2$ gives $\{3, 4, 6\}$ as required. Now 3 and 6 differ by 3, so they had to arise as $1 + i, 4 + i$ for some $i$, as 1 and 4 are the only two numbers in the starting block which differ by 3. Clearly the key property possessed by $\{1, 2, 4\}$ which enables it to produce a balanced design with $\lambda = 1$ is that every non-zero number $\pmod 7$ occurs exactly once as a difference between two of its elements: $1 = 2 - 1, 2 = 4 - 2, 3 = 4 - 1, 4 = 1 - 4, 5 = 2 - 4, 6 = 1 - 2$.

**Definition 2.1.1** A $(v, k, \lambda)$ *difference set* (mod $v$) or a *cyclic* $(v, k, \lambda)$ *difference set* is a set $D = \{d_1, \ldots, d_k\}$ of distinct elements of $Z_v$ such that each non-zero $d \in Z_v$ can be expressed in the form $d = d_i - d_j$ in precisely $\lambda$ ways.

**Example 2.1.1** $\{1, 2, 4\}$ is a cyclic $(7, 3, 1)$ difference set.

**Example 2.1.2** $\{1,3,4,5,9\}$ is a cyclic $(11,5,2)$ difference set. The differences are $5 - 4 = 4 - 3 = 1$, $5 - 3 = 3 - 1 = 2$, $4 - 1 = 1 - 9 = 3$, etc.

As indicated above, difference sets are a very valuable tool in the construction of symmetric designs. They appear to have been first studied by Kirkman in 1857; he considered the case $\lambda = 1$ and displayed the following examples in addition to that of Example 2.1.1.

**Example 2.1.3** (a) $\{1,2,4,10\}$ is a $(13,4,1)$ difference set in $Z_{13}$.

(b) $\{1,2,5,15,17\}$ is a $(21,5,1)$ difference set in $Z_{21}$.

(c) $\{1,2,7,19,23,30\}$ is a $(31,6,1)$ difference set in $Z_{31}$.

Kirkman stated that he could not find a $(43,7,1)$ difference set, and asserted (correctly) that probably none exists, and he pointed out how these difference sets lead to $(n^2 + n + 1, n + 1, 1)$ designs for $n = 2,3,4,5$. In a previous paper he had constructed such designs for all prime values of $n$ by a different method. The fact that one existed for $n = 4$, where $n$ is not a prime, came as a surprise to him, but as has already been mentioned it is now known that these finite projective planes exist for all prime powers $n$.

The parameters of a difference set are related, just as for block designs.

**Lemma 2.1.1** *If a cyclic $(v,k,\lambda)$ difference set exists then $\lambda(v-1) = k(k-1)$.*

**Proof** There are $k(k-1)$ differences arising from the $k$ elements of the difference set, and they represent each of the $v - 1$ non-zero integers (mod $v$), $\lambda$ times each.

The importance of difference sets in the construction of symmetric designs is now formalized.

**Definition 2.1.2** If $D = \{d_1, \ldots, d_k\}$ is a $(v,k,\lambda)$ difference set (mod $v$) then the set $D + a = \{d_1 + a, \ldots, d_k + a\}$ is called a *translate* of $D$.

**Lemma 2.1.2** *Each translate of a $(v,k,\lambda)$ difference set is also a $(v,k,\lambda)$ difference set.*

**Proof** $(d_i + a) - (d_j + a) = m \Leftrightarrow d_i - d_j = m$, so the number of representations of $m$ as a difference of elements of $D + a$ is equal to the number of representations as a difference of elements of $D$.

Developing a starting block cyclically just corresponds to forming the translates.

**Theorem 2.1.3** *If $D = \{d_1, \ldots, d_k\}$ is a cyclic $(v, k, \lambda)$ difference set then the translates $D, D + 1, \ldots, D + (v - 1)$ are the blocks of a symmetric $(v, k, \lambda)$ design.*

**Proof** The translates are certainly all of size $k$. It therefore remains only to show that, if $a, b \in Z_v$, then $a$ and $b$ occur together in precisely $\lambda$ translates. Now $a = d_i + (a - d_i)$ for each $i$, so $a$ occurs in the translates $D + (a - d_i)$. Similarly, $b$ occurs in the translates $D + (b - d_i)$. Thus, $a, b$ occur together in a translate $D + d$ precisely when $d = a - d_i = b - d_j$ for some $i, j$. But $a - d_i = b - d_j \Leftrightarrow a - b = d_i - d_j$, and there are $\lambda$ pairs $i, j$ for which this holds. So $a, b$ occur together in exactly $\lambda$ translates.

**Example 2.1.4** (a) $\{1, 2, 4, 10\}$ (mod 13) yields a finite projective plane of order 3. (b) $\{1, 3, 4, 5, 9\}$ (mod 11) yields a $(11, 5, 2)$ design.

Theorem 2.1.3 shows the importance of difference sets. The next problem is how to construct them.

*Complementary difference sets*

Consider, for example, $D = \{1, 2, 4, 10\}$ (mod 13). The complement of $D$ in $Z_{13}$ is $\bar{D} = \{3, 5, 6, 7, 8, 9, 11, 12, 13\}$, and it is routine to check that $\bar{D}$ is a $(13, 9, 6)$ difference set. It is now shown that in general $\bar{D} = Z_v - D$ is always a difference set.

**Lemma 2.1.4** *Let $D$ be a $(v, k, \lambda)$ difference set (mod $v$). Then any non-zero $d$ (mod $v$) occurs*

  (i) *$\lambda$ times as a difference $e - f, e, f \in D$;*
  (ii) *$k - \lambda$ times as a difference $e - f, e \in D, f \in \bar{D}$;*
  (iii) *$k - \lambda$ times as a difference $e - f, e \in \bar{D}, f \in D$;*
  (iv) *$v - 2k + \lambda$ times as a difference $e - f, e, f \in \bar{D}$.*

**Proof** (i) is immediate. Next note that for each $e \in D$, there is a unique representation of $d$ as $d = e - x$ (take $x = e - d!$). For $\lambda$ choices of $e$ we have $x \in D$ (by (i)), so for $k - \lambda$ choices of $e$ we have $x \in \bar{D}$; this proves (ii), and the proof of (iii) is similar. Finally, $d$ has $v$ representations as a difference of members of $Z_v$; so $v = \lambda + (k - \lambda) + (k - \lambda) + $ number of representations with $e, f \in \bar{D}$.

The following is now immediate from (iv).

**Theorem 2.1.5** *If $D$ is a $(v, k, \lambda)$ difference set (mod $v$) then $\bar{D}$ is a $(v, v - k, v - 2k + \lambda)$ difference set (mod $v$), called the complementary difference set.*

**Example 2.1.5** From Example 2.1.1 we obtain the fact that $\{3, 5, 6, 7\}$ is a $(7, 4, 2)$ difference set.

*Difference sets in groups other than $Z_v$*

Instead of considering elements of $Z_v$, consider an arbitrary additive abelian group $G$.

**Definition 2.1.3** A $(v, k, \lambda)$ *difference set in an additive abelian group $G$ of order $v$* is a set $D = \{d_1, \ldots, d_k\}$ of distinct elements of $G$ such that each non-zero element $g$ of $G$ has exactly $\lambda$ representations as $g = d_i - d_j$.

**Example 2.1.6** The group $Z_4 \oplus Z_4$ consists of ordered pairs $(i, j)$ where $i, j \in Z_4$. Addition is defined mod 4 by $(i, j) + (k, l) = (i + k, j + l)$. The set

$$D = \{(0, 0), (0, 2), (1, 0), (3, 0), (1, 1), (3, 3)\}$$

is a $(16, 6, 2)$ difference set. For example, $(3, 1)$ has two representations $(3, 1) = (3, 3) - (0, 2) = (0, 2) - (1, 1)$, and similarly $(2, 2) = (3, 3) - (1, 1) = (1, 1) - (3, 3)$.

Any difference set in any additive abelian group will lead to a symmetric design: simply take as blocks all the translates $D + g$, $g \in G$, and the proof of Theorem 2.1.3 goes through. Thus from Example 2.1.6 we obtain a $(16, 6, 2)$ design. Often in our applications we shall take $G$ to be the additive group of a finite field. It should be pointed out that not every symmetric design can be obtained from a difference set in some group. For example, a $(31, 10, 3)$ design will be constructed in Exercises 2, number 22 which is not cyclic; but it it arose from a difference set in a group $G$ then $G$, being of prime order, would have to be $Z_{31}$. See Example 2.5.4.

*Quadratic residue difference sets*

One of the most important constructions of difference sets makes use of the *quadratic residues* or *squares* in GF$(q)$. Recall from section 1.6 that these are precisely the even powers of a primitive element $\theta$ of GF$(q)$. The following theorem is due to Paley (1933).

**Theorem 2.1.6** *Let* $q = p^\alpha \equiv 3$ (mod 4). *Then the non-zero squares in* GF$(q)$ *form a* $(q, \frac{1}{2}(q - 1), \frac{1}{4}(q - 3))$ *difference set.*

**Proof** Let $q = 4t + 3$ and let $\theta$ be a primitive element of GF$(q)$. Let $Q$ denote the set of non-zero squares; then $Q = \{\theta^2, \theta^4, \ldots, \theta^{4t+2} = 1\}$. Further, by Theorem 1.6.2, $-1 \notin Q$, and $-Q = \{\theta, \theta^3, \ldots, \theta^{4t+1}\}$.

Suppose that 1 can be written as a difference of elements of $Q$ as

$$1 = \theta^{2a_1} - \theta^{2b_1} = \cdots = \theta^{2a_r} - \theta^{2b_r}.$$

Then any $\theta^{2s} \in Q$ can be written as

$$\theta^{2s} = \theta^{2(a_1+s)} - \theta^{2(b_1+s)} = \cdots = \theta^{2(a_r+s)} - \theta^{2(b_r+s)},$$

so that every representation of 1 gives rise to a representation of $\theta^{2s}$. Conversely, every representation of $\theta^{2s}$ gives rise to a representation of 1 (this time *divide* by $\theta^{2s}$). So every member of $Q$ has the same number of representations as a difference of elements of $Q$. Further, if $q \notin Q$, then $-q \in Q$ and every representation $-q = \theta^{2a} - \theta^{2b}$ gives a corresponding representation $q = \theta^{2b} - \theta^{2a}$. Thus every element of $-Q$ has the same number of representations as every element of $Q$. Thus $Q$ is indeed a difference set. Clearly $k = |Q| = 2t + 1 = \frac{1}{2}(q - 1)$, so that finally $\lambda(v - 1) = k(k - 1)$ gives $\lambda = t = \frac{1}{4}(q - 3)$.

**Corollary 2.1.7** *If* $q = p^\alpha \equiv 3$ (mod 4) *then there exists a symmetric* $(q, \frac{1}{2}(q - 1), \frac{1}{4}(q - 3))$ *design*

**Example 2.1.7** Take $q = p = 11$. Then $Q = \{1, 4, 9, 5, 3\}$ is a $(11, 5, 2)$ difference set giving rise to a $(11, 5, 2)$ design.

Corollary 2.1.7 gives an infinite family of symmetric designs since there are infinitely many primes congruent to 3 (mod 4). These are all *Hadamard designs*, with parameters $(4m - 1, 2m - 1, m - 1)$ for some $m$. Such designs will be studied further in Chapter 3.

**Corollary 2.1.8** *If* $q = p^\alpha \equiv 3$ (mod 4), *the quadratic non-residues in* GF($q$), *together with* 0, *form a* $(q, \frac{1}{2}(q + 1), \frac{1}{4}(q + 1))$ *difference set which yields a* $(q, \frac{1}{2}(q + 1), \frac{1}{4}(q + 1))$ *design.*

**Proof** This follows immediately from Theorem 2.1.5 and Corollary 2.1.7.

**Example 2.1.8** Take $q = 19$ so that $Q = \{1, 4, 9, 16, 6, 17, 11, 7, 5\}$. So $\{2, 3, 8, 10, 12, 13, 14, 15, 18, 19\}$ is a $(19, 10, 5)$ difference set which yields a $(19, 10, 5)$ design.

A more complicated analogous result for $q \equiv 1$ (mod 4) will be presented in the next section.

## 2.2 Difference systems

It is clear that the designs obtained from difference sets in a group of order $v$ are always symmetric, since $b = v$. Can the method of differences be adapted to provide a means of constructing non-symmetric designs?

Consider the sets $D_1 = \{1, 2, 5\}$ and $D_2 = \{1, 3, 9\}$ in $Z_{13}$. The differences arising from $D_1$ are $1, 12, 4, 9, 3, 10$ and those arising from $D_2$ are $2, 11, 8, 5, 6, 7$. So between them $D_1$ and $D_2$ give each non-zero element of $Z_{13}$ as a difference exactly once.

**Definition 2.2.1** Let $D_1, \ldots, D_t$ be sets of size $k$ in an additive abelian group $G$ of order $v$ such that the differences arising from the $D_i$ give each non-zero element of $G$ exactly $\lambda$ times. Then $D_1, \ldots, D_t$ are said to form a $(v, k, \lambda)$ *difference system* in $G$.

Note that $D_1, \ldots, D_t$ need not be disjoint.

**Example 2.2.1** (a) $\{1, 2, 5\}$ and $\{1, 3, 9\}$ form a $(13, 3, 1)$ difference system in $Z_{13}$. (b) $\{0, 1, 7, 11\}$, $\{0, 2, 3, 14\}$, $\{0, 4, 6, 9\}$ form a $(19, 4, 2)$ difference system in $Z_{19}$.

Just as for difference sets, the parameters are related.

**Lemma 2.2.1** *If $D_1, \ldots, D_t$ form a $(v, k, \lambda)$ difference system then*

$$tk(k - 1) = \lambda(v - 1).$$

**Proof** Each of the $t$ sets $D_i$ gives $k(k - 1)$ differences, so there are $tk(k - 1)$ differences altogether. These represent the $v - 1$ non-zero elements of $G$, each $\lambda$ times.

Difference systems generate block designs, just as difference sets did.

**Theorem 2.2.2** *Let $D_1, \ldots, D_t$ form a $(v, k, \lambda)$ difference system in the additive abelian group $G = \{g_0, \ldots, g_{v-1}\}$. Then the sets $D_i + g_j$, $1 \leqslant i \leqslant t$, $0 \leqslant j \leqslant v - 1$, are the blocks of a $(v, vt, kt, k, \lambda)$ design.*

**Proof** The sets $D_i + g_j$ are all of size $k$, and clearly there are $tv$ of them. So it remains to establish balance.

Let $D_i = \{d_{i1}, \ldots, d_{ik}\}$, $1 \leqslant i \leqslant t$, so that

$$D_i + g_j = \{d_{i1} + g_j, \ldots, d_{ik} + g_j\}.$$

Let $a, b$ be any two elements of $G$. Now $a = d_{ih} + (a - d_{ih})$ so $a$ occurs in the translate $D_i + (a - d_{ih})$, $1 \leqslant h \leqslant k$. Similarly, $b$ occurs in the translates $D_i + (b - d_{ih})$. Thus $a, b$ occur together in a translate $D_i + g_j$ precisely when $g_j = a - d_{ih} = b - d_{il}$ for some $h, l$. But $a - d_{ih} = b - d_{il} \Leftrightarrow a - b = d_{ih} - d_{il}$, and since there are precisely $\lambda$ choices of $i, h, l$ for which this holds it follows that $a, b$ occur together in exactly $\lambda$ translates. Finally, the value of $r$ follows from eqn (1.2).

**Example 2.2.2** It follows from Example 2.2.1(a) that there exists a $(13, 26, 6, 3, 1)$ design, i.e. a Steiner triple system of order 13.

Note that the design obtained from Theorem 2.2.2 is symmetric if and only if $t = 1$, i.e. if and only if the difference system is a difference set. Clearly the theorem gives a powerful method of constructing designs; it was used to good effect by Bose in his 1939 paper. The method had however been used considerably earlier; for example, Anstice used it in 1853 to construct $(v, 3, 1)$ designs whenever $v$ is a prime of the form $6m + 1$. Bose extended this to all prime powers of that form.

**Theorem 2.2.3** *Let $q = p^\alpha = 6m + 1$, and let $\theta$ be a primitive element of $GF(q)$. Then the sets $D_i = \{0, \theta^i, \theta^{m+i}\}$, $0 \leqslant i \leqslant m - 1$, form a $(6m - 1, 3, 1)$ difference system.*

**Proof** We have $\theta^{6m} = 1$ and $\theta^{3m} = -1$. Also,

$$0 = \theta^{3m} + 1 = (\theta^m + 1)(\theta^{2m} - \theta^m + 1)$$

where $\theta^m + 1 \neq \theta$; so $\theta^{2m} = \theta^m - 1$. Now the differences arising from $D_i$ are $\theta^i$, $-\theta^i$, $\theta^{m+i}$, $-\theta^{m+i}$, $\theta^i(\theta^m - 1)$, $-\theta^i(\theta^m - 1)$, i.e. $\theta^i$, $\theta^{3m+i}$, $\theta^{m+i}$, $\theta^{4m+i}$, $\theta^{2m+i}$, $\theta^{5m+i}$. As $i$ takes values from 0 to $m - 1$ these differences give all the powers of $\theta$ from $\theta^0$ to $\theta^{6m-1}$, i.e. all the non-zero elements of $GF(q)$, each exactly once.

**Corollary 2.2.4** *If $q = p^\alpha = 6m + 1$, there exists a $(6m + 1, 3, 1)$ design, i.e. a Steiner triple system of order $6m + 1$.*

**Example 2.2.3** Take $q = 13$, $m = 2$, $\theta = 2$; then $D_0 = \{0, 1, 4\}$ and $D_1 = \{0, 2, 8\}$ yield a Steiner triple system of order 13 whose blocks are $\{i, 1 + i, 4 + i\}$ and $\{i, 2 + i, 8 + i\}$, $0 \leqslant i \leqslant 12 \pmod{13}$. This is just Example 2.2.1(a) (why?).

Another difference system can be obtained by using quadratic residues and non-residues in $GF(q)$ where $q = p^\alpha \equiv 1 \pmod 4$; this parallels Theorem 2.1.6.

**Theorem 2.2.5** *If $q = p^\alpha = 4m + 1$ then the set $Q$ of quadratic residues and the set $R$ of (non-zero) quadratic non-residues in $GF(q)$ form a $(4m + 1, 2m, 2m - 1)$ difference system.*

**Proof**  Let $\theta$ be a primitive element of GF($q$) so that $\theta^{4m} = 1$. We then have $Q = \{\theta^2, \theta^4, \ldots, \theta^{4m} = 1\}$ and $R = \{\theta, \theta^3, \ldots, \theta^{4m-1}\}$. Since $\theta^{2m} = -1$, $-1 \in Q$, so $Q = -Q$ and $R = -R$.

As in the proof of Theorem 2.1.6, all elements of $Q$ have the same number (say $\lambda_1$) of representations as a difference of elements of $Q$. A similar argument shows that all elements of $R$ have the same number (say $\lambda_2$) of representations as a difference of elements of $Q$. But, since $R = \theta Q$, to every representation of $r \in R$ as a difference $q_1 - q_2$ of elements of $Q$, there corresponds a representation of $\theta r \in Q$ as a difference $\theta q_1 - \theta q_2$ of elements of $R$; so every element of $Q$ arises $\lambda_2$ times as a difference of elements of $R$. Similarly every element of $R$ arises exactly $\lambda_1$ times as a difference of elements of $R$. So, altogether, each non-zero element of GF($q$) occurs $\lambda_1 + \lambda_2$ times as a difference arising from $Q$ or $R$; so $Q$ and $R$ do indeed form a difference system. Since $v = 4m + 1$ and $k = 2m$, Lemma 2.2.1 finally gives $4m(2m - 1) = \lambda \cdot 4m$, i.e. $\lambda = 2m - 1$.

**Corollary 2.2.6**  *For each* $q = p^\alpha = 4m + 1$ *there exists a* $(4m + 1, 8m + 2, 4m, 2m, 2m - 1)$ *design.*

**Example 2.2.4**  Take $q = 17$. Take $Q = \{1, 4, 9, 16, 8, 2, 15, 13\}$ and $R = \{3, 5, 6, 7, 10, 11, 12, 14\}$ yield a $(17, 34, 16, 8, 7)$ design.

Difference systems yield block designs because the sets of a difference system possess two essential properties: they are all of the same size, and between them they give every non-zero element as a difference the same number ($\lambda$) of times. If the sets are now permitted to be of different sizes, but the balance ($\lambda$) property is still retained, the method will yield balance designs with blocks of different sizes, i.e. *pairwise balanced designs*.

**Example 2.2.5**  The sets $\{0, 1, 3, 7\}$, $\{0, 1, 3, 9\}$, and $\{0, 1, 5\}$ yield every element of $Z_{11}$ as a difference exactly three times (e.g. $4 = 7 - 3 = 5 - 1 = 0 - 7$), and so the translates of these sets form a PBD$(11, \{3, 4\}, 3)$ consisting of 22 blocks of size 4 and 11 blocks of size 3.

If a new symbol $\infty$ is adjoined to every one of the 3-element blocks in this example, then all the blocks will be of size 4, and any pair of elements including $\infty$ will also occur in exactly three blocks; for any $a$ will occur in the blocks $\{\infty, a, 1 + a, 5 + a\}$, $\{\infty, a - 1, a, a + 4\}$, and $\{\infty, 6 + a, 7 + a, a\}$ Thus a $(12, 4, 3)$BIBD is obtained. Note that this method works because $\lambda$ is equal to the size of the small block in the difference system, which is itself one less than the size of all the other blocks. In general this method of adjoining $\infty$ will produce a BIBD only if $\lambda = k - 1$ where the system has one block of size $k - 1$ and all other blocks of size $k$.

**Example 2.2.6** $\{0, 1, 2, 4\}$ and $\{3, 4, 6\}$ in $Z_7$ yield a PBD$(7, \{3, 4\}, 3)$ and hence a $(8, 4, 3)$BIBD with blocks $\{i, 1+i, 2+i, 4+i\}$, $\{\infty, 3+i, 4+i, 6+i\}$, $0 \leqslant i \leqslant 6$.

As an application of this idea, we have the following theorem.

**Theorem 2.2.7** *If* $q = p^\alpha \equiv 3 \pmod 4$ *then there exists a* PBD$(q, \{\frac{1}{2}(q-1), \frac{1}{2}(q+1), \frac{1}{2}(q-1))$ *and a* $(q+1, \frac{1}{2}(q+1), \frac{1}{2}(q-1))$ *design.*

**Proof** Let $Q$ and $R$ denote the sets of non-zero quadratic residues and non-residues respectively in GF$(q)$ and let $D_1 = \{0\} \cup Q$, $D_2 = R$, so that $|D_1| = \frac{1}{2}(q+1)$ and $|D_2| = \frac{1}{2}(q-1)$. As in the proof of Theorem 2.1.6, the differences of elements of $Q$ give every non-zero member of GF$(q)$ exactly $\frac{1}{4}(q-3)$ times; so $D_1$ gives all of these differences, as well as the differences $q - 0$, $0 - q$ ($q \in Q$), i.e. all the elements of $Q$ and $-Q = R$ once more. Further, since $R = -Q$, $R$ also gives all non-zero elements of GF$(q)$ as differences $\frac{1}{4}(q-3)$ times. So altogether each non-zero member of GF$(q)$ arises as a difference $2 \cdot \frac{1}{4}(q-3) + 1 = \frac{1}{2}(q-1)$ times. So $D_1, D_2$ generate a PBD$(q, \{\frac{1}{2}(q-1), \frac{1}{2}(q+1)\}, \frac{1}{2}(q-1))$. Adjoining $\infty$ to each translate of $D_2$ now yields a $(q+1, \frac{1}{2}(q+1), \frac{1}{2}(q-1))$ design.

**Example 2.2.7** Take $q = 11$. Then $D_1 = \{0, 1, 3, 4, 5, 9\}$ and $D_2 = \{2, 6, 7, 8, 10\}$ generate a PBD$(11, \{5, 6\}, 5)$, and the blocks

$$\{i, 1+i, 3+i, 4+i, 5+i, 9+i\}, \{\infty, 2+i, 6+i, 7+i, 8+i, 10+i\},$$
$$0 \leqslant i \leqslant 10,$$

form a $(12, 6, 5)$BIBD.

## 2.3 Difference systems for tournaments

The 'classical' method of constructing a league schedule for $2n$ teams, described in section 1.2, yields a schedule in which the first round games are

$$\infty \text{ v } 1, \quad 2 \text{ v } (2n-1), \quad 3 \text{ v } (2n-2), \ldots, \quad n \text{ v } (n+1),$$

and in which further rounds are obtained by adding 1 (mod $2n - 1$) to each number except $\infty$. The reason why this works is that the pairs $\{2, 2n-1\}$, $\{3, 2n-2\}, \ldots, \{n, n+1\}$ form a difference system in $Z_{2n-1}$: they give rise to the differences $\pm 2, \pm 4, \pm 6, \ldots, \pm(2n-2)$, i.e. $2, 4, 6, \ldots, 2n-2$; $2n-3, 2n-5, \ldots, 1$, i.e. each non-zero member of $Z_{2n-1}$, once each. Thus if $a, b \in Z_{2n-1}$, $a$ and $b$ will play each other in one of the games derived from the pair in the difference system whose members differ by $a - b$.

**Definition 2.3.1**  (a) A *starter* in an abelian group $G$ of order $2n - 1$ is a set of $n - 1$ unordered pairs $\{x_1, y_1\}, \ldots, \{x_{n-1}, y_{n-1}\}$ of elements of $G$ such that:

(i) $x_1, y_1, \ldots, x_{n-1}, y_{n-1}$ are precisely all the non-zero elements of $G$;

(ii) $\pm(x_1 - y_1), \ldots, \pm(x_{n-1} - y_{n-1})$ are also precisely the non-zero elements of $G$.

(b) The pairs $\{x_1, y_1\}, \ldots, \{x_{2n-2}, y_{2n-2}\}$ form a *2-fold starter* in $G$ if

(i) $x_1, y_1, \ldots, x_{2n-2}, y_{2n-2}$ are the non-zero elements of $G$, each occurring twice.

(ii) $\pm(x_1 - y_1), \ldots, \pm(x_{2n-2} - y_{2n-2})$ are the non-zero elements of $G$, each occurring twice.

**Example 2.3.1**  The pairs $\{1, 2\}, \{4, 8\}, \{5, 10\}, \{9, 7\}, \{3, 6\}$ form a starter in $Z_{11}$ since the differences are $\pm 1$, $\pm 4$, $\pm 5$, $\pm 2$, $\pm 3$. They therefore yield a cyclic league schedule for 12 teams:

| | | | | | |
|---|---|---|---|---|---|
| Day 1: | $\infty$ v 0 | 1 v 2 | 4 v 8 | 5 v 10 | 9 v 7 | 3 v 6 |
| Day 2: | $\infty$ v 1 | 2 v 3 | 5 v 9 | 6 v 0 | 10 v 8 | 4 v 7 |
| Day 3: | $\infty$ v 2 | 3 v 4 | 6 v 10 | 7 v 1 | 0 v 9 | 5 v 8 |
| ... | | | | | | |

Difference systems can also be used to construct whist tournaments; these were briefly introduced in section 1.4 as resolvable doubles schedules.

**Definition 2.3.2**  A *whist tournament*. Wh($4n$), for $4n$ players is a schedule of games each involving two players against two others, such that:

(i) the games are arranged in $4n - 1$ rounds, each of $n$ games;

(ii) each player plays in exactly one game in each round;

(iii) each player partners every other player exactly once;

(iv) each player opposes every other player exactly twice.

In section 1.4 a schedule for four players was exhibited; each round had just one game. Note the cyclic nature of that tournament: 4 played the role of $\infty$, and the others moved cyclically mod 3. Similar cyclic schedules have been known for other multiples of 4 for some time. For example, in the first volume of *Whist*, dated 1890–1891, Safford listed schedules for all $4n \leqslant 40$, most of which are cyclic. One of his examples is the following.

**Example 2.3.2** A cyclic Wh(12). The players are $\infty, 0, 1, \ldots, 9, 10$.

| Round 1: | $\infty, 0$ v $4,5$ | $1,10$ v $2,8$ | $3,7$ v $6,9$ |
|---|---|---|---|
| Round 2: | $\infty, 1$ v $5,6$ | $2,0$ v $3,9$ | $4,8$ v $7,10$ |
| Round 3: | $\infty, 2$ v $6,7$ | $3,1$ v $4,10$ | $5,9$ v $8,0$ |
| Round 4: | $\infty, 3$ v $7,8$ | $4,2$ v $5,0$ | $6,10$ v $9,1$ |
| Round 5: | $\infty, 4$ v $8,9$ | $5,3$ v $6,1$ | $7,0$ v $10,2$ |
| Round 6: | $\infty, 5$ v $9,10$ | $6,4$ v $7,2$ | $8,1$ v $0,3$ |
| Round 7: | $\infty, 6$ v $10,0$ | $7,5$ v $8,3$ | $9,2$ v $1,4$ |
| Round 8: | $\infty, 7$ v $0,1$ | $8,6$ v $9,4$ | $10,3$ v $2,5$ |
| Round 9: | $\infty, 8$ v $1,2$ | $9,7$ v $10,5$ | $0,4$ v $3,6$ |
| Round 10: | $\infty, 9$ v $2,3$ | $10,8$ v $0,6$ | $1,5$ v $4,7$ |
| Round 11: | $\infty, 10$ v $3,4$ | $0,9$ v $1,7$ | $2,6$ v $5,8.$ |

Note the cyclic structure (mod 11). Clearly the fact that this is a valid Wh(12) depends on a clever choice of games in the first round. Certainly $\infty$ will partner each other player once and oppose each other player twice, but what conditions are necessary to ensure balance between the other players? Since each pair must be partners exactly once, the pairs (apart from the one including $\infty$) in the first round must form a difference system, just as in the football league case. This is clearly satisfied above since the differences between pairs of partners are $\pm 1$, $\pm 2$, $\pm 5$, $\pm 4$, $\pm 3$. But, further, each pair must oppose each other twice, so the pairs of *opponents* in the first round games must form a difference system with $\lambda = 2$! This again is easily checked in the above example, since the pairs of opponents are

$$\{0,4\}, \{0,5\}, \{1,2\}, \{1,8\}, \{10,2\},\{10,8\}, \{3,6\}, \{3,9\}, \{7,6\}, \{7,9\}$$

and these give differences $\pm 4$, $\pm 5$, $\pm 1$, $\pm 4$, $\pm 3$, $\pm 2$, $\pm 3$, $\pm 5$, $\pm 1$, $\pm 2$ i.e. each non-zero element of $Z_{11}$ twice each.

The following lemma is now clear.

**Lemma 2.3.1** *The games $\infty, 0$ v $b_1, d_1$; $a_2, c_2$ v $b_2, d_2$; $\ldots$; $a_n, c_n$ v $b_n, d_n$, where the $a_i, b_i, c_i, d_i$ are the non-zero elements of an additive group $G$, can be taken as the first round games of a cyclic Wh($4n$) if*

(a) *the pairs $\{a_i, c_i\}$ and $\{b_i, d_i\}$ form a starter, and*
(b) *the pairs $\{a_i, b_i\}$, $\{b_i, c_i\}$, $\{c_i, d_i\}$, $\{d_i, a_i\}$ form a 2-fold starter.*

**Example 2.3.3** A Wh(16) can be obtained by developing (mod 15) the following first round games:

$$\infty, 0 \text{ v } 5,10 \quad 1,2 \text{ v } 4,8 \quad 3,11 \text{ v } 12,14 \quad 6,9 \text{ v } 7,13.$$

For the differences between partners are $\pm 5$, $\pm 1$, $\pm 4$, $\pm 7$, $\pm 2$, $\pm 3$, $\pm 6$, and the differences between opponents are $\pm 5$, $\pm 5$, $\pm 3$, $\pm 7$, $\pm 2$, $\pm 6$, $\pm 6$, $\pm 4$, $\pm 1$, $\pm 3$, $\pm 1$, $\pm 7$, $\pm 2$, $\pm 4$.

It does not appear to be easy to construct difference systems for Wh($4n$); the proof of the existence of a Wh($4n$) for all $n \geqslant 1$ given in Chapter 11 will depend largely on other methods. However, we can use difference systems to construct many similar whist tournaments for $4n + 1$ players, where one player sits out in each round.

**Definition 2.3.3** A *whist tournament* Wh($4n + 1$) for $4n + 1$ players is a schedule of games each involving two players against two others, such that:

  (i) the games are arranged in $4n + 1$ rounds, each of $n$ games;
 (ii) each player plays in one game in all but one of the rounds;
(iii) each players partners every other player exactly once;
 (v) each player opposes every other player exactly twice.

**Example 2.3.4** The schedule for five players given in section 1.4 is a Wh(5).

Parallel to Lemma 2.3.1, we have the following means of constructing cyclic Wh($4n + 1$).

**Lemma 2.3.2** *The games* $a_i, c_i$ v $b_i, d_i$, $1 \leqslant i \leqslant n$, *where the* $a_i, b_i, c_i, d_i$ *are the non-zero elements of the abelian group $G$ of order $4n + 1$, can be taken as the first round games of a cyclic* Wh($4n + 1$) *if conditions* (a) *and* (b) *of the previous Lemma hold.*

The following theorem was proved by Baker (1975a).

**Theorem 2.3.3** *Let* $q = p^\alpha = 4n + 1$. *Then a* Wh($4n + 1$) *exists.*

**Proof** Let $\theta$ be a primitive element of GF($q$), so that $\theta^{4n} = 1$ and $\theta^{2n} = -1$. Consider the following games, which involve each non-zero element of GF($q$) exactly once:

$$\theta^i, \theta^{2n+i} \text{ v } \theta^{n+i}, \theta^{3n+i} \quad (0 \leqslant i \leqslant n - 1).$$

The differences between partners are $\pm \theta^i(\theta^{2n} - 1)$ and $\pm \theta^{n+i}(\theta^{2n} - 1)$, i.e. $\pm 2\theta^i$ and $\pm 2\theta^{n+i}$, i.e. $2\theta^i$, $2\theta^{2n+i}$, $2\theta^{n+i}$, $2\theta^{3n+i}$. As $i$ takes values $0, \ldots, n - 1$, these differences are twice the non-zero elements of GF($q$), i.e. all the nonzero elements. Also, the differences between opponents are

$\pm \theta^i(\theta^n - 1)$, $\pm \theta^i(\theta^{3n} - 1)$, $\pm \theta^{n+i}(\theta^n - 1)$ and $\pm \theta^{2n+i}(\theta^n - 1)$. Writing $\theta^{3n} - 1$ as $\theta^{3n} - \theta^{4n} = -\theta^{3n}(\theta^n - 1) = \theta^n(\theta^n - 1)$, these are $\theta^n - 1$ times $\pm \theta^i$, $\pm \theta^{n+i}$, $\pm \theta^{n+i}$, $\pm \theta^i$, i.e. every non-zero element of GF($q$) twice. Thus the games can be taken as the first round games of a Wh($4n + 1$). Other rounds are obtained by adding a fixed element of GF($q$) to each entry.

**Example 2.3.5** A Wh(13) tournament.
Take $q = 13$, $\theta = 2$. The games of the first round are

$$1, 2^6 \text{ v } 2^3, 2^9 \quad 2, 2^7 \text{ v } 2^4, 2^{10} \quad 2^2, 2^8 \text{ v } 2^5, 2^{11}$$

i.e.

$$1, 12 \text{ v } 8, 5 \quad 2, 11 \text{ v } 3, 10 \quad 4, 9 \text{ v } 6, 7.$$

So the Wh(13) is:

| | | | |
|---|---|---|---|
| Round 1: | 1, 12 v 8, 5 | 2, 11 v 3, 10 | 4, 9 v 6, 7 |
| Round 2: | 2, 13 v 9, 6 | 3, 12 v 4, 11 | 5, 10 v 7, 8 |
| Round 3: | 3, 1 v 10, 7 | 4, 13 v 5, 12 | 6, 11 v 8, 9 |
| Round 4: | 4, 2 v 11, 8 | 5, 1 v 6, 13 | 7, 12 v 9, 10 |
| Round 5: | 5, 3 v 12, 9 | 6, 2 v 7, 1 | 8, 13 v 10, 11 |
| Round 6: | 6, 4 v 13, 10 | 7, 3 v 8, 2 | 9, 1 v 11, 12 |
| Round 7: | 7, 5 v 1, 11 | 8, 4 v 9, 3 | 10, 2 v 12, 13 |
| Round 8: | 8, 6 v 2, 12 | 9, 5 v 10, 4 | 11, 3 v 13, 1 |
| Round 9: | 9, 7 v 3, 13 | 10, 6 v 11, 5 | 12, 4 v 1, 2 |
| Round 10: | 10, 8 v 4, 1 | 11, 7 v 12, 6 | 13, 5 v 2, 3 |
| Round 11: | 11, 9 v 5, 2 | 12, 8 v 13, 7 | 1, 6 v 3, 4 |
| Round 12: | 12, 10 v 6, 3 | 13, 9 v 1, 8 | 2, 7 v 4, 5 |
| Round 13: | 13, 11 v 7, 4 | 1, 10 v 2, 9 | 3, 8 v 5, 6. |

## 2.4 Pure and mixed differences

The difference systems introduced in section 2.2 are not sufficiently powerful for our purposes. They enable some non-symmetric designs to be constructed, but not enough. However, it is possible to alter the method slightly to obtain a much more powerful tool, which was used as early as 1852 by Anstice.

The basic idea is really quite simple. Instead of taking the elements of an additive group as the elements of the design, we take several copies of each element, distinguished by a subscript. As an introductory example, consider the following (10, 3, 2)BIBD given by Bose in his 1939 paper, where the elements are of the form $x_i$ where $x \in Z_5$ and $i = 1$ or 2; thus there are two copies of each of the five elements of $Z_5$.

**Example 2.4.1**   A $(10, 30, 9, 3, 2)$BIBD. The blocks are:

$$\{0_2, 1_2, 2_2\} \; \{1_1, 4_1, 0_2\} \; \{2_1, 3_1, 0_2\} \; \{1_1, 4_1, 2_2\} \; \{2_1, 3_1, 2_2\} \; \{0_1, 0_2, 2_2\}$$
$$\{1_2, 2_2, 3_2\} \; \{2_1, 0_1, 1_2\} \; \{3_1, 4_1, 1_2\} \; \{2_1, 0_1, 3_2\} \; \{3_1, 4_1, 3_2\} \; \{1_1, 1_2, 3_2\}$$
$$\{2_2, 3_2, 4_2\} \; \{3_1, 1_1, 2_2\} \; \{4_1, 0_1, 2_2\} \; \{3_1, 1_1, 4_2\} \; \{4_1, 0_1, 4_2\} \; \{2_1, 2_2, 4_2\}$$
$$\{3_2, 4_2, 0_2\} \; \{4_1, 2_1, 3_2\} \; \{0_1, 1_1, 3_2\} \; \{4_1, 2_2, 0_2\} \; \{0_1, 1_1, 0_2\} \; \{3_1, 3_2, 0_2\}$$
$$\{4_2, 0_2, 1_2\} \; \{0_1, 3_1, 4_2\} \; \{1_1, 2_1, 4_2\} \; \{0_1, 3_1, 1_2\} \; \{1_1, 2_1, 1_2\} \; \{4_1, 4_2, 1_2\}.$$

Note that the 30 blocks are obtained cyclically from the six blocks in the first row, leaving the subscripts unchanged in the cyclic development. What is special about the choice of six 'initial' blocks which makes this method work? Again, it is a matter of differences. Each pair of elements has to occur twice in a block. Thus, considering those elements of fixed suffix, the *pure* differences arising from pairs of elements in the initial blocks with that suffix must give every non-zero element of $Z_5$ exactly twice. For example, the differences arising from those elements in the six initial blocks with suffix 1, i.e. the *pure* $(1.1)$ *difference*, are $\pm 3$, $\pm 1$, $\pm 3$, $\pm 1$, i.e. each non-zero element of $Z_5$ twice. Similarly, for suffix 2: the pure $(2, 2)$ differences are $\pm 1$, $\pm 1$, $\pm 2$, $\pm 2$, i.e. every non-zero element twice. But as well as this, the *mixed* $(i, j)$ *differences*, between elements of suffix $i$ and those of suffix $j$, where $i \neq j$, must give every element of $Z_5$ twice, including 0. Thus, the mixed $(1, 2)$ differences are $1 - 0 = 1$, $4 - 0 = 4$, $2 - 0 = 2$, $3 - 0 = 3$, $1 - 2 = 4$, $4 - 2 = 2$, $2 - 2 = 0$, $3 - 2 = 1$, $0 - 0 = 0$, $0 - 2 = 3$, i.e. each element of $Z_5$ twice; and the mixed $(2, 1)$ differences, being their negatives, also give each element twice.

**Definition 2.4.1**   Let $G$ be an additive abelian group, and let $H$ consist of $t$ element $x_1, \ldots, x_t$ corresponding to each element $x$ of $G$. Then the $k$-subsets $D_1, \ldots, D_s$ of $H$ are said to form a *mixed difference system* if there exists an integer $\lambda$ such that, for each $j \leqslant t$, every non-zero element of $G$ occurs $\lambda$ times as a pure $(j, j)$ difference of elements of the $D_i$, and, for every $i, j \leqslant t$, $i \neq j$, every element of $G$ occurs $\lambda$ times as a mixed $(i, j)$ difference.

**Example 2.4.2**   Take $G = Z_3$ and $t = 3$, and consider

$$\{1_1, 2_1, 0_2\}, \quad \{1_2, 2_2, 0_3\}, \quad \{1_3, 2_3, 0_1\}, \quad \{0_1, 0_2, 0_3\}.$$

The $(1, 1)$ differences are $1 - 2 = 2$, $2 - 1 = 1$, i.e. all non-zero elements of $Z_3$.
The $(2, 2)$ differences are $1 - 2 = 2$, $2 - 1 = 1$ also.
The $(3, 3)$ differences are $1 - 2 = 2$, $2 - 1 = 1$ again.

The $(1, 2)$ differences are $1 - 0$, $2 - 0$, $0 - 0$, i.e. every element of $Z_3$.
The $(2, 1)$ differences are the negatives of the $(1, 2)$ differences.
The $(2, 3)$ differences are $2 - 0$, $1 - 0$, $0 - 0$, i.e. every element once.
The $(3, 2)$ differences are their negatives.
The $(1, 3)$ differences are $0 - 1$, $0 - 2$, $0 - 0$, i.e. every element once.
The $(3, 1)$ differences are their negatives.

Thus the four given sets form a mixed difference system in $Z_3$. If they are developed cyclically (mod 3), the following 12 blocks are obtained which form a $(9, 12, 4, 3, 1)$ design, i.e. a Steiner triple system of order 9:

$$\{1_1, 2_1, 0_2\} \quad \{1_2, 2_2, 0_3\} \quad \{1_3, 2_3, 0_1\} \quad \{0_1, 0_2, 0_3\}$$
$$\{2_1, 0_1, 1_2\} \quad \{2_2, 0_2, 1_3\} \quad \{2_3, 0_3, 1_1\} \quad \{1_1, 1_2, 1_3\}$$
$$\{0_1, 1_1, 2_2\} \quad \{0_2, 1_2, 2_3\} \quad \{0_3, 1_3, 2_1\} \quad \{2_1, 2_2, 2_3\}.$$

**Theorem 2.4.1** *Let $D_1, \ldots, D_s$ form a mixed difference system in $H$, where $H$ consists of elements $x_1, \ldots, x_t$ for each $x$ in an additive group $G$. Then the sets $D_j + g$ $(g \in G)$ defined by $x_j + g = (x + g)_j$ form a BIBD.*

**Proof** The blocks are all of size $k = |D_i|$, and we have to check balance. The elements $x_i, y_i$ will occur together in the same block precisely when $x = u + g$, $y = v + g$ for some $g \in G$, and some pair $u, v$ of elements of $G$, with $u_i, v_i$ occurring together in the same set of the difference system. But $x = u + g$ and $y = v + g$ for some $g \Leftrightarrow u - v = x - y$, and the difference $x - y$ occurs precisely $\lambda$ times as a pure $(i, i)$ difference, so that there are $\lambda$ choices for $u, v$. A similar argument holds for mixed differences.

**Example 2.4.3** Take $G = Z_8$, $t = 2$, and consider

$$\{1_1, 7_1, 0_2\}, \{2_1, 6_1, 0_2\}, \{3_1, 4_1, 0_2\}, \{2_1, 5_1, 0_2\}, \{0_1, 0_2, 4_2\},$$
$$\{0_2, 1_2, 2_2\}, \{0_1, 7_1, 0_2\}, \{1_1, 6_1, 0_2\}, \{3_1, 5_1, 0_2\}, \{0_2, 2_2, 5_2\}.$$

The $(1, 1)$ differences are $\pm 2$, $\pm 4$, $\pm 1$, $\pm 3$, $\pm 1$, $\pm 3$, $\pm 2$, i.e. each non-zero element of $Z_8$ twice. Similarly for the $(2, 2)$ differences. The $(1, 2)$ differences are $1, 7, 2, 6, 3, 4, 2, 5, 0, 4, 0, 7, 1, 6, 3, 5$, i.e. every element twice; similarly for the $(2, 1)$ differences. Thus the 10 sets form a mixed difference system, and their translates form a $(16, 3, 2)$ design.

Even Theorem 2.4.1 is not quite general enough. A further extension is to include some infinite elements $\infty$ such that $\infty + g = g$ for all $g \in G$.

**Example 2.4.4** Consider the sets $\{\infty, 1_1, 1_2\}$, $\{2_1, 4_1, 3_2\}$, $\{3_1, 7_1, 5_2\}$, $\{5_1, 6_1, 2_2\}$, $\{4_2, 6_2, 7_2\}$, where we have taken $G = Z_7$ and $t = 2$. The $(1, 1)$

differences are $\pm 2$, $\pm 4$, $\pm 1$, i.e. each non-zero element of $Z_7$ once. The $(2, 1)$ differences are $0$, $\pm 1$, $\pm 2$, $\pm 3$, i.e. every element once. Similarly for the other differences. So, without $\infty$, every pair will occur together once when the sets are developed (mod 7). But clearly $\infty$ will occur once with each element also; so we obtain a $(15, 3, 1)$ design with blocks:

$$\{\infty, 1_1, 1_2\} \quad \{2_1, 4_1, 3_2\} \quad \{3_1, 7_1, 5_2\} \quad \{5_1, 6_1, 2_2\} \quad \{4_2, 6_2, 7_2\}$$
$$\{\infty, 2_1, 2_2\} \quad \{3_1, 5_1, 4_2\} \quad \{4_1, 1_1, 6_2\} \quad \{6_1, 7_1, 3_2\} \quad \{5_2, 7_2, 1_2\}$$
$$\{\infty, 3_1, 3_2\} \quad \{4_1, 6_1, 5_2\} \quad \{5_1, 2_1, 7_2\} \quad \{7_1, 1_1, 4_2\} \quad \{6_2, 1_2, 2_2\}$$
$$\{\infty, 4_1, 4_2\} \quad \{5_1, 7_1, 6_2\} \quad \{6_1, 3_1, 1_2\} \quad \{1_1, 2_1, 5_2\} \quad \{7_2, 2_2, 3_2\}$$
$$\{\infty, 5_1, 5_2\} \quad \{6_1, 1_1, 7_2\} \quad \{7_1, 4_1, 2_2\} \quad \{2_1, 3_1, 6_2\} \quad \{1_2, 3_2, 4_2\}$$
$$\{\infty, 6_1, 6_2\} \quad \{7_1, 2_1, 1_2\} \quad \{1_1, 5_1, 3_2\} \quad \{3_1, 4_1, 7_2\} \quad \{2_2, 4_2, 5_2\}$$
$$\{\infty, 7_1, 7_2\} \quad \{1_1, 3_1, 2_2\} \quad \{2_1, 6_1, 4_2\} \quad \{4_1, 5_1, 1_2\} \quad \{3_2, 5_2, 6_2\}.$$

This is just Example 1.2.1 in disguise.

Having seen some particular examples, we now turn to some more general mixed difference systems. First, a 1939 example due to Bose.

**Theorem 2.4.2** *The sets*

$$\{0_1, 0_2, 0_3\}, \{1_i, (2u)_i, 0_{i+1}\}, \{2_i, (2u-1)_i, 0_{i+1}\}, \ldots, \{u_i, (u+1)_i, 0_{i+1}\},$$
$$1 \leqslant i \leqslant 3,$$

*where the elements are* mod $2u + 1$ *and the suffices are* mod 3, *form a mixed difference system.*

**Proof** The $(i, i)$ differences are $\pm(2u - 1)$, $\pm(2u - 3)$, $\ldots$, $\pm 1$, i.e. all the non-zero elements of $Z_{2u+1}$ once. The $(1, 2)$ differences are $0, 1, 2u$, $2, 2u - 1, \ldots, u, u + 1$, i.e. every element once. Similarly for the other differences.

**Corollary 2.4.3** *A Steiner triple system of order* $6u + 3$ *exists for all* $u \geqslant 0$.

**Proof** By Theorem 2.4.1, the sets of the difference system of Theorem 2.4.2 yield a BIBD; it has $v = 3(2u + 1) = 6u + 3$, $k = 3$ and $\lambda = 1$.

The Steiner triple systems obtained by the above construction are not in general resolvable. An infinite family of resolvable ones is now constructed; this construction is due to Ray-Chaudhuri and Wilson (1971).

**Theorem 2.4.4** *Let* $q = p^\alpha = 6m + 1$. *Then a resolvable Steiner triple system of order* $3q$ *exists.*

**Proof** Let $\theta$ be a primitive element of GF($q$), so that $\theta^{6m} = 1$, $\theta^{3m} = -1$, and $\theta^{2m} + 1 = \theta^m$. Take three copies of each element GF($q$), and consider the sets

$$A = \{0_1, 0_2, 0_3\},$$

$$B_{ij} = \{\theta_j^i, \theta_j^{i+2m}, \theta_j^{i+4m}\}, \quad 1 \leqslant i \leqslant m, 1 \leqslant j \leqslant 3,$$

$$C_{ij} = \{\theta_j^{i+m}, \theta_{j+1}^{i+3m}, \theta_{j+2}^{i+5m}\}, \quad 1 \leqslant i \leqslant m, 1 \leqslant j \leqslant 3 \ (\text{mod } 3),$$

$$D_{ij} = \{\theta_j^i, \theta_{j+1}^{i+2m}, \theta_{j+2}^{i+4m}\}, \quad 1 \leqslant j \leqslant m, 1 \leqslant j \leqslant 3 \ (\text{mod } 3).$$

We claim that these $9m + 1$ sets form a mixed difference system with $\lambda = 1$. The $(j, j)$ differences are $\pm\theta^i(\theta^{2m} - 1)$, $\pm\theta^{i+2m}(\theta^{2m} - 1)$, $\pm\theta^i(\theta^{4m} - 1)$, i.e. $\theta^{2m} - 1$ times $\pm\theta^i$, $\pm\theta^{i+2m}$, $\pm\theta^i(\theta^{2m} + 1)$, i.e. $\theta^{2m} - 1$ times $\pm\theta^i$, $\pm\theta^{i+2m}$, $\pm\theta^{i+m}$, i.e. each non-zero element of GF($q$) exactly once. Now consider the mixed differences. The $(2, 1)$ differences are $0$, $\theta^{i+m}(\theta^{2m} - 1)$, $\theta^{i+3m}(\theta^{2m} - 1)$, $\theta^{i+m}(1 - \theta^{4m})$, $\theta^i(\theta^{2m} - 1)$, $\theta^{i+2m}(\theta^{2m} - 1)$, $\theta^i(1 - \theta^{4m})$, i.e. $0$, $\theta^{i+m}(\theta^{2m} - 1)$, $\theta^{i+3m}(\theta^{2m} - 1)$, $\theta^{i+5m}(\theta^{2m} - 1)$, $\theta^i(\theta^{2m} - 1)$, $\theta^{i+2m}(\theta^{2m} - 1)$, $\theta^{i+4m}(\theta^{2m} - 1)$, i.e. each element once. Similarly for the other mixed differences.

Thus the sets form a mixed difference system with $\lambda = 1$, and so the translates form a Steiner triple system of order $3q$. Further, it is resolvable. The sets $A, B_{ij}, C_{ij}$ ($1 \leqslant i \leqslant m$, $1 \leqslant j \leqslant 3$) form one resolution class, and the translates give a further $6m$ classes. Finally, each $D_{ij}$ with its translates give a further resolution class; so we obtain a further $3m$ classes, giving $9m + 1$ in all.

**Example 2.4.5** Take $q = 7$, $m = 1$, $\theta = 3$. Then the sets

$$A = \{0_1, 0_2, 0_3\}$$

| | | |
|---|---|---|
| $B_{11} = \{3_1, 6_1, 5_1\}$ | $B_{12} = \{3_2, 6_2, 5_2\}$ | $B_{13} = \{3_3, 6_3, 5_3\}$ |
| $C_{11} = \{2_1, 4_2, 1_3\}$ | $C_{12} = \{2_2, 4_3, 1_1\}$ | $C_{13} = \{2_3, 4_1, 1_2\}$ |
| $D_{11} = \{3_1, 6_2, 5_3\}$ | $D_{12} = \{3_2, 6_3, 5_1\}$ | $D_{13} = \{3_3, 6_1, 5_2\}$ |

form a mixed difference system whose translates are the blocks of a resolvable $(21, 3, 1)$ design. This is displayed fully in Example 7.1.2.

This completes the presentation of the work involving differences which will be needed in the constructions of various types of design in later chapters. The final section of this chapter is devoted to a further look at existence problems for difference sets; in this study, the ideal of a *multiplier* is important.

## 2.5 Multipliers of difference sets

Having seen the usefulness of difference sets in constructing symmetric designs, it is natural to turn to the existence problem: for which sets of parameters $v, k, \lambda$ satisfying $\lambda(v - 1) = k(k - 1)$ do difference sets exist?

A powerful method of both ruling out potential sets of parameters and constructing difference sets for other sets of parameters involves the use of *multipliers*. The idea of a multiplier was elaborated by M. Hall about a hundred years after Kirkman introduced the idea in 1857. Kirkman observed that if $D = \{1, 2, 4\}$ in $Z_7$, then $2D = \{2d : d \in D\} = \{2, 4, 1\} = D$, and $4D = \{4, 1, 2\} = D$, while, in $Z_{13}$, if $E = \{1, 2, 4, 10\}$ then $3F = \{3, 6, 12, 4\} = E + 2$.

**Definition 2.5.1** The integer $t$ is a *multiplier* of a difference set $D = \{d_1, \ldots, d_k\}$ in $Z_v$, if $tD = \{td : d \in D\}$ is a translate of $D$, i.e. if $tD = D + s$ for some $s \in Z_v$. If $tD = D$, i.e. if $s = 0$, then $t$ is said to *fix D*.

**Example 2.5.1** $D = \{1, 3, 4, 5, 9\}$ in $Z_{11}$ has 3 a multiplier which fixes $D$ since $3D = \{3, 9, 1, 4, 5\} = D$.

Clearly 1 is always a multiplier of a difference set in $Z_v$; it is called the *trivial* multiplier. Every known difference set in any $Z_v$ has a non-trivial multiplier, but it has not yet been *proved* that such a non-trivial multiplier must always exist. However, an extremely important result is the following. (See, for example, Hall (1986) for a proof.)

**Theorem 2.5.1** (The multiplier theorem of Hall and Ryser, 1951.) *If* $D = \{d_1, \ldots, d_k\}$ *is a* $(v, k, \lambda)$ *difference set in* $Z_v$ *and if* $p$ *is a prime divisor of* $n = k - \lambda$ *with* $p > \lambda$ *and* $(p, v) = 1$, *then* $p$ *is a multiplier of* $D$.

**Example 2.5.1** (again) Here $(v, k, \lambda) = (11, 5, 2)$, so $n = 3$. The prime $p = 3$ satisfies the conditions of the theorem and hence is a multiplier, as was noted above.

Note that if $t$ is a multiplier of $D$ then $tD = D + s$ for some $s$, so that $t(D + a) = (D + a) + s + (t - 1)a$; so $t$ is also a multiplier of every translate of $D$. Note further that if $(t - 1, v) = 1$ then $a$ can be chosen so that $a(t - 1) \equiv -s \pmod{v}$, and, for this $a$, $t(D + a) = D + a$, i.e. $t$ fixes $D + a$. For example, $D = \{1, 2, 4, 10\}$ is a difference set in $Z_{13}$ with $3D = D + 2$; solving $a(t - 1) \equiv -s \pmod{v}$, i.e. $2a \equiv -3 \pmod{13}$ gives $a \equiv 12$, and $D + 12 = \{0, 1, 3, 9\}$ has the property that $3(D + 12) = D + 12$. So a translate has been found which is fixed by the multiplier 3. We can in fact do better than this.

**Theorem 2.5.2** *If $D$ is a $(v, k, \lambda)$ difference set in $Z_v$ with $(v, k) = 1$, then there exists a translate of $D$ which is fixed by every multiplier of $D$.*

**Proof** Consider the translate $D + a$. If $D = \{d_1, \ldots, d_k\}$ then the sum of the elements of $D + a$ is $\sum_i d_i + ka$, and the congruence $ka \equiv -\sum_i d_i$ (mod $v$) has a unique solution for $a$ (mod $v$) since $(k, v) = 1$. Thus there is a value of $a$ such that the sum of the elements of $D + a$ is $0$ (mod $v$). Now let $t$ be any multiplier of $D$; $t$ is then also a multiplier of $D + a$, and $t(D + a) = D + b$ for some $b$. Thus

$$0 \equiv t0 \equiv t\left(\sum_i d_i + ka\right) \equiv t\sum_i (d_i + a) \equiv \sum_i d_i + kb.$$

Since $\sum_i d_i + ka \equiv 0$ has a *unique* solution (mod $v$) it now follows that $b = a$, i.e. $t$ does indeed fix $D + a$.

**Example 2.5.2** Take $D = \{1, 2, 4, 10\}$ in $Z_{13}$. Then 3 is a multiplier of $D$. The equation $ka \equiv -\sum_i d_i$ (mod $v$) is $4a \equiv 9$ (mod 13) which has solution $a = 12$. So $D + 12$ will be fixed by 3:

$$3(D + 12) = 3\{0, 1, 3, 9\} = \{0, 3, 9, 1\} = D + 12.$$

Theorems 2.5.1 and 2.5.2 can be used to find difference sets.

**Example 2.5.3** Suppose we want to find a $(37, 9, 2)$ difference set. Since 37 is prime, the difference set, if it exists, must exist in $Z_{37}$, the only group of order 37. The parameters certainly satisfy $k(k - 1) = \lambda(v - 1)$. Since $n = k - \lambda = 7$, take $p = 7$; by Theorem 2.5.1, 7 must be a multiplier. Now in view of Theorem 2.5.2 we can suppose that the difference set we are looking for is in fact fixed by 7. So, if $a \in D$, then $7a, 7^2 a, 7^3 a, \ldots$ are all in $D$. But the order of 7 (mod 37) is 9, so $D$ must be

$$D = \{a, 7a, 7^2 a, \ldots, 7^8 a\}$$

for some $a$. If we take $a = 1$ we find that

$$\{1, 7, 12, 10, 33, 9, 26, 34, 16\}$$

is indeed a difference set.

On the other hand, the theorems can also be used to establish non-existence results.

**Example 2.5.4** The parameters $(v, k, \lambda) = (31, 10, 3)$ certainly satisfy $k(k - 1) = \lambda(v - 1)$, but it will now be shown that no $(31, 10, 3)$ difference set exists. Since 31 is prime we consider the group $Z_{31}$. If there is a $(31, 10, 3)$ difference set $D$, then $n = 10 - 3 = 7$ will be a multiplier, and we can again suppose that $D$ is fixed by 7, so that $7D = D$. Thus if $c \in D$ then $7c, 7^2c, 7^3c, \ldots$ are all in $D$. But the order of 7 (mod 31) is greater than 10, so no $(31, 10, 3)$ difference set can exist.

**Example 2.5.5** For a long time the question of the existence or non-existence of a $(111, 11, 1)$ design, i.e. of a finite projective plane of order 10, was one of the tantalizing open problems in mathematics. However, it has now been shown that no such design exists: see Lam, Thiel, and Swiercz (1989). Clearly the proof of this result is far beyond the scope of this book, but we can more easily establish the non-existence of a $(111, 11, 1)$ difference set in $Z_{111}$.

Suppose then that $D$ is a $(111, 11, 1)$ difference set in $Z_{111}$. Then $n = k - \lambda = 10$, and 2 and 5 are both multiplier. By Theorem 2.5.2 we can suppose that $D = 2D = 5D$ so that, if $a \in D$, $a \neq 0$, then $a, 2a, 4a$, and $5a$ are also in $D$. But then $a$ has two representations as a difference of elements of $D : a = 2a - a = 5a - 4a$. Since $\lambda = 1$ we must therefore have $2a \equiv 5a$ and $a \equiv 4a$ (mod 111), i.e. $3a \equiv 0$, whence $a = 37$ or 74. This proves that the only possible elements of $D$ are $0, 37, 74$ whereas $D$ is supposed to have 11 elements!

## 2.6   Exercises 2

1. Verify that the following are difference sets (each of which will yield a finite projective plane).
   (a) $\{0, 1, 7, 19, 23, 44, 47, 49\}$ in $Z_{57}$, giving a $(57, 8, 1)$ design.
   (b) $\{1, 2, 4, 8, 16, 32, 64, 55, 37\}$ in $Z_{73}$, giving a $(73, 9, 1)$ design.
   (c) $\{0, 1, 3, 9, 27, 49, 56, 61, 77, 81\}$ in $Z_{91}$, giving a $(91, 10, 1)$ design.
2. Verify that
   (a) $\{0, 1, 2, 4, 5, 8, 10\}$ is a $(15, 7, 3)$ difference set in $Z_{15}$;
   (b) $\{(0, 0), (1, 0), (4, 0), (0, 1), (2, 1), (3, 1), (0, 2)\}$ is a $(15, 7, 3)$ difference set in $Z_5 \oplus Z_3$.
3. Verify that $\{(1, 1), (2, 2), (3, 3), (4, 4), (5, 5), (0, 1), (0, 2), (0, 3), (0, 4), (0, 5), (1, 0), (2, 0), (3, 0), (4, 0), (5, 0)\}$ is a $(36, 15, 6)$ difference set in $Z_6 \oplus Z_6$. (M. Hall)
4. Construct a $(19, 9, 4)$ difference set and a $(23, 11, 5)$ difference set.
5. Verify that the fourth powers (mod 37) form a $(37, 9, 2)$ difference set. Show also that the squares (mod 37) can be used to construct a $(37, 18, 17)$ difference system.
6. Show that if $\{d_1, \ldots, d_k\}$ is a difference set in $Z_v$ and $(n, v) = 1$, then $\{nd_1, \ldots, nd_k\}$ is also a difference set in $Z_v$.
7. Suppose that a symmetric $(v, k, \lambda)$ design $D$ on $\{1, \ldots, v\}$ is obtained from a difference set $\{d_1, \ldots, d_k\}$ in $Z_v$. The incidence matrix $A$ of $D$

therefore has 1s in positions $(i, d_j + i - 1)$, $1 \leqslant i \leqslant v$, $1 \leqslant j \leqslant k$, where the second component is interpreted mod $v$. Show that the dual design, i.e. the design whose incidence matrix is $A'$, is isomorphic to $D$. (Bhat and Shrikhande, 1970)

8. Verify that
   (a) $\{0, 1, 3, 12\}$, $\{0, 1, 5, 13\}$, $\{0, 4, 6, 9\}$ form a $\{19, 4, 2\}$ difference system (mod 19);
   (b) $\{0, 6, 11, 15, 18, 20, 21\}$, $\{0, 5, 7, 8, 9, 13, 19\}$ form a $(22, 7, 4)$ difference system (mod 22);
   (c) $\{(0, 0), (0, 1), (1, 0), (4, 4)\}$, $\{(0, 0), (2, 0), (0, 2), (3.3)\}$ form a $(25, 4, 1)$ difference system in $Z_5 \oplus Z_5$;
   (d) $\{0, 1, 3, 24\}$, $\{0, 10, 18, 30\}$, $\{0, 4, 26, 32\}$ form a $(37, 4, 1)$ difference system (mod 37).
   
   In each case write down the parameters of the block design obtained from it.

9. Verify that $\{0, 1, 3, 7, 25, 38\}$, $\{0, 5, 20, 32, 46, 75\}$, $\{0, 8, 17, 47, 57, 80\}$ yield a $(91, 6, 1)$ design. (Mills, 1975)

10. Verify that $\{0, 7, 10, 11, 23\}$, $\{0, 5, 14, 20, 22\}$ form a difference system in $Z_{41}$ which yields a $(41, 5, 1)$ design.

11. Verify that the pairs $\{1, n - 1\}, \{2, n - 2\}, \ldots, \{\frac{1}{2}(n - 1), \frac{1}{2}(n + 1)\}$ form a difference system (called a *patterned starter*) in $Z_n$ if $n$ is odd. Tie up with section 1.2.

12. Show that if $q = 6m + 1$ is a prime power and $\theta$ is a primitive element of $GF(q)$ then the sets $D_i = \{0, \theta^i, \theta^{2m+i}, \theta^{4m+i}\}$, $0 \leqslant i \leqslant m - 1$, yield a $(6m + 1, 4, 2)$ design. In particular, there exists a $(25, 4, 2)$ design. (Bose, 1942)

13. If $q = 4m + 1$ is a prime power and $\theta$ is a primitive element of $GF(q)$, then the sets $D_i = \{\theta^i, \theta^{m+i}, \theta^{2m+i}, \theta^{3m+i}\}$, $0 \leqslant i \leqslant m - 1$, yield a $(4m + 1, 4, 3)$ design.

14. The sets $\{0, 2i - 1, 4i - 1\}$, $\{0, 2m + 2i - 1, 4i - 3\}$, $1 \leqslant i \leqslant m$, form a $(6m + 1, 3, 2)$ difference system (mod $6m + 1$). (Heffter, 1897)

15. The sets $\{0, i, 2m + 3 - i\}$, $1 \leqslant i \leqslant m + 1$; $\{0, 2i, 3m + 3 + i\}$, $1 \leqslant i \leqslant m$; $\{\infty, 0, 3m + 2\}$ yield a $(6m + 6, 3, 2)$ design.

16. Let $q = 12m + 1$ be a prime power and let $\theta$ be a primitive element of $GF(q)$ such that $\theta^{4m} - 1$ is an *odd* power of $\theta$. Let $D_i = \{0, \theta^{2i}, \theta^{4m+2i}, \theta^{8m+2i}\}$, $0 \leqslant i \leqslant m - 1$. Noting that $\theta^{4m} + 1 = \theta^{2m}$, prove that the $D_i$ form a difference system giving rise to a $(12m + 1, 4, 1)$ design. In particular,
   (i) $q = 13$, $\theta = 2$ give a $(13, 4, 1)$ design;
   (ii) $q = 37$, $\theta = 2$ fail since $\theta^{12} - 1 = \theta^{10}$;
   (iii) $q = 73$, $\theta = 5$ give a $(73, 4, 1)$ design;
   (iv) $q = 25$, $\theta$ satisfying $\theta^2 + 2\theta + 3 = 0$ over $Z_5$ give a $(25, 4, 1)$ design. (Bose, 1939)

17. Construct a Wh(20) by starting with: $\infty, 0$ v $11, 12$; $3, 9$ v $16, 1$; $4, 14$ v $13, 18$; $6, 8$ v $5, 2$; $7, 15$ v $10, 17$.

18. Use Theorem 2.3.3 to construct a Wh(17) and a Wh(9). (See Example 11.1.3)

19. Use mixed differences in $Z_3 \oplus Z_3$ to construct a $(28, 4, 1)$ design as follows. Take $t = 3$, and write $(x, y)$ as $xy$. Then the following sets form a mixed difference system:
$\{00_3, 10_3, 21_3, 11_1)$, $\{00_1, 10_1, 21_1, 11_2\}$, $\{00_2, 10_2, 21_2, 11_3\}$,
$\{00_1, 01_1, 10_2, 12_3\}$, $\{00_2, 01_2, 10_3, 12_1\}$, $\{00_3, 01_3, 10_1, 12_2\}$,
$\{\infty, 00_1, 00_2, 00_3\}$. (K. Kishen; see Bose, 1939.)

20. Verify that $\{0_1, 3_1, 9_1, 10_1\}$, $\{0_1, 0_2, 2_2, 7_2\}$, $\{0_1, 0_2, 9_2, 10_2\}$, $\{0_1, 2_1, 5_2, 8_2\}$, $\{0_1, 3_1, 4_2, 7_2\}$, $\{0_1, 4_1, 3_2, 9_2\}$, $\{0_1, 5_1, 2_2, 6_2\}$ (mod 11) yield a $(22, 4, 2)$ design.

21. Verify that (mod 24) the blocks $\{0_1, 1_1, 3_1, 7_1, 8_2\}$, $\{0_2, 1_2, 3_2, 7_2, 8_1\}$, $\{0_1, 5_1, 13_1, 15_2\}$, $\{0_2, 5_2, 13_2, 15_1\}$, $\{0_1, 9_1, 3_2, 13_2\}$, $\{0_2, 9_2, 3_1, 13_1\}$, $\{0_1, 12_1, 0_2, 12_2\}$ form a mixed difference system which yields a PBD(48, {4, 5}, 1). (Brouwer)

22. Verify that the following blocks form a mixed difference system (mod 7) which yields a $(31, 10, 3)$ design.
$\{1_1, 6_1, 2_2, 5_2, 3_3, 4_3, 3_4, 5_4, 6_4, \infty_1\}$
$\{2_1, 5_1, 3_2, 4_2, 1_3, 6_3, 3_4, 5_4, 6_4, \infty_2\}$
$\{3_1, 4_1, 1_2, 6_2, 2_3, 5_3, 3_4, 5_4, 6_4, \infty_3\}$
$\{1_1, 2_1, 4_1, 1_2, 2_2, 4_2, 1_3, 2_3, 4_3, 0_4\}$
$\{0_1, 1_1, 2_1, 3_1, 4_1, 5_1, 6_1, \infty_1, \infty_2, \infty_3\}$
$\{0_2, 1_2, 2_2, 3_2, 4_2, 5_2, 6_2, \infty_1, \infty_2, \infty_3\}$
$\{0_3, 1_3, 2_3, 3_3, 4_3, 5_3, 6_3, \infty_1, \infty_2, \infty_3\}$ (Hall, 1986)

23. Let $q = 4m + 1$ be a prime power, and let $\theta$ be a primitive element of GF($q$). Show that the following sets form a mixed difference system which yields a resolvable $(3q + 1, 4, 1)$ design.
$A = \{0_1, 0_2, 0_3, \infty\}$;
$B_{ij} = \{\theta_i^i, \theta_i^{i+2m}, \theta_{j+1}^{i+m}, \theta_{j+1}^{i+3m}\}$, $0 \leqslant i \leqslant m - 1$, $1 \leqslant j \leqslant 3$ (mod 3).

24. Find a translate of $\{1, 2, 5, 15, 17\}$ (mod 21) which is fixed by every multiplier.

25. Verify that $D = \{1, 2, 4, 10\}$ and $E = \{1, 2, 5, 7\}$ are both $(13, 4, 1)$ difference sets (mod 13). Find $t$ such that $tD = E$. Use $E$ (mod 14) to construct a GDD (14, 4, 2).

26. Construct designs from the following;
    (1) $\{0, 2, 5, 6, 20, 31, 41\}$ (mod 48);
    (2) $\{1, 6, 8, 14, 38, 48, 49, 52\}$ (mod 63);
    (3) $\{0, 1, 6\}$, $\{0, 2, 10\}$, $\{0, 4, 11\}$, $\{0, 3, 12\}$ (mod 26);
    (4) $\{0, 1, 4\}$, $\{0, 2, 8\}$ (mod 13);
    (5) $\{0, 1, 4\}$, $\{0, 2, 8\}$ (mod 15).

27. Show that the set of multipliers of a difference set form a group.

28. Use multipliers to construct (i) $(73, 9, 1)$, (ii) $(91, 10, 1)$, (iii) $(19, 9, 4)$ difference sets.

29. Use multipliers to show that no $(56, 11, 2)$ difference set in $Z_{56}$ exists.

30. Show that no $(79, 13, 2)$ difference set exists. (Note that the order of 11 mod 79 is 39.)

31. Show that if $D$ is a cyclic $(n^2 + n + 1, n + 1, 1)$ difference set then $n$ is a multiplier of $D$. Illustrate with $n = 4$.

32. Find a multiplier for the difference set of 2(a).

33. *A product result for difference families.* Recall that Theorem 2.2.3 produces a $(v, 3, 1)$ difference system whenever $v$ is a prime power, $v \equiv 1 \pmod 6$. Thus, for example, $(7, 3, 1)$ and $(13, 3, 1)$ difference systems exist. This example shows how to produce a $(91, 3, 1)$ difference system from them $(91 = 7 \times 13)$.

  (i) A $(7, 3, 1)$ difference system is $\{0, 1, 3\}$; a $\{13, 3, 1\}$ difference system is $\{0, 3, 4\}$, $\{0, 5, 7\}$. Consider the triples $\{0, 1 + 7j, 3 + 14j\}$, $0 \leqslant j \leqslant 12$; $\{0, 21, 28\}$, $\{0, 35, 49\}$: verify that they form a $(91, 3, 1)$ difference system.

  (ii) More generally, if $\{0, a_1^{(i)}, \ldots, a_{k-1}^{(i)}\}$, $1 \leqslant i \leqslant t$, is a $(v, k, 1)$ difference system in $Z_v$ and $\{0, b_1^{(i)}, \ldots, b_{k-1}^{(i)}\}$, $1 \leqslant i \leqslant s$, is a $(w, k, 1)$ difference system in $Z_w$, where $(w, (k-1)!) = 1$, then the sets $\{0, a_1^{(i)} + hv, a_2^{(i)} + 2hv, \ldots, a_{k-1}^{(i)} + (k-1)hv\}$, $0 \leqslant h \leqslant w - 1$, $1 \leqslant i \leqslant t$, and $\{0, vb_1^{(i)}, \ldots, vb_{k-1}^{(i)}\}$, $1 \leqslant i \leqslant s$, form a $(vw, k, 1)$ difference system. (Colbourn and Colbourn, 1980).

# 3

# Symmetric designs

## 3.1 The extreme cases

In this chapter we look at symmetric designs, studying in particular Hadamard designs and finite projective planes. These are the 'extreme' symmetric designs in the sense of Theorem 3.1.2 below.

Recall that Fisher's inequality asserts that, if a $(v, b, r, k, \lambda)$ design exists, then necessarily $b \geqslant v$. The designs with $b = v$ (and hence $r = k$) have the additional property (Theorem 1.1.4) that each pair of blocks intersect in $\lambda$ elements; the symmetries between properties involving blocks and elements lead us to call such designs *symmetric*. Recall also that the complementary design of a symmetric $(v, k, \lambda)$ design turns out to be a symmetric $(v, v - k, v - 2k + \lambda)$ design provided that $v - 2k + \lambda > 0$ (Corollary 1.1.7); also, $n = k - \lambda$ is unchanged by complementation.

**Definition 3.1.1** A symmetric $(v, k, \lambda)$ design with $k < v - 1$ is called a *non-trivial* design.

In other words, a *trivial* symmetric design is one with $k = v - 1$. In such a design each block omits just one element; thus, since each element has to occur in a block equally often, the design simply consists of all $(v - 1)$-subsets $\mu$ times, say, where $vk = vr = bk = v\mu k$, i.e. $\mu = 1$. Thus the trivial design consists of each $(v - 1)$-subset once, and $\lambda = k(k - 1)/(v - 1) = k - 1 = v - 2$, so that $v - 2k + \lambda = v - 2(v - 1) + (v - 2) = 0$.

**Lemma 3.1.1** $v - 2k + \lambda > 0$ *for a non-trivial symmetric* $(v, k, \lambda)$ *design.*

**Proof** $\lambda(v - 1) = k(k - 1)$ where $k < v - 1$; so $\lambda < k - 1$. But $\lambda(v - k) = \lambda(v - 1) + \lambda(1 - k) = k(k - 1) - \lambda(k - 1) = (k - 1)(k - \lambda)$; so, if $\lambda < k - 1$, $v - k > k - \lambda$, i.e. $v - 2k + \lambda > 0$.

**Theorem 3.1.2** *Suppose that a non-trivial symmetric* $(v, k, \lambda)$ *design exists, and let* $n = k - \lambda$. *Then*

$$4n - 1 \leqslant v \leqslant n^2 + n + 1$$

**Proof** Let $\lambda' = v - 2k + \lambda$. Then $\lambda' > 0$ by the lemma, and $\lambda + \lambda' = v - 2n$.

Also $\lambda\lambda' = \lambda(v - 2k + \lambda) = \lambda v - 2k\lambda + \lambda^2 = k(k - 1) + \lambda - 2k\lambda + \lambda^2$
$= (k - \lambda)^2 - k + \lambda = n^2 - n = n(n - 1)$.

But $(\lambda + \lambda')^2 \geqslant 4\lambda\lambda'$; so $(v - 2n)^2 \geqslant 4n(n - 1)$. Observe that $4n(n - 1)$ is clearly not a perfect square; so in fact the inequality just proved can be strengthened to a strict inequality: $(v - 2n)^2 > 4n(n - 1)$, so $(v - 2n)^2 \geqslant 4n^2 - 4n + 1 = (2n - 1)^2$. Now $v - 2n = v - 2k + 2\lambda > 0$, so it follows that $v - 2n \geqslant 2n - 1$, i.e. $v \geqslant 4n - 1$.

To prove the other inequality, start with $v - 2k + \lambda \geqslant 1$; then $0 \leqslant v - 2n - \lambda - 1$. Multiplying throughout this inequality by $\lambda - 1$ gives
$$0 \leqslant (\lambda - 1)(v - 2n - \lambda) - (\lambda - 1),$$
i.e. $0 \leqslant \lambda(v - 2n - \lambda) - (v - 2n - 1) = \lambda(v - 2k + \lambda) - (v - 2n - 1)$
$= \lambda\lambda' - (v - 2n - 1)$,
i.e. $0 \leqslant n^2 - n - (v - 2n - 1)$ since $\lambda\lambda' = n(n - 1)$,
i.e. $v \leqslant n^2 + n + 1$.

Having obtained these upper and lower bounds for $v$ in terms of $n = k - \lambda$, the obvious thing to do is to investigate when these bounds can be attained.

**Theorem 3.1.3** *Any symmetric design with* $v = n^2 + n + 1$, $n = k - \lambda$, *is either a finite projective plane of order $n$ or its complement.*

**Proof** Suppose that $v = n^2 + n + 1$. We shall show that $\lambda = 1$ or $\lambda = v - 2n - 1$. Start with

$$v - 2n - 1 = n^2 - n = \lambda\lambda' = \lambda(v - 2n - \lambda) = \lambda(v - 2n - 1) - \lambda(\lambda - 1),$$

which yields $(\lambda - 1)(v - 2n - 1 - \lambda) = 0$, i.e. $\lambda = 1$ or $\lambda = v - 2n - 1$ as required. If $\lambda = 1$ then $k(k - 1) = \lambda(v - 1) = v - 1 = n(n + 1)$, so that $k = n + 1$, and the design is a finite projective plane. If $\lambda = v - 2n - 1$ then $\lambda' = 1$ and so the complementary design is a finite projective plane.

So one extreme of Theorem 3.1.2 corresponds to finite projective planes and their complements. Finite projective planes are in fact the only symmetric designs with $\lambda = 1$; for if $\lambda = 1$ then $k = n + \lambda = n + 1$ and $v - 1 = k(k - 1) = n^2 + n$.

**Theorem 3.1.4** *Any symmetric design with* $v = 4n - 1$, $n = k - \lambda$, *is either a* $(4n - 1, 2n - 1, n - 1)$ *design or its complement.*

**Proof** If $v = 4n - 1$ then $k(k - 1) = \lambda(4n - 2) = 2\lambda(2n - 1) = 2(k - n)(2n - 1)$, so that

$$0 = k^2 - k - k(4n - 2) + 2n(2n - 1)$$
$$= k^2 - k(4n - 1) + 2n(2n - 1) = (k - 2n)(k - 2n + 1);$$

thus $k = 2n$ or $2n - 1$. If $k = 2n - 1$ then $\lambda(v - 1) = k(k - 1)$ yields $\lambda = n - 1$ and we have a $(4n - 1, 2n - 1, n - 1)$ design. If $k = 2n$ then $\lambda = n$ and we have a $(4n - 1, 2n, n)$ design.

**Definition 3.1.2** A $(4n - 1, 2n - 1, n - 1)$ design is called a *Hadamard design of order* $n$.

For example, a $(7, 3, 1)$ design is both a finite projective plane of order 2 and a Hadamard design of order 2; the lower and upper bounds of Theorem 3.1.2 coincide when $n = 2$. The reason for the name 'Hadamard design' is explained in the next section.

When do finite projective planes and Hadamard design exist? It is conjectured that a Hadamard design of order $n$ exists for all $n \geqslant 2$; at the time of writing the smallest open case is $n = 107$. Several methods of constructing Hadamard designs will be given in this chapter. It is also conjectured that a finite projective plane of order $n$ exists if and only if $n$ is a prime power; see Section 3.5 for further discussion.

## 3.2 Hadamard designs and matrices

This section will be concerned with the existence of Hadamard designs. Their existence is related to that of certain matrices.

**Definition 3.2.1** A square $m \times m$ matrix $H$ all of whose entries are $\pm 1$ is called a *Hadamard matrix of order* $m$ if $H'H = mI$.

**Example 3.2.1**

$$\begin{bmatrix} 1 & 1 \\ 1 & -1 \end{bmatrix} \text{ and } \begin{bmatrix} 1 & 1 & -1 & -1 \\ 1 & -1 & -1 & 1 \\ 1 & 1 & 1 & 1 \\ 1 & -1 & 1 & -1 \end{bmatrix}$$

are, respectively, Hadamard matrices of orders 2 and 4.

Hadamard matrices were first studied in the nineteenth century. They were studied by Sylvester in 1867 in connection with a tiling problem; among other things Sylvester saw how to construct Hadamard matrices of orders $2^n$. Then in 1893 Hadamard discovered that if $A = (a_{ij})$ is an $m \times m$ matrix with $|a_{ij}| \leq 1$ for all $i, j$, then the determinant $|A|$ can be at most $m^{m/2}$, this value being attained precisely when $A'A = mI$.

There are several elementary remarks to be made about Definition 3.2.1. The condition $H'H = mI$ essentially asserts (consider non-diagonal entries) that the scalar product of the $i$th row of $H'$ and the $j$th column of $H$ is zero whenever $i \neq j$, i.e. the scalar product of any two different columns of $H$ is zero, i.e. the columns of $H$ are *orthogonal*. Further, $H'H = mI$ asserts that $H$ is invertible with inverse $H^{-1} = (1/m)H'$, so that $H$ and $H'$ must commute; thus $HH' = mI$, and any two columns of $H'$, i.e. any two rows of $H$, are orthogonal. This establishes the following lemma.

**Lemma 3.2.1** *An $m \times m$ $(+1, -1)$ matrix $H$ is a Hadamard matrix precisely when the rows of $H$ are pairwise orthogonal (or, equivalently, when the columns of $H$ are pairwise orthogonal).*

Next observe that the property $H'H = mI$ is unaffected if any row or column is multiplied by $-1$. So, given a Hadamard matrix, we can obtain from it a Hadamard matrix of the same size whose first row and first column consist entirely of +1s. Such a Hadamard matrix is said to be *normalized*. For example, the second matrix in Example 3.2.1 can be normalized by multiplying the last two columns by $-1$.

**Lemma 3.2.2** *Let $H$ be a Hadamard matrix of order $m$ in which the first row consists entirely of $+1$s. Then every other row has $\frac{1}{2}m + 1$s and $\frac{1}{2}m - 1$s. Further, if $m > 2$, each pair of rows apart from the first have $+1$s together in $\frac{1}{4}m$ columns and $-1$s together in $\frac{1}{4}m$ columns.*

**Proof** Let $R_1$ denote the first row, which consists entirely of +1s. Let $R_2$ be any other row and suppose it has $x$ +1s and $y$ −1s. The scalar product of $R_1$ and $R_2$ is then $x - y$ and so, since the rows are pairwise orthogonal, it follows that $x - y = 0$, i.e. $x = y$. But $x + y = m$, so $x = y = \frac{1}{2}m$.

If $m > 2$, let $R_3$ be any other row. Suppose there are $\lambda$ columns in which $R_2$ and $R_3$ both have +1; then there will be $\frac{1}{2}m - \lambda$ columns in which $R_2$ has +1 and $R_3$ has −1. Suppose also there that are $\mu$ columns in which both $R_2$ and $R_3$ have −1; then there are $\frac{1}{2}m - \mu$ columns in which $R_2$ has −1 and $R_3$ has +1. Since $R_3$ must have $\frac{1}{2}m + 1$s, it follows that $\lambda + (\frac{1}{2}m - \mu) = \frac{1}{2}m$; so $\lambda = \mu$. Further, since $R_2$

and $R_3$ are orthogonal, $0 = 1 \cdot 1$ ($\lambda$ times) $+ 1 \cdot (-1)(\frac{1}{2}m - \lambda$ times) $+ (-1) \cdot (-1)(\mu$ times) $+ (-1) \cdot 1(\frac{1}{2}m - \mu$ times), i.e.

$$0 = \lambda - (\tfrac{1}{2}m - \lambda) + \mu - (\tfrac{1}{2}m - \mu) = 2\lambda + 2\mu - m = 4\lambda - m;$$

so $\lambda = \mu = \frac{1}{4}m$.

As an immediate consequence we have the following important theorem.

**Theorem 3.2.3** *If a Hadamard matrix of order $m$ exists then $m = 2$ or $m = 4n$ for some positive integer $n$.*

It is conjectured that a Hadamard matrix of order $m$ does in fact exist whenever $m$ is a multiple of 4. This conjecture is equivalent to the one for Hadamard designs at the end of section 3.1, in view of the following result which establishes the connection between Hadamard designs and matrices.

**Theorem 3.2.4** *A Hadamard $(4n - 1, 2n - 1, n - 1)$ design exists if and only if a Hadamard matrix of order $4n$ exists.*

**Proof** Suppose first that a Hadamard matrix of order $4n$ exists. Take a normalized Hadamard matrix $H$ of this order, remove the first row and column, and in the resulting matrix replace every $-1$ by 0. The matrix $A$ thus obtained is a $(0, 1)$ matrix; interpret it as the incidence matrix of a design. By Lemma 3.2.2, each row of $A$ has $2n - 1$ 1s, so each block in the design has size $2n - 1$. Further, since $H'$ satisfies the conditions of Lemma 3.2.2, any two rows of $H'$ (i.e., any two columns of $H$) have $+1$s together $n$ times, so any two columns of $A$ have 1s together $n - 1$ times. Thus in the design of which $A$ is the incidence matrix, any two elements occur together in $n - 1$ blocks. Thus $A$ is the incidence matrix of a $(4n - 1, 2n - 1, n - 1)$ design.

Conversely, suppose that a $(4n - 1, 2n - 1, n - 1)$ design exists. Take its incidence matrix $A$, replace every 0 by $-1$ and add a further row and column of $+1$s. This gives a $(+1, -1)$ matrix $H$ of order $4n$. Consider the scalar product of the $i$th and $j$th columns, $i \neq j$. These columns have $+1$s together $1 + (n - 1) = n$ times since $\lambda = n - 1$ in the design. Also, each column contains $r + 1 = 2n + 1$s, so there are $2(2n - n) = 2n$ places where the two columns differ, and the scalar product of the two columns is therefore $2n(+1) + 2n(-1) = 0$. It now follows from Lemma 3.2.1 that $H$ is a Hadamard matrix.

**Example 3.2.2** Start with the seven-point plane with incidence matrix $A$

below to obtain the Hadamard matrix $H$ of order 8. In the second of these matrices, $+, -$ stand for $+1, -1$ respectively.

$$A = \begin{bmatrix} 1 & 1 & 0 & 1 & 0 & 0 & 0 \\ 0 & 1 & 1 & 0 & 1 & 0 & 0 \\ 0 & 0 & 1 & 1 & 0 & 1 & 0 \\ 0 & 0 & 0 & 1 & 1 & 0 & 1 \\ 1 & 0 & 0 & 0 & 1 & 1 & 0 \\ 0 & 1 & 0 & 0 & 0 & 1 & 1 \\ 1 & 0 & 1 & 0 & 0 & 0 & 1 \end{bmatrix}$$

$$H = \begin{bmatrix} + & + & + & + & + & + & + & + \\ + & + & + & - & + & - & - & - \\ + & - & + & + & - & + & - & - \\ + & - & - & + & + & - & + & - \\ + & - & - & - & + & + & - & + \\ + & + & - & - & - & + & + & - \\ + & - & + & - & - & - & + & + \\ + & + & - & + & - & - & - & + \end{bmatrix}.$$

The equivalence of matrices and designs as expressed by Theorem 3.2.4 provides a two-prolonged attack on the problem of constructing Hadamard matrices or designs. One method of constructing Hadamard designs is to construct $(4n - 1, 2n - 1, n - 1)$ difference sets and then use Theorem 2.1.3. Thus, for example, Corollary 2.1.7 shows that $(4n - 1, 2n - 1, n - 1)$ designs exist whenever $4n - 1$ is a prime power.

**Theorem 3.2.5** *If $q \equiv 3$ (mod 4) is a prime power then there exists a Hadamard matrix of order $q + 1$.*

This result was first proved in the case when $q$ is a prime by Scarpis in 1898; it was then proved for prime powers by Paley in 1933.

**Example 3.2.3** The squares (mod 11) are $1, 3, 4, 5, 9$ and they form a $(11, 5, 2)$ difference set whose translates are the blocks of a cyclic $(11, 5, 2)$ symmetric design. The first row of the incidence matrix of the design is

$$1 \quad 0 \quad 1 \quad 1 \quad 1 \quad 0 \quad 0 \quad 0 \quad 1 \quad 0 \quad 0$$

and the matrix yields a Hadamard matrix (of order 12) as in the previous example.

Scarpis also showed that, if $q \equiv 1$ (mod 4) is a prime, then a Hadamard matrix of order $2(q + 1)$ exists. This result was also generalized by Paley to prime powers; this will be our Theorem 3.3.5.

## 3.3 Constructing Hadamard Matrices

Although it has not yet been shown that Hadamard matrices of order $4n$ exist for all $n \geqslant 1$, many infinite families of Hadamard matrices have been constructed. One such family was given in Theorem 3.2.5. It is the aim of the section to describe some of the other constructions.

It has already been remarked that Sylvester showed the existence of Hadamard matrices of orders $2^m$ for all $m \geqslant 1$, thereby establishing the first infinite family. There is a very simple construction of such matrices. Let

$$H_0 = [1], \qquad H_1 = \begin{bmatrix} 1 & 1 \\ 1 & -1 \end{bmatrix}$$

and, in general, for $n \geqslant 1$ define

$$H_n \begin{bmatrix} H_{n-1} & H_{n-1} \\ H_{n-1} & -H_{n-1} \end{bmatrix}.$$

**Theorem 3.3.1** *$H_n$ is a Hadamard matrix of order $2^n$.*

**Proof** Proceed by induction on $n$. $H_n$ is certainly a square $(+1, -1)$ matrix of order $2^n$. Consider any two rows $R_i, R_j$ of $H_n, i > j$. Let $\mathbf{x}_i$ denote the first half of $R_i$ and $\mathbf{y}_i$ the second half; similarly for $\mathbf{x}_j$ and $\mathbf{y}_j$. Then, if $i \neq j + 2^{n-1}$, $\mathbf{x}_i \cdot \mathbf{x}_j = 0$ and $\mathbf{y}_i \cdot \mathbf{y}_j = 0$, so that $R_i$ and $R_j$ are orthogonal. If $i = j + 2^{n-1}$ then $\mathbf{x}_i = \mathbf{x}_j$ and $\mathbf{y}_i = -\mathbf{y}_j$, so that the scalar product of $R_i$ and $R_j$ is

$$\mathbf{x}_i \cdot \mathbf{x}_j + \mathbf{y}_i \cdot \mathbf{y}_j = \mathbf{x}_i \cdot \mathbf{x}_i - \mathbf{y}_i \cdot \mathbf{y}_i = 2^{n-1} - 2^{n-1} = 0.$$

**Example 3.3.1** $H_3$ is the following Hadamard matrix of order 8.

| + | + | + | + | + | + | + | + |
|---|---|---|---|---|---|---|---|
| + | − | + | − | + | − | + | − |
| + | + | − | − | + | + | − | − |
| + | − | − | + | + | − | − | + |
| + | + | + | + | − | − | − | − |
| + | − | + | − | − | + | − | + |
| + | + | − | − | − | − | + | + |
| + | − | − | + | − | + | + | −. |

**Example 3.3.2** Take the right-hand half of $H_6$ and replace each $-1$ by 0. Equivalently, take the rows of $H_5$ and replace each $-1$ by 0 (thus obtaining 32 $(0, 1)$ sequences of length 32) and then also take their complements. This gives 64 $(0, 1)$ sequences of length 32, each containing 16 1s and 16 0s (except for one sequence with 32 1s and one with 32 0s).

Any two of these sequences differ in 16 or 32 places. These 64 sequences were taken as the codewords of the Mars Mariner 9 code, used in sending photographs of Mars back to Earth in 1971. Since any two codewords differ in at least 16 places, seven 'errors' can be made in a codeword and the resulting sequence is still closer to the original codeword than to any other.

This example touches on a very important area of combinatorics, namely the study of error-correcting codes. We make no attempt to do justice to this topic in this book, referring the reader to one of the many textbooks devoted to the subject, such as Hill (1986).

The construction of Theorem 3.3.1 is a special case of a more general one.

**Definition 3.3.1** If $A = (a_{ij})$ and $B$ are square matrices of orders $m, n$ respectively, then the *Kronecker product* $A \otimes B$ is defined to be the $mn \times mn$ matrix

$$\begin{bmatrix} a_{11}B & a_{12}B & \cdots & a_{1m}B \\ a_{21}B & a_{22}B & \cdots & a_{2m}B \\ \vdots & & & \\ a_{m1}B & a_{m2}B & \cdots & a_{mm}B \end{bmatrix}.$$

For example, $H_n$ is just the Kronecker product $H_1 \otimes H_{n-1}$.

**Theorem 3.3.2** *If $A$ and $B$ are Hadamard matrices then so is $A \otimes B$.*

**Proof** $C = A \otimes B$ is certainly a $(+1, -1)$ matrix. Further,

$$CC' = \begin{bmatrix} a_{11}B & \cdots & a_{1m}B \\ \vdots & & \\ a_{m1}B & \cdots & a_{mm}B \end{bmatrix} \begin{bmatrix} a_{11}B' & \cdots & a_{m1}B' \\ \vdots & & \\ a_{1m}B' & \cdots & a_{mm}B' \end{bmatrix}$$

so that the $(i, j)$th block of $CC'$ is

$$(a_{i1}B \cdots a_{im}B) \cdot (a_{j1}B' \cdots a_{jm}B') = a_{i1}a_{j1}BB' + \cdots + a_{im}a_{jm}BB'$$

$$= (a_{i1}a_{j1} + \cdots + a_{im}a_{jm})nI_n$$

$$= \begin{cases} 0 & \text{if } i \neq j; \\ mnI_n & \text{if } i = j. \end{cases}$$

Thus

$$CC' = \begin{bmatrix} mnI_n & & 0 \\ & \ddots & \\ 0 & & mnI_n \end{bmatrix} = mnI_{mn}.$$

**Corollary 3.3.3** *If there exist Hadamard matrices of orders $m, n$ then there exists a Hadamard matrix of order mn.*

Using this result and Theorem 3.2.5 we can now construct Hadamard matrices of orders 4, $8 = 2 \times 4$, $12 = 11 + 1$, $16 = 4 \times 4$, $20 = 19 + 1$, $24 = 23 + 1$, $28 = 27 + 1$, $32 = 2 \times 16$, $40 = 2 \times 20$, $44 = 43 + 1$, $48 = 47 + 1$, $56 = 2 \times 28$, $60 = 59 + 1$, $64 = 2 \times 32$, $68 = 67 + 1$, $72 = 71 + 1$, $80 = 79 + 1$, $84 = 83 + 1$, $88 = 2 \times 44$, $96 = 2 \times 48$. The missing values $4n \leqslant 100$ are $4n = 36, 52, 76, 92, 100$. All of these except 92 are of the form $2(q + 1)$ where $q$ is a prime power, $q \equiv 1 \pmod 4$, and can be dealt with by the next theorem. Note however that there are other ways of dealing with 36; for example see Exercises 3, number 7. The case $4n = 92$ will then be dealt with using a method introduced by Williamson in 1944.

The proof of the next construction of Hadamard matrices involves both Kronecker products and the properties of squares in finite fields. Let $Q$ and $R$ denote respectively the sets of non-zero squares and non-squares in $\mathrm{GF}(q)$, and define the quadratic character $\chi$ as follows.

**Definition 3.3.2**   $\chi$ is defined on $\mathrm{GF}(q)$ by

$$\chi(0) = 0$$
$$\chi(x) = 1 \qquad \text{if } x \in Q$$
$$\chi(x) = -1 \qquad \text{if } x \in R.$$

Thus, if $\theta$ is a primitive element of $\mathrm{GF}(q)$, $\chi(\theta^i) = (-1)^i$, and, in general, $\chi(ab) = \chi(a)\chi(b)$.

**Lemma 3.3.4**   *Let $q \equiv 1 \pmod 4$ and $\mathrm{GF}(q) = \{a_1, a_2, \ldots, a_{q-1}, a_q = 0\}$. Let $\chi$ be the quadratic character of $\mathrm{GF}(q)$ and define the $q \times q$ matrix $S = (s_{ij})$ by $s_{ij} = \chi(a_i - a_j)$. Then*

  (i) *$S$ is a symmetric matrix;*
 (ii) *each row of $S$ has $\frac{1}{2}(q - 1)$ +1s and $\frac{1}{2}(q - 1)$ −1s, and hence has sum 0;*
(iii) *$SS' = qI_q - J_q$;*
 (iv) *if*

$$K = \begin{bmatrix} 0 & 1 & 1 & \cdots & 1 \\ 1 & & & & \\ \vdots & & & S & \\ 1 & & & & \end{bmatrix}$$

  *then $KK' = qI_{q+1}$.*

**Proof**

(i) Since $q \equiv 1 \pmod 4$ it follows from Theorem 1.6.2 that $\chi(x) = \chi(-x)$ for all $x \in \mathrm{GF}(q)$. Thus $s_{ij} = \chi(a_i - a_j) = \chi(a_j - a_i) = s_{ji}$.

(ii) For fixed $i$, $\chi(a_i - a_j) = 1$ precisely when $a_i - a_j \in Q$, i.e., when $a_i = a_j + y$ for some $y \in Q$. There are $\frac{1}{2}(q - 1)$ choices of $y$ and hence $\frac{1}{2}(q - 1)$ choices of $j$.

(iii) First of all, the $(i, i)$th entry in $SS'$ is $\sum_{j=1}^{q} s_{ij}^2 = \sum_{j=1}^{q} \chi^2(a_i - a_j) = q - 1$. Next, if $i \neq j$, the $(i, j)$th entry of $SS'$ is the product of the $i$th and $j$th rows of $S$, i.e.

$$\sum_{h=1}^{q} s_{ih} s_{jh} = \sum_{h=1}^{q} \chi(a_i - a_h)\chi(a_j - a_h) = \sum_{l=1}^{q} \chi(a_l)\chi(a_l + a), \quad (3.1)$$

where $a = a_j - a_i$; for, as $a_h$ varies over $\mathrm{GF}(q)$ so does $a_l = a_j - a_h$. Now each $a_l + a$ $(l \neq q)$ can be written as $c_l a_l$ for some $c_l$, namely $c_l = (a_l + a)a_l^{-1}$, and $c_l \neq c_k$ if $l \neq k$. Thus (3.1) is

$$\sum_{l=1}^{q} \chi(a_l)\chi(a_l)\chi(c_l) = \sum_{l=1}^{q-1} \chi(c_l).$$

Since $c_l$ takes every value except 1, this is just $\sum_{i=1}^{q} \chi(a_i) - 1 = -1$. Thus the diagonal entries in $SS'$ are $q - 1$ and the off-diagonal entries are $-1$. So $SS' = qI - J$ as required.

(iv) We have

$$KK' = \begin{bmatrix} 0 & 1 & \cdots & 1 \\ 1 & & & \\ \vdots & & S & \\ 1 & & & \end{bmatrix} \begin{bmatrix} 0 & 1 & \cdots & 1 \\ 1 & & & \\ \vdots & & S' & \\ 1 & & & \end{bmatrix}$$

$$= \begin{bmatrix} q & 0 & \cdots & 0 \\ 0 & & & \\ \vdots & & SS' + J & \\ 0 & & & \end{bmatrix} \quad \text{by (ii)}$$

$$= qI_{q+1} \quad \text{by (iii)}.$$

We can now prove the following result of Paley (1933).

**Theorem 3.3.5** *If $q \equiv 1$ (mod 4) is a prime power then there exists a Hadamard matrix of order $2(q + 1)$.*

**Proof** Take

$$H = \begin{bmatrix} K+I & K-I \\ K-I & -K-I \end{bmatrix}$$

where $K$ is as in the lemma.

**Example 3.3.3** Take $q = 5$. Then

$$K = \begin{bmatrix} 0 & 1 & 1 & 1 & 1 & 1 \\ 1 & 0 & 1 & -1 & -1 & 1 \\ 1 & 1 & 0 & 1 & -1 & -1 \\ 1 & -1 & 1 & 0 & 1 & -1 \\ 1 & -1 & -1 & 1 & 0 & 1 \\ 1 & 1 & -1 & -1 & 1 & 0 \end{bmatrix},$$

yielding the following Hadamard matrix of order 12.

```
+  +  +  +  +  +  -  +  +  +  +  +
+  +  +  -  -  +  +  -  +  -  -  +
+  +  +  +  -  -  +  +  -  +  -  -
+  -  +  +  +  -  +  -  +  -  +  -
+  -  -  +  +  +  +  -  -  +  -  +
+  +  -  -  +  +  +  +  -  -  +  -
-  +  +  +  +  +  -  -  -  -  -  -
+  -  +  -  -  +  -  -  -  +  +  -
+  +  -  +  -  -  -  -  -  -  +  +
+  -  +  -  +  -  -  +  -  -  -  +
+  -  -  +  -  +  -  +  +  -  -  -
+  +  -  -  +  -  -  -  +  +  -  -.
```

As already mentioned, Theorem 3.3.5 and the earlier results enable us to construct Hadamard matrices of orders $4n$ for all $4n \leqslant 100$, except 92. We now describe a method due to Williamson which will deal with this case. The method is interesting from the number theoretic point of view, for it mimics the classical result of Lagrange that every integer can be expressed as a sum of four squares (see Hardy and Wright, 1979). For example,

$$35 = 5^2 + 3^2 + 1^2 + 0^2, \quad 15 = 3^2 + 2^2 + 1^2 + 1^2.$$

In general, if $m$ is a positive integer there exist integers $a, b, c, d$ such that

$$m = a^2 + b^2 + c^2 + d^2.$$

If we now define the matrix $H$ by

$$H = \begin{bmatrix} -a & b & c & d \\ b & a & d & -c \\ c & -d & a & b \\ d & c & -b & a \end{bmatrix}$$

then

$$HH' = (a^2 + b^2 + c^2 + d^2)I_4 = mI_4. \tag{3.2}$$

Williamson's idea was to replace the integers $a, b, c, d$ by $n \times n$ matrices $A, B, C, D$. Let

$$H = \begin{bmatrix} -A & B & C & D \\ B & A & D & -C \\ C & -D & A & B \\ D & C & -B & A \end{bmatrix}. \tag{3.3}$$

Then, if $A, B, C, D$ are symmetric and commute with each other, it is easy to check that

$$HH' = I_4 \otimes (A^2 + B^2 + C^2 + D^2).$$

If $A, B, C, D$ can be chosen so that

$$A^2 + B^2 + C^2 + D^2 = 4nI_n$$

then it will follow that $HH' = 4nI_{4n}$, i.e. $H$ is a Hadamard matrix of order $4n$.

It might appear that just too much is being demanded of $A, B, C, D$ for the method to be useful, but in special cases it is indeed useful.

**Example 3.3.4**  Take

$$A = \begin{bmatrix} 1 & 1 & 1 \\ 1 & 1 & 1 \\ 1 & 1 & 1 \end{bmatrix}, \quad B = C = D = \begin{bmatrix} 1 & -1 & -1 \\ -1 & 1 & -1 \\ -1 & -1 & 1 \end{bmatrix}.$$

Then $A, B, C, D$ are symmetric and commute with each other; further, $A^2 = 3A$ and $B^2 = C^2 = D^2 = 4I - A$ so that $A^2 + B^2 + C^2 + D^2 = 12I_3$. Thus a Hadamard matrix of order 12 is obtained from eqn (3.3).

For larger values of $n$, Williamson simplified his search by looking for *circulant* matrices. Any circulant matrix

$$\begin{bmatrix} a_1 & a_2 & \cdots & a_n \\ a_n & a_1 & \cdots & a_{n-1} \\ \vdots & & & \\ a_2 & a_3 & \cdots & a_1 \end{bmatrix}$$

can be written as $a_1 I + a_2 K + a_3 K^2 + \cdots + a_n K^{n-1}$, where $K$ is the permutation matrix

$$K = \begin{bmatrix} 0 & 1 & 0 & 0 & \cdots & 0 \\ 0 & 0 & 1 & 0 & \cdots & 0 \\ 0 & 0 & 0 & & \cdots & 0 \\ \vdots & & & & & \\ 1 & 0 & 0 & 0 & \cdots & 0 \end{bmatrix}.$$

Since polynomials in $K$ clearly commute, the required commutative property is automatically satisfied. The symmetric property requires $a_2 = a_n$, $a_3 = a_{n-1}, \ldots$. Williamson dealt with the case $n = 43$, but here we present the case $n = 23$.

**Example 3.3.5** (Baumert *et al.*, 1962) Consider $n = 23$ in order to construct a Hadamard matrix of order 92. A solution is provided by choosing $A, B, C, D$ to be the $23 \times 23$ circulant matrices whose first rows are given by:

```
A:  + + − − − + − − − + − + + − + − − − + − − − +
B:  + − + + − + + − − + + + + + + − − + + − + + −
C:  + + + − − − + + − + − + + − + − + + − − − + +
D:  + + + − + + + − + − − − − − − + − + + + − + +.
```

This method has been used to find Hadamard matrices of order 92, 116, 172. Then in 1972 Turyn used the method for infinitely many orders: details can be found in Hall (1986).

Another possible method of constructing Hadamard matrices is worth mentioning because of its use of symmetric designs which are not Hadamard designs. The idea here is to take the incidence matrix $A$ of a symmetric design and replace each 0 by $-1$. When is the resulting matrix Hadamard? Observe that replacing each 0 by $-1$ is equivalent to forming the matrix $H = 2A - J$.

**Lemma 3.3.6** *If $A$ is the incidence matrix of a symmetric $(v, k, \lambda)$ design then $H = 2A - J$ is Hadamard if and only if $v = 4(k - \lambda)$.*

**Proof** $HH' = (2A - J)(2A' - J) = 4AA' - 2AJ - 2JA' + J^2$

$$= 4((k - \lambda)I + \lambda J) - 2kJ - 2kJ + vJ$$
$$= 4(k - \lambda)I + (v - 4(k - \lambda))J$$
$$= vI \text{ if and only if } v = 4(k - \lambda).$$

**Example 3.3.6** Take the $(4,3,2)$ design consisting of the four blocks $\{b, c, d\}, \{a, c, d\}, \{a, b, d\}, \{a, b, c\}$. Then

$$A = \begin{bmatrix} 0 & 1 & 1 & 1 \\ 1 & 0 & 1 & 1 \\ 1 & 1 & 0 & 1 \\ 1 & 1 & 1 & 0 \end{bmatrix}$$

yields the Hadamard matrix

$$H = \begin{bmatrix} -1 & 1 & 1 & 1 \\ 1 & -1 & 1 & 1 \\ 1 & 1 & -1 & 1 \\ 1 & 1 & 1 & -1 \end{bmatrix}$$

which, incidentally, is the only circulant Hadamard matrix known.

It is straightforward (Exercises 3, number 7) to show that the condition $v = 4(k - \lambda)$ is satisfied precisely when $v = 4m^2$, $k = 2m^2 \pm m$, $\lambda = m^2 \pm m$, for some integer $m$. In the case $m = 3$ a Hadamard matrix of order 36 can be obtained from a $(36, 15, 6)$ design. Conversely, a $(36, 15, 6)$ design can be constructed by first constructing a Hadamard matrix.

**Example 3.3.7** (Spence, 1971) In the Williamson construction, take $A, B, C, D$ to be the circulant matrices with first rows as follows:

$$
\begin{array}{llllllllllll}
A: & + & - & - & - & + & + & - & - & - \\
B: & + & - & - & + & - & - & + & - & - \\
C: & + & - & + & - & - & - & - & + & - \\
D: & + & + & - & - & - & - & - & - & + \, .
\end{array}
$$

Then the matrix $H$ in eqn (3.3) has $15 + 1$s in each row and column, so that $HJ = -6J = JH$. Let $K = \frac{1}{2}(H + J)$; then $K$ is a $(0, 1)$ matrix such that

$$KK' = \tfrac{1}{4}(H + J)(H' + J) = \tfrac{1}{4}(36I - 6J - 6J + 36J)$$
$$= 9I + 6J,$$

and

$$JK = \tfrac{1}{2}(-6J + 36J) = 15J = KJ.$$

So $K$ is the incidence matrix of a $(36, 15, 6)$ design.

Clearly not every Hadamard matrix of order $4m$ yields a symmetric design; the Hadamard matrix has to be *regular*, i.e. have constant row sums. But if the matrix is regular, the method does indeed work (see Exercises 3, number 8).

## 3.4  Finite projective planes

It has been noted that symmetric designs with $\lambda = 1$ have parameters $v = n^2 + n + 1$, $k = n + 1$ for some integer $n \geq 2$, and are called *finite projective planes*. Kirkman established the existence of such designs for all prime values of $n$ (see Theorem 4.5.4), and in the early twentieth century Veblen and Bussey (1906) showed how to construct such designs whenever $n$ is a prime power. These designs, derived from projective geometry, were shown to be cyclic by Singer in 1939. Kirkman, it will be recalled, had presented $(q^2 + q + 1, q + 1, 1)$ difference sets for several prime values of $q$, each yielding a cyclic plane (Example 2.1.3). Instead of giving the usual projective geometry construction, we now present the proof due to Bose and Chowla (1962) of the existence of these difference sets for all prime powers $q$.

**Theorem 3.4.1**  *If $q$ is a prime power, there exists a cyclic $(q^2 + q + 1, q + 1, 1)$ difference set.*

**Proof**  Let $q = p^\alpha$ and let $\theta$ be a primitive element of $GF(q^3)$. Then inside $GF(q^3)$ the elements $0, \theta^u, \theta^{2u}, \ldots, \theta^{(q-1)u}$, where $u = (q^3 - 1)/(q - 1) = q^2 + q + 1$, constitute a copy $G$ of $GF(q)$.

For each $b_i$ in $G$, $1 \leq i \leq q$, there is a unique $a_i$, $0 \leq a_i \leq q^3 - 1$, such that $\theta + b_i = \theta^{a_i}$. We first of all prove that these $a_i$ are all distinct (mod $q^2 + q + 1$). Suppose that $\theta + b_1 = \theta^{a_1}$ and $\theta + b_2 = \theta^{a_2}$, where $a_1 = m(q^2 + q + 1) + a_2$. Then

$$\theta^{a_1} = \theta^{a_2} \theta^{m(q^2 + q + 1)} = \theta^{a_2} w \quad \text{for some } w \in G.$$

Thus $\theta + b_1 = w(\theta + b_2)$, where $b_1, b_2, w \in G$. But this forces $\theta$ to be in $G$, contradicting the fact that $\theta$ generates $GF(q^3)$.

So all the $a_i$ are distinct (mod $q^2 + q + 1$). Further, they are all non-zero; for if $\theta + b_i = \theta^0 = 1$, then again $\theta$ would be in $G$.

We now put $a_{q+1} = 0$ and consider

$$D = \{a_1, \ldots, a_{q+1}\}.$$

We claim that $D$ is a cyclic $(q^2 + q + 1, q + 1, 1)$ difference set. To prove this, suppose that $a_i - a_j \equiv a_k - a_l \not\equiv 0$, $a_i \neq a_k$. Then $a_i + a_l \equiv a_j + a_k$. If none of these $a_i$ is 0, we have

$$(\theta + b_i)(\theta + b_l) = (\theta + b_j)(\theta + b_k)g \qquad (3.4)$$

for some $g \in G$, and if this is not an identity then $\theta$ satisfies an equation of degree $\leqslant 2$ over $G$ and hence cannot generate $GF(q^3)$. So eqn (3.4) is an identity, $g = 1$, $b_i + b_l = b_j + b_k$ and $b_i b_l = b_j b_k$. Thus $b_i(b_j + b_k - b_i) = b_j b_k$, whence $(b_i - b_k)(b_j - b_i) = 0$. Thus $b_i = b_k$ or $b_j = b_i$, a contradiction. Finally, if $a_i = 0$ then $\theta + b_l = (\theta + b_j)(\theta + b_k)g$ and again $\theta$ satisfies a quadratic equation over $G$, giving a contradiction.

**Example 3.4.1** Take $q = 3$, $q^3 = 27$; take $\theta$ satisfying $\theta^3 + 2\theta + 1 = 0$ (see Section 1.6), i.e. $\theta^3 = \theta + 2$ over $GF(3)$. Since $\theta + 0 = \theta^1$, $\theta + 1 = \theta^9$ and $\theta + 2 = \theta^3$, we obtain $D = \{0, 1, 3, 9\}$ as a $(13, 4, 1)$ difference set.

**Example 3.4.2** Take $q = 4$, $q^3 = 64$. Take $\theta$ satisfying $\theta^6 = \theta + 1$. Here $GF(4) = \{0, \theta^{21}, \theta^{42}, \theta^{63} = 1\}$ and, by using $\theta^6 = \theta + 1$, we can show that $\theta^{21} = \theta^5 + \theta^4 + \theta^3 + \theta + 1$, so that $\theta + \theta^{21} = \theta^5 + \theta^4 + \theta^3 + 1 = \theta^{60}$. Similarly we get $\theta + \theta^{42} = \theta^{29}$, $\theta + 0 = \theta^1$, $\theta + 1 = \theta^6$. Since $60 \equiv 18$ and $29 \equiv 8$ (mod 21) we obtain the $(21, 5, 1)$ difference set $D = \{0, 1, 6, 8, 18\}$.

### 3.5 The Bruck–Chowla–Ryser theorem

In this section we consider further the question of the existence of symmetric $(v, k, \lambda)$ designs. For such a design to exist the condition $\lambda(v - 1) = k(k - 1)$ must be satisfied, but this necessary condition is not in itself sufficient to ensure the existence of a design. This is apparent from the following result, first proved independently by Schutzenberger (1949) and Shrikhande (1950).

**Theorem 3.5.1** *If a symmetric $(v, k, \lambda)$ design exists with $v$ even, then $k - \lambda$ must be a perfect square.*

**Proof** Let $A$ be the incidence matrix so that, as in the proof of Fisher's inequality, $|A|^2 = k^2(k - \lambda)^{v-1}$. Thus $|A| = k(k - \lambda)^{(v-1)/2}$. Since $|A|$ and $k$ are integers, $(k - \lambda)^{(v-1)/2}$ must be rational. Since $v - 1$ is odd, this can happen only if $k - \lambda$ is a perfect square.

**Example 3.5.1** No $(22, 7, 2)$ design can exist, although $2(22 - 1) = 7 \cdot 6$, since 22 is even and $7 - 2 = 5$ is not a square.

Thus certain potential sets of parameters $(v, k, \lambda)$ with $v$ even can be ruled out. Odd values of $v$ are much harder to deal with, but the following result is of great importance. It is the combined work of Bruck and Ryser (1949) and Chowla and Ryser (1950).

**Theorem 3.5.2** (Bruck–Chowla–Ryser theorem)  *Suppose that a symmetric $(v, k, \lambda)$ design exists with $v$ odd. Then the equation*

$$z^2 = (k - \lambda)x^2 + (-1)^{(v-1)/2}\lambda y^2$$

*has a non-trivial integer solution $x, y, z$.*

**Example 3.5.2**  Suppose a $(29, 8, 2)$ design exists. Then there is a non-trivial integer solution of the equation

$$z^2 = 6x^2 + 2y^2.$$

Suppose we take such a solution and cancel out any common factor. Since $z$ must be even, put $z = 2w$ so that $2w^2 = 3x^2 + y^2$; thus $y^2 + w^2 \equiv 0 \pmod{3}$. But $0^2 \equiv 0$, $1^2 \equiv 2^2 \equiv 1 \pmod 3$, so $y^2 + w^2 \equiv 0 \pmod 3$ only if $y \equiv w \equiv 0 \pmod 3$. Thus $y = 3t$, $w = 3u$, so $6u^2 = x^2 + 3t^2$. So $x$ is also a multiple of 3, a contradiction. So no $(29, 8, 2)$ design can exist.

We do not give a proof of the Bruck–Chowla–Ryser theorem, but refer the reader to Anderson (1990) or Hall (1986). But we close this section by considering what BCR has to tell us about finite projective planes. If a $(n^2 + n + 1, n + 1, 1)$ design exists, then the equation

$$z^2 = nx^2 + (-1)^{n(n+1)/2}y^2$$

must have a non-trivial integer solution. Now $\frac{1}{2}n(n + 1)$ is odd if $n \equiv 1$ or 2 (mod 4) and is even if $n \equiv 0$ or 3 (mod 4). Thus in the case of $n \equiv 0$ or 3, the equation to be solvable is

$$z^2 = nx^2 + y^2,$$

which *does* have the non-trivial solution $y = z = 1$, $x = 0$. So no information is obtained in this case. However, if $n \equiv 1$ or 2 then the equation is $y^2 + z^2 = nx^2$, i.e.

$$n = \left(\frac{y}{x}\right)^2 + \left(\frac{z}{x}\right)^2,$$

i.e. $n$ must be expressible as the sum of two rational squares. This condition can be re-expressed in terms of the prime factors of $n$, as we now show.

**Lemma 3.5.3**  *Let $n = p_1 \cdots p_r m^2$, where the $p_i$ are distinct primes. Then, if $n = x^2 + y^2$ for rational $x, y$, then $p_i \equiv 1 \pmod 4$ for each $i$.*

**Proof**  Let $x = u/v$, $y = w/z$ be expressed in simplest form. Then $p_1 \cdots p_r m^2 v^2 z^2 = (uz)^2 + (vw)^2$. Consider any $i$, and note that $p_i$ does not divide either of $uz$ or $vw$ since otherwise $p_i$ divides the right-hand side, and hence the left, an even number of times. So $(uz)^2 + (vw)^2 \equiv 0$ (mod $p_i$) where neither of $uz$ and $vw$ is zero (mod $p_i$). Thus $(uz)^2$, which is clearly a square, has the property that its negative is also a square. By Theorem 1.6.2 this only happens if $p_i \equiv 1$ (mod 4).

**Corollary 3.5.4** *If $n \equiv 1$ or 2 (mod 4) and n is divisible by a prime $p \equiv 3$ (mod 4) to an odd power, then no finite projective plane of order n exists.*

**Example 3.5.3** There are no finite projective planes of orders 6, 14, 21, 22, 30.

Note finally that the converse of BCR is not true; for example, no $(111, 11, 1)$ design exists.

## 3.6 Exercises 3

1. Check that, in Chapters 1 and 2, finite projective planes of orders $2, 3, 4, 5, 7, 8, 9$ and Hadamard designs of orders $2, 3, 4, 5, 6$ have been exhibited.
2. Use a $(15, 7, 3)$ design to construct a Hadamard matrix of order 16.
3. A Hadamard matrix $H$ is called *skew-Hadamard* if $H = S + I$ for some skew symmetric matrix $S$. Show that $SS' = (n - 1)I$ must hold, where $n$ is the order of $H$. Show that if $H = S + I$ is skew-Hadamard of order $n$ then
$$K = \begin{bmatrix} S+I & S+I \\ S-I & -S+I \end{bmatrix}$$
   is skew-Hadamard of order $2n$.
4. Verify that the Kronecker product satisfies the following:
   (i) $(A + B) \otimes C = A \otimes C + B \otimes C$;
   (ii) $A \otimes (B + C) = A \otimes B + A \otimes C$;
   (iii) $(A \otimes B)(C \otimes D) = AC \otimes BD$;
   (iv) $(A \otimes B)' = A' \otimes B'$;
   (v) $(A \otimes B) \otimes C = A \otimes (B \otimes C)$.
5. (i) Write down all of the 16 $(+1, -1)$ sequences of length 4. Show that they can be grouped to form four Hadamard matrices of order 4.
   (ii) Use induction to show that the $2^{2^n}$ $(+1, -1)$ sequences of length $2^n$ can be grouped to form $2^{2^n - n}$ Hadamard matrices of order $2^n$. (Paley 1933).
6. Show that the existence of a Hadamard matrix of order $4n$ implies the existence of designs corresponding to each of the following sets of parameters $(v, k, \lambda)$: (i) $(4n - 1, 2n, n)$, (ii) $(2n - 1, n - 1, n - 2)$, (iii) $(2n, n, n - 1)$.

7. (i) If $A$ is a square $(0,1)$ matrix and $B$ is obtained from $A$ by replacing each 0 by $-1$, then $B = 2A - J$.

   (ii) If $A$ is the incidence matrix of a symmetric $(v, k, \lambda)$ design then $B = 2A - J$ is Hadamard if and only if $v = 4(k - \lambda)$: see Lemma 3.3.6. Show that this condition $v = 4(k - \lambda)$ holds only if $v = 4m^2$, $k = 2m^2 \pm m$, $\lambda = m^2 \pm m$ for some integer $m$. (Use Theorem 3.5.1.)

   (iii) Use Exercises 1 number 19 to show how to construct a Hadamard matrix of order 36.

8. A Hadamard matrix $H$ of order $4n$ *is regular* if all of its row and column sums are equal, i.e. if $HJ = JH = tJ$ for some integer $t$.

   (i) Write down an example of a regular Hadamard matrix of order 4.

   (ii) Show that if each $-1$ is replaced by 0, so that $A = \frac{1}{2}(H + J)$ is obtained, then $AA' = nI + (n + \frac{1}{2}t)J$ and $JA = AJ = (2n + \frac{1}{2}t)J$. Deduce that $A$ is the incidence matrix of a symmetric

   $$(4n, 2n + \tfrac{1}{2}t, n + \tfrac{1}{2}t)$$

   design. Using eqn (1.1), deduce that $4n = t^2$, so that $n$ has to be a perfect square: $n = m^2$ gives a $(4m^2, 2m^2 + m, m^2 + m)$ design.

9. Verify that the methods described in the text enable Hadamard matrices of all orders $4n \leqslant 200$, except $116, 156, 172, 188$, to be constructed.

10. By writing $A, B, C, D$ as polynomials in $K$, verify that, in Example 3.3.5, $A^2 + B^2 + C^2 + D^2 = 92I$.

11. How many of the $2^n$ binary sequences of length $n$ can be chosen to form a collection $C$ such that any two sequences in $C$ differ in at least $\frac{1}{2}n$ places? Corrádi and Katai showed that

$$|C| \leqslant \begin{cases} n + 1 & \text{if } n \text{ is odd;} \\ n + 2 & \text{if } n \equiv 2 \pmod 4; \\ 2n & \text{if } n \equiv 0 \pmod 4. \end{cases}$$

Use Hadamard matrices to show that these are best possible bounds. (Corrádi and Katai, 1969).

12. Define a *symmetric conference matrix* $N$ of order $n \equiv 2 \pmod 4$ to be a $(+1, -1)$ matrix $N$ of order $n$ such that $N = R + I$ where $R$ is symmetric, $RR' = (n-1)I$, and $RJ = 0$.

   (i) Verify that the matrix $K + I$, where $K$ is as in Lemma 3.3.4, is a symmetric conference matrix (s.c.m.).

   (ii) Show that if $N = R + I$ is an s.c.m. of order $n$ then

   $$H = \begin{bmatrix} -R - I & R - I \\ R - I & R + I \end{bmatrix}$$

   is a symmetric Hadamard matrix of order $2n$.

(iii) If an s.c.m. is normalized to the form

$$I + \begin{bmatrix} 0 & 1 & \cdots & 1 \\ 1 & & & \\ \cdot & & W & \\ 1 & & & \end{bmatrix},$$

then $W$ is called the *core*. Verify that $WW' = (n-1)I - J$ and that $WJ = JW = 0$.

(iv) Let $N = R + I$ be an s.c.m. of order $n$ with core $W$. Let

$$X = \begin{bmatrix} J & J \\ J & -J \end{bmatrix} \quad \text{and} \quad Y = \begin{bmatrix} W+I & W-I \\ W-I & -W-I \end{bmatrix}$$

so that $XY' + YX' = 0$. Then $H = I_n \otimes X + R \otimes Y$ is a symmetric Hadamard matrix of order $2n(n-1)$.

(v) Deduce from (i) and (iv) that, whenever $q \equiv 1 \pmod 4$ is a prime power, there exists a symmetric Hadamard matrix of order $2q(q+1)$. (J. S. Wallis, 1972).

13. (i) Suppose that in a round robin tournament of $n$ teams, every team gets beaten by at least $m$ other teams (for some fixed value of $m$). Show that $n \geqslant 2m + 1$, with equality if and only if each team is beaten by exactly $m$ teams.

(ii) Suppose that in a round robin tournament, for every pair of teams we can find $m$ other teams which beat them both. Show that each team is beaten by at least $2m + 1$ teams, so that $n \geqslant 4m + 3$. Consider the extreme case where $n = 4m + 3$, and let $S_i$ denote the set of teams which beat the $i$th team. Show that $|S_i| = 2m + 1$ and $|S_i \cap S_j| = m$ whenever $i \neq j$. Deduce that a skew-Hadamard matrix of order $4m + 4$ then exists. (Szekeres and Szekeres, 1965)

14. *Symmetric designs with* $v = 2^m$. Suppose that a symmetric $(v, k, \lambda)$ design exists with $v = 2^m$ for some integer $m$. Assume $k \leqslant \frac{1}{2}v$.

(i) Since $v$ is even, $n = k - \lambda$ must be a square; $n = 2^{2s}n_1^2$ where $n_1$ is odd. Use $k(k-1) = \lambda(v-1)$ to show that $k = 2^s k_1$, $\lambda = 2^s \lambda_1$ where $k_1 \lambda_1$ are odd. Deduce that $\lambda_1 2^{m-2} = (k_1 - n_1)(k_1 + n_1)$ and hence that $k_1 + n_1$ is divisible by $2^{m-s-1}$.

(ii) Deduce that $k_1 + n_1 = 2^{m-s-1}$, $k_1 - n_1 = 2\lambda_1$, so that $n = 2^{m-s}$.

(iii) Thus $m = 2t$ for some $t$, and the design is a $(2^{2t}, 2^{2t-1} \pm 2^{t-1}, 2^{2t-2} \pm 2^{t-1})$ design.

(iv) Such designs will in fact exist if, by Exercises 3, number 8, a regular Hadamard matrix of order $2^{2t}$ exists. That such a matrix exists will be shown later, using orthogonal Latin squares (Exercises 4, number 13). (Mann, 1965)

15. (i) Suppose that the restriction of a finite projective plane $\Pi$ of order $n$ to $m^2 + m + 1$ of its element is a finite projective plane $\Pi_0$ of

order $m$ ($\Pi_0$ is then a *subplane* of $\Pi$). Show that the number of points of $\Pi$ not on a line of $\Pi_0$ is $(n - m)(n - m^2)$, so that either $n = m^2$ or there is a point of $\Pi$ not on any line of $\Pi_0$. Deduce that $n = m^2$ or $n \geqslant m^2 + m$. Any subplane of order $\sqrt{n}$ is called a *Baer subplane*; if $n = q^2$, a Baer subplane always exists.

(ii) Find a Baer subplane in a FPP of order 4.

(iii) Let $\Pi$ be a plane of order $q^2$ and $\Pi_0$ a Baer subplane. Choose any point $P \in \Pi_0$; let $L_i'$ denote the set of $q^2 - q$ points of the line $L_i$ of $\Pi$ which do not lie in $\Pi_0$, and let $L^*$ denote any line of $\Pi$ through $P$ but containing no other point of $\Pi_0$. Let $X = L^* \cup L_0 \cup \cdots \cup L_q$ so that $|X| = q^3 + q^2 - q + 1$. Then the non-trivial intersections of lines of $\Pi$ with $X$ form a PBD($q^3 + q^2 - q + 1$, $\{q + 1, q^2 + 1, q^2 - q + 1\}, 1$) in which there is just one block of size $q^2 + 1$. (This will be used in section 11.5.)

16. Show that the following cannot be the parameters $(v, k, \lambda)$ of a BIBD: $(43, 7, 1)$, $(22, 7, 2)$, $(29, 8, 2)$, $(46, 10, 2)$, $(34, 12, 4)$, $(67, 12, 2)$, $(53, 13, 3)$, $(92, 14, 2)$, $(106, 15, 2)$, $(43, 15, 5)$, $(137, 17, 2)$, $(103, 18, 3)$, $(72, 20, 5)$.

17. Show that none of the following is ruled out by Theorem 3.5.1 or BCR as possible parameters of a symmetric design: $(16, 6, 2)$, $(37, 9, 2)$, $(25, 9, 3)$, $(31, 10, 3)$, $(56, 11, 2)$, $(45, 12, 3)$. (These are all the possibilities with $k \leqslant 12$ apart from Hadamard designs and FPPs; designs exist in each case.)

# 4

# Orthogonal Latin squares

## 4.1 Early results

The main aim of this chapter is to present an account of the work of Bose, Shrikhande, Parker and others which culminated in the 1960 proof that two mutually orthogonal Latin squares (MOLS) of order $n$ exist whenever $n \neq 2$ or 6. The concept of orthogonality of Latin squares was introduced in section 1.3; it was shown there that, if $N(n)$ denotes the largest value of $m$ for which $m$ MOLS of order $n$ exist, then $N(n) \leq n - 1$ for each $n$, and $N(n) = n - 1$ if and only if a finite projective plane of order $n$ exists. It therefore follows from the results of Chapter 3 that $N(q) = q - 1$ whenever $q$ is a prime power.

**Theorem 4.1.1** *If $q$ is a prime power, then there exist $q - 1$ MOLS of order $q$.*

Since this has been established indirectly via Theorems 1.3.5 and 3.4.1, there is strictly no need for a further proof. However, a particularly elegant constructive one is readily available. This proof was published independently by Bose (1938) and Stevens (1939), but, remarkably, the construction is present in E.H. Moore's long 1896 paper.

**Proof** Let $\mathrm{GF}(q) = \{\lambda_1, \lambda_2, \ldots, \lambda_{q-1}, \lambda_q = 0\}$, and define $q \times q$ arrays $A_1, \ldots, A_{q-1}$ by taking the $(i, j)$th entry of $A_k$ to be $\lambda_i \lambda_k + \lambda_j$, $1 \leq k \leq q - 1$. First check that each $A_k$ is a Latin square. No two entries in the $i$th row of $A_k$ are equal, for

$$\lambda_i \lambda_k + \lambda_j = \lambda_i \lambda_k + \lambda_J \Rightarrow \lambda_j = \lambda_J \Rightarrow j = J;$$

and no two entries in the $j$th column of $A_k$ are equal since

$$\lambda_i \lambda_k + \lambda_j = \lambda_I \lambda_k + \lambda_j \Rightarrow \lambda_i \lambda_k = \lambda_I \lambda_k \Rightarrow \lambda_i = \lambda_I$$

(since $\lambda_k^{-1}$ exists) $\Rightarrow i = I$.

So each $A_k$ is certainly a Latin square. Further, the squares are mutually orthogonal. For suppose, for some $k \neq K$, that

$$\lambda_i \lambda_k + \lambda_j = \lambda_I \lambda_k + \lambda_J \quad \text{and} \quad \lambda_i \lambda_K + \lambda_j = \lambda_I \lambda_K + \lambda_J.$$

Subtraction of one equation from the other gives $\lambda_i(\lambda_k - \lambda_K) = \lambda_I(\lambda_k - \lambda_K)$, whence $\lambda_i = \lambda_I$ since $\lambda_k \neq \lambda_K$; thus $i = I$. It now follows that $\lambda_j = \lambda_J$, so that $j = J$.

**Example 4.1.1** Two MOLS of order 3. Take $\lambda_1 = 1$, $\lambda_2 = 2$ and $\lambda_3 = 0$ to obtain

$$\begin{bmatrix} 2 & 0 & 1 \\ 0 & 1 & 2 \\ 1 & 2 & 0 \end{bmatrix} \quad \text{and} \quad \begin{bmatrix} 0 & 1 & 2 \\ 2 & 0 & 1 \\ 1 & 2 & 0 \end{bmatrix},$$

which, on reversing the order of the rows, are equivalent to

$$\begin{bmatrix} 1 & 2 & 3 \\ 3 & 1 & 2 \\ 2 & 3 & 1 \end{bmatrix} \quad \text{and} \quad \begin{bmatrix} 1 & 2 & 3 \\ 2 & 3 & 1 \\ 3 & 1 & 2 \end{bmatrix}. \tag{4.1}$$

**Example 4.1.2** Three MOLS of order 4. Here we use $GF(4) = \{0, 1, x, x^2\}$, where $x^2 = x + 1$ (as in Example 1.6.2). With $\lambda_1 = 1$, $\lambda_2 = x$, $\lambda_3 = x^2$, $\lambda_4 = 0$ we obtain

$$\begin{bmatrix} 0 & x^2 & x & 1 \\ x^2 & 0 & 1 & x \\ x & 1 & 0 & x^2 \\ 1 & x & x^2 & 0 \end{bmatrix}, \quad \begin{bmatrix} x^2 & 0 & 1 & x \\ x & 1 & 0 & x^2 \\ 0 & x^2 & x & 1 \\ 1 & x & x^2 & 0 \end{bmatrix}, \quad \begin{bmatrix} x & 1 & 0 & x^2 \\ 0 & x^2 & x & 1 \\ x^2 & 0 & 1 & x \\ 1 & x & x^2 & 0 \end{bmatrix}$$

which, on reversing the order of rows and replacing $1, x, x^2, 0$ by $1, 2, 3, 4$, are just the squares exhibited in Example 1.3.4.

Note that if the $A_i$ are constructed as in Theorem 4.1.1, each has last row $\lambda_1, \lambda_2, \ldots, \lambda_q$; further, the other rows are permuted cyclically from one square to the next, the $i$th row of $A_k$ being the $(i-1)$th row of $A_{k+1}$ $(1 < i < q)$.

**Example 4.1.3** Four MOLS of order 5. For $1 \leqslant i \leqslant 4$, take $\lambda_i = 2^i$ since 2 is a primitive root (mod 5); $2^1 = 2$, $2^2 = 4$, $2^3 = 3$, $2^4 = 1$. The squares

| | | | | | | | | | | | | | | | | | | | |
|---|---|---|---|---|---|---|---|---|---|---|---|---|---|---|---|---|---|---|---|
| 1 | 3 | 2 | 0 | 4 | | 0 | 2 | 1 | 4 | 3 | | 3 | 0 | 4 | 2 | 1 | | 4 | 1 | 0 | 3 | 2 |
| 0 | 2 | 1 | 4 | 3 | | 3 | 0 | 4 | 2 | 1 | | 4 | 1 | 0 | 3 | 2 | | 1 | 3 | 2 | 0 | 4 |
| 3 | 0 | 4 | 2 | 1 | | 4 | 1 | 0 | 3 | 2 | | 1 | 3 | 2 | 0 | 4 | | 0 | 2 | 1 | 4 | 3 |
| 4 | 1 | 0 | 3 | 2 | | 1 | 3 | 2 | 0 | 4 | | 0 | 2 | 1 | 4 | 3 | | 3 | 0 | 4 | 2 | 1 |
| 2 | 4 | 3 | 1 | 0 | | 2 | 4 | 3 | 1 | 0 | | 2 | 4 | 3 | 1 | 0 | | 2 | 4 | 3 | 1 | 0 |

are obtained. If we now relabel 2, 4, 3, 1, 0 by 1, 2, 3, 4, 5 and take the rows in reverse order we obtain

| 1 2 3 4 5 | 1 2 3 4 5 | 1 2 3 4 5 | 1 2 3 5 5 |
|---|---|---|---|
| 2 4 5 3 1 | 4 3 1 5 2 | 5 1 4 2 3 | 3 5 2 1 4 |
| 3 5 2 1 4 | 2 4 5 3 1 | 4 3 1 5 2 | 5 1 4 2 3 |
| 5 1 4 2 3 | 3 5 2 1 4 | 2 4 5 3 1 | 4 3 1 5 2 |
| 4 3 1 5 2 | 5 1 4 2 3 | 3 5 2 1 4 | 2 4 5 3 1. |

The converse of Theorem 4.1.1, that if $N(n) = n - 1$ then $n$ is a prime power, was stated by Wernicke in 1910, but his proof was fallacious and the truth or falsity is still undecided. It is, of course, equivalent to the statement that a finite projective plane of order $n$ exists if and only if $n$ is a prime power. If a plane of order 12 were to exist, there would be 11 MOLS of order 12, whereas the best known to date is that $N(12) \geqslant 5$ (see Exercises 4, number 14).

Two further results which appeared early on in the study of MOLS must now be mentioned. In 1900 Tarry published a proof that $N(6) = 1$. His proof was very long and tedious, appearing as it did long before relationships between MOLS and other combinatorial structures were appreciated. A more recent proof due to Stinson (1984) is more accessible. The other result, which plays an important role in the study of MOLS, shows that MOLS of order $m$ and MOLS of order $n$ can be combined to produce MOLS of order $mn$. Since we already know how to construct MOLS of prime power orders, the usefulness of this result is apparent. The theorem is usually known as MacNeish's theorem (1922), but again the result goes back to Moore's 1896 paper.

**Theorem 4.1.2** (Moore–MacNeish)   $N(mn) \geqslant \min(N(m), N(n))$.

**Proof**  Let $A^{(1)}, \ldots, A^{(s)}$ be $s$ MOLS of order $m$ on $\{0, 1, \ldots, m - 1\}$ and let $B^{(1)}, \ldots, B^{(s)}$ be $s$ MOLS of order $n$ on $\{0, 1, \ldots, n - 1\}$. It will be shown how to construct $s$ MOLS $C^{(1)}, \ldots, C^{(s)}$ of order $mn$ on $\{0, 1, \ldots, mn - 1\}$.

If $A$ is any Latin square of order $m$ and $B$ is any of order $n$, define the product of $A$ and $B$ to be

$$C = A \times B = \begin{bmatrix} B + a_{11}nJ & B + a_{12}nJ & \cdots & B + a_{1m}nJ \\ \vdots & & & \vdots \\ B + a_{m1}nJ & B + a_{m2}nJ & \cdots & B + a_{mm}nJ \end{bmatrix}.$$

Then $C$ is an $mn \times mn$ array, and, since the $a_{ij}$ take values $0, \ldots, m - 1$ and the entries of $B$ are $0, \ldots, n - 1$, the entries in $C$ are the numbers 0 up to $n - 1 + n(m - 1) = mn - 1$. Further, $C$ is a Latin square. For consider any row of $C$. Since the $a_{ij}$ in that row are each of $0, \ldots, m - 1$ once, and the entries from $B$ are $0, \ldots, n - 1$ once, the entries in that row of $C$ are precisely every possible number of the form $an + b$ with $0 \leqslant a \leqslant m - 1$ and $0 \leqslant b \leqslant n - 1$, i.e. every number from 0 to $mn - 1$. A similar argument holds for the columns.

So let $C^{(t)}$ denote the product of $A^{(t)}$ and $B^{(t)}$, $1 \leqslant t \leqslant s$. We show that $C^{(1)}, \ldots, C^{(s)}$ are MOLS. Suppose that

$$c_{ij}^{(r)} = c_{IJ}^{(r)} \quad \text{and} \quad c_{ij}^{(t)} = c_{IJ}^{(t)}. \tag{4.2}$$

To find what $c_{ij}$ is, write $i$ as $i = (k - 1)n + l$, $1 \leqslant l \leqslant n$, and $j = (g - 1)n + h$, $1 \leqslant h \leqslant n$; then $c_{ij}^{(r)} = b_{lh}^{(r)} + na_{kg}^{(r)}$. Similarly write $I = (K - 1)n + L$, $J = (G - 1)n + H$. Then (4.2) becomes

$$b_{lh}^{(r)} + na_{kg}^{(r)} = b_{LH}^{(r)} + na_{KG}^{(r)}, \qquad b_{lh}^{(t)} + na_{kg}^{(t)} = b_{LH}^{(t)} + na_{KG}^{(t)}.$$

Since each integer has a unique representation of the form $an + b$, it follows that $b_{lh}^{(r)} = b_{LH}^{(r)}$, $b_{lh}^{(t)} = b_{LH}^{(t)}$, $a_{kg}^{(r)} = a_{KG}^{(r)}$, $a_{kg}^{(t)} = a_{KG}^{(t)}$. Since $B^{(r)}$ and $B^{(t)}$ are orthogonal $l = L$ and $h = H$; since $A^{(r)}$ and $A^{(t)}$ are orthogonal, $k = K$ and $g = G$; so finally $i = I$ and $j = J$.

Since $N(p^\alpha) = p^\alpha - 1$ we now have the following result.

**Corollary 4.1.3** *If $n = p_1^{\alpha_1}, \ldots, p_r^{\alpha_r}$ then $N(n) \geqslant (\min_i p_i^{\alpha_i}) - 1$.*

**Example 4.1.4** Let $n = 12m + 7$. Then the smallest prime power which could possibly divide $n$ is 5; so $N(n) \geqslant 5 - 1 = 4$.

Note that if $n$ is odd then each $p_i^{\alpha_i}$ is at least 3, so that $N(n) \geqslant 2$. Also, if $n$ is a multiple of 4, then each $p_i^{\alpha_i}$ is again at least 3 so that $N(n) \geqslant 2$. This establishes the following.

**Theorem 4.1.4** *$N(n) \geqslant 2$ for all $n \not\equiv 2$ (mod 4), $n \geqslant 1$.*

The above results were essentially what was known about MOLS before 1940. By this time there were two conjectures:

(i) *Euler's conjecture*: $N(n) = 1$ if $n \equiv 2$ (mod 4);
(ii) *MacNeish's conjecture*: if $n = p_1^{\alpha_1} \cdots p_r^{\alpha_r}$ then $N(n) = (\min_i p_i^{\alpha_i}) - 1$.

If (ii) were correct, it would not only imply the truth of Euler's conjecture but it would also imply that if $n$ is not a prime power then no finite projective plane of order $n$ exists. However, both conjectures have turned out to be false. In the case of (i), it has already been mentioned that $N(n) \geqslant 2$ for all $n > 6$, $n \equiv 2$ (mod 4). The first such $n$ actually to be dealt with was $n = 22$ (Bose and Shrikhande, 1959), but shortly afterwards Parker showed that if $q = p^\alpha \equiv 3$ (mod 4) then $N(\frac{1}{2}(3q - 1)) \geqslant 2$. Taking $q = 7$ gives $N(10) \geqslant 2$. This result also destroyed the MacNeish conjecture, although it had already been shown to be false by Parker (1959*b*) who had shown that $N(21) \geqslant 4$. All these developments are described in this chapter.

## 4.2 Orthogonal arrays

Before proceeding with the study of MOLS it is convenient to exhibit an alternative representation of a set of MOLS as an orthogonal array. Consider the two MOLS of order 3 given by (4.1). From them the following array is obtained.

$$
\begin{array}{ccccccccc}
1 & 1 & 1 & 2 & 2 & 2 & 3 & 3 & 3 \\
1 & 2 & 3 & 1 & 2 & 3 & 1 & 2 & 3 \\
1 & 2 & 3 & 3 & 1 & 2 & 2 & 3 & 1 \\
1 & 2 & 3 & 2 & 3 & 1 & 3 & 1 & 2.
\end{array}
\qquad (4.3)
$$

The first two rows label the positions in the squares, the first row giving the row number and the second row giving the column number. In the third row are placed the corresponding entries of the first Latin square, and in the fourth row are placed the entries of the second square.

**Definition 4.2.1** An *orthogonal array* $\mathrm{OA}(s, n)$ on $n$ symbols is an $s \times n^2$ array such that any two rows give, in their vertical pairs, each ordered pair of symbols exactly once.

Clearly, expression (4.3) is an $\mathrm{OA}(4, 3)$; the third and fourth rows have the required property since the corresponding squares are orthogonal; the first and third because in a Latin square each element occurs just once in each row.

**Theorem 4.2.1** *An $\mathrm{OA}(s, n)$ exists if and only if $N(n) \geqslant s - 2$.*

**Proof** Given $s - 2$ MOLS of order $n$, construct an array as described above. If $i \geqslant 1$, row 1 and row $i + 2$ give each ordered pair once since each row of the $i$th square contains each element once. Similarly for rows 2 and $i + 2$. For rows $i + 2$ and $j + 2$, the orthogonality of the $i$th and $j$th squares is precisely the property of the array.

Conversely, given an $\mathrm{OA}(s, n)$, reorder the columns so that the first two rows are

$$
\begin{array}{ccccccccccccc}
1 & 1 & \cdots & 1 & 2 & 2 & \cdots & 2 & \cdots & n & n & \cdots & n \\
1 & 2 & \cdots & n & 2 & 2 & \cdots & n & \cdots & 1 & 2 & \cdots & n.
\end{array}
$$

Then the squares $A_k$, $1 \leqslant k \leqslant s - 2$, defined by taking the $(i, j)$th entry of $A_k$ the entry in the $(k + 2)$th row under the pair $\binom{i}{j}$, are clearly orthogonal.

**Example 4.2.1**   The following OA(5, 4) corresponds to the three MOLS of order 4 exhibited in Example 1.3.4.

$$
\begin{array}{cccccccccccccccc}
1 & 1 & 1 & 1 & 2 & 2 & 2 & 2 & 3 & 3 & 3 & 3 & 4 & 4 & 4 & 4 \\
1 & 2 & 3 & 4 & 1 & 2 & 3 & 4 & 1 & 2 & 3 & 4 & 1 & 2 & 3 & 4 \\
1 & 2 & 3 & 4 & 2 & 1 & 4 & 3 & 3 & 4 & 1 & 2 & 4 & 3 & 2 & 1 \\
1 & 2 & 3 & 4 & 4 & 3 & 2 & 1 & 2 & 1 & 4 & 3 & 3 & 4 & 1 & 2 \\
1 & 2 & 3 & 4 & 3 & 4 & 1 & 2 & 4 & 3 & 2 & 1 & 2 & 1 & 4 & 3.
\end{array}
$$

To show that $N(n) \geqslant 2$, it is sufficient to construct an OA(4, $n$). Parker was one of the first to make use of orthogonal arrays, using them to deal with $n = 10$ in particular (see Exercises 4, number 5). A simpler construction of suitable OAs, given by Bose, Shrikhande, and Parker in 1960, is now described. It cleverly uses the idea of a difference system; in this case it is the patterned difference system of Exercises 2, number 11.

**Theorem 4.2.2**   *If* $N(m) \geqslant 2$, *then* $N(3m + 1) \geqslant 2$.

**Proof**   Consider the following $4 \times 4m$ array $A_0$:

$$
A_0 = \begin{bmatrix}
0 & 0 & \cdots & 0 & 1 & \cdots & m & 2m & \cdots & m+1 & x_1 & \cdots & x_m \\
1 & 2 & \cdots & m & 0 & \cdots & 0 & x_1 & \cdots & x_m & 2m & \cdots & m+1 \\
2m & 2m-1 & \cdots & m+1 & x_1 & \cdots & x_m & 0 & \cdots & 0 & 1 & \cdots & m \\
x_1 & x_2 & \cdots & x_m & 2m & \cdots & m+1 & 1 & \cdots & m & 0 & \cdots & 0
\end{bmatrix}.
$$

Here the $x_i$ are $m$ symbols distinct from $0, \ldots, 2m$. Let $A_i$ be the array obtained from $A_0$ by adding $i$ to each numerical entry (mod $2m + 1$, reducing to $0, 1, \ldots, 2m$) and leaving each $x_i$ unchanged. Further, since $N(m) \geqslant 2$, there exists an OA(4, $m$) on $x_1, \ldots, x_m$ which we denote by $A^*$. If also

$$
E = \begin{bmatrix}
0 & 1 & \cdots & 2m \\
0 & 1 & \cdots & 2m \\
0 & 1 & \cdots & 2m \\
0 & 1 & \cdots & 2m
\end{bmatrix}
$$

we claim that the array

$$
D = \begin{bmatrix} E & A_0 & A_1 & \cdots & A_{2m} & A^* \end{bmatrix}
$$

is an OA(4, 3m + 1). Certainly, the number of columns is $2m + 1 + (2m + 1)4m + m^2 = (3m + 1)^2$. It is also clear that in any two rows the ordered pair $(x_i, n)$ will occur exactly once for each $i$ and each $n$, as will the pair $(n, x_i)$, and the parts $E$ and $A^*$ will give every ordered pair of the form $(n, n)$ or $(x_i, x_j)$ exactly once. So it remains to show that if $0 \leqslant u \leqslant 2m$ and $0 \leqslant v \leqslant 2m$, $u \neq v$, the pair $(u, v)$ occurs once. As an example, consider the second and third rows. The differences between corresponding numbers in these two rows in $A_0$ are $2m - 1, 2m - 3, \ldots, 3, 1$ and their negatives (mod $2m + 1$), i.e. each of $1, 2, \ldots, 2m$ precisely once. Thus the ordered pair $(u, v)$ will arise in the second and third rows of $D$ from the pair in $A_0$ differing by $u - v$, with a suitable number added to each. A similar argument applies to other pairs of rows.

**Corollary 4.2.3** $N(12t + 10) \geqslant 2$ *for all integers* $t$.

**Proof** Take $m = 4t + 3$ in the theorem.

In particular, $N(10) \geqslant 2$ and $N(22) \geqslant 2$. Here for example is the Graeco-Latin square obtained by Parker (1959$a$) to show that $N(10) \geqslant 2$.

**Example 4.2.2**

| | | | | | | | | | |
|----|----|----|----|----|----|----|----|----|----|
| 00 | 47 | 18 | 76 | 29 | 93 | 85 | 34 | 61 | 52 |
| 86 | 11 | 57 | 28 | 70 | 39 | 94 | 45 | 02 | 63 |
| 95 | 80 | 22 | 67 | 38 | 71 | 49 | 56 | 13 | 04 |
| 59 | 96 | 81 | 33 | 07 | 48 | 72 | 60 | 24 | 15 |
| 73 | 69 | 90 | 82 | 44 | 17 | 58 | 01 | 35 | 26 |
| 68 | 74 | 09 | 91 | 83 | 55 | 27 | 12 | 46 | 30 |
| 37 | 08 | 75 | 19 | 92 | 84 | 66 | 23 | 50 | 41 |
| 14 | 25 | 36 | 40 | 51 | 62 | 03 | 77 | 88 | 99 |
| 21 | 32 | 43 | 54 | 65 | 06 | 10 | 89 | 97 | 78 |
| 42 | 53 | 64 | 05 | 16 | 20 | 31 | 98 | 79 | 87 |

It is interesting to note that the following Latin square of order 10 has the remarkable property that it is orthogonal to its transpose, thus giving an independent verification that $N(10) \geqslant 2$.

**Example 4.2.3**

| | | | | | | | | | |
|----|----|----|----|----|----|----|----|----|----|
| 0 | 2 | 8 | 6 | $\infty$ | 7 | 1 | 5 | 4 | 3 |
| 5 | 1 | 3 | 0 | 7 | $\infty$ | 8 | 2 | 6 | 4 |
| 7 | 6 | 2 | 4 | 1 | 8 | $\infty$ | 0 | 3 | 5 |
| 4 | 8 | 7 | 3 | 5 | 2 | 0 | $\infty$ | 1 | 6 |
| 2 | 5 | 0 | 8 | 4 | 6 | 3 | 1 | $\infty$ | 7 |
| $\infty$ | 3 | 6 | 1 | 0 | 5 | 7 | 4 | 2 | 8 |
| 3 | $\infty$ | 4 | 7 | 2 | 1 | 6 | 8 | 5 | 0 |
| 6 | 4 | $\infty$ | 5 | 8 | 3 | 2 | 7 | 0 | 1 |
| 1 | 7 | 5 | $\infty$ | 6 | 0 | 4 | 3 | 8 | 2 |
| 8 | 0 | 1 | 2 | 3 | 4 | 5 | 6 | 7 | $\infty$ |

Note the structure: if we ignore the last row and column, the rest of the square is obtained by cyclically developing the first row mod 9, with $\infty + 1 = \infty$. The last row and column complete the Latin square.

A Latin square orthogonal to its transpose is called a *self-orthogonal Latin square* (SOLS). In 1976 Brayton, Coppersmith, and Hoffman showed remarkably that a SOLS of order $n$ exists for all $n$ except $1, 2, 3, 6$; this will be discussed in Chapter 5. Meanwhile note the following observation which they made about the OA used in the proof of Theorem 4.2.2: if $(a, b, c, d)'$ is a column of $A_0$ then so is $(b, a, d, c)'$. This tells us that if the $(a, b)$th entry in the first square is $c$ and in the second square is $d$, then the $(b, a)$th entry in the first square is $d$ and in the second square is $c$. But this is just the condition for self-orthogonality. So if $A^*$ also has this property, i.e. if a SOLS of order $m$ exists, there also exists a SOLS of order $3m + 1$.

**Theorem 4.2.4** *If a SOLS of order $m$ exists then a SOLS of order $3m + 1$ also exists.*

Note that this theorem cannot be used to establish the existence of a SOLS of order 10, since none of order 3 exists. The square in Example 4.2.3 was obtained differently. Note also that the following example of a SOLS of order 14 enables us to assert that $N(14) \geqslant 2$.

**Example 4.2.4** A SOLS of order 14.
Follow the ideas of Example 4.2.3, cyclically developing

$$0 \quad 8 \quad 3 \quad 12 \quad 9 \quad 2 \quad 5 \quad 10 \quad 6 \quad 11 \quad 1 \quad 4 \quad \infty$$

(mod 13).

We close this section by using an OA(4, 26) to show that $N(26) \geqslant 2$.

**Example 4.2.5** An OA(4, 26). First note that

$$
\begin{array}{ccccccc}
1 & 7 & 6 & 5 & 4 & 3 & 2 \\
3 & 2 & 1 & 7 & 6 & 5 & 4 \\
5 & 4 & 3 & 2 & 1 & 7 & 6 \\
7 & 6 & 5 & 4 & 3 & 2 & 1 \\
2 & 1 & 7 & 6 & 5 & 4 & 3 \\
4 & 3 & 2 & 1 & 7 & 6 & 5 \\
6 & 5 & 4 & 3 & 2 & 1 & 7
\end{array}
$$

is a SOLS of order 7. Now let

$$
A_0 = \begin{bmatrix}
0 & x_1 & x_2 & x_3 & x_4 & x_5 & x_6 & x_7 & 6 & 1 & \cdots & 3 & 15 & \cdots & 1 & 0 & \cdots \\
3 & 15 & 10 & 7 & 8 & 12 & 9 & 6 & 1 & 0 & \cdots & 0 & x_1 & \cdots & 6 & 1 & \cdots \\
1 & 0 & 0 & 0 & 0 & 0 & 0 & 0 & 0 & x_1 & \cdots & 6 & 1 & \cdots & 3 & 15 & \cdots \\
6 & 1 & 2 & 4 & 6 & 7 & 8 & 10 & 3 & 15 & \cdots & 1 & 0 & \cdots & 0 & x_1 & \cdots
\end{bmatrix},
$$

where the numerical entries are considered mod 19. We check that the numerical differences between any two rows are precisely $1, \ldots, 18$. In view of the cyclic structure of $A_0$ it is sufficient to do this for rows 1 and 2 and for rows 1 and 3. But the differences between rows 1 and 2 are

$$16, 5, 1, 2, 4, 6, 7, 8, 10 \quad \text{and their negatives (mod 19),}$$

while the differences between rows 1 and 3 are

$$18, 6, 16, 14, 8, 3, 2, 5, 1, 15, 17, 4, 9, 12, 11, 7, 10, 13,$$

i.e. $1, \ldots, 18$ in each case.

It follows that if $A_i$ is obtained from $A_0$ by adding $i$ (mod 19) to each numerical entry, then $A_0 A_1 \ldots A_{18}$ has the property that the vertical pairs in any two rows give every ordered pair of distinct number once, and every pair involving a number and an $x_i$ once. Thus if $B$ denotes an OA(4,7) on $x_1, \ldots, x_7$ corresponding to SOLS of order 7 displayed above, and if $E$ is a $4 \times 19$ array with $(i, i, i, i)'$ as its $(i+1)$th column, $0 \leqslant i \leqslant 18$, then

$$\begin{bmatrix} A_0 & A_1 & \cdots & A_{18} & B & E \end{bmatrix}$$

is an OA(4, 26).

## 4.3 Using pairwise balanced designs

One of the first uses Parker made of orthogonal arrays was to disprove MacNeish's conjecture. If that conjecture had been correct, then we would have had $N(21) = \min(3 - 1, 7 - 1) = 2$; however, Parker showed that $N(21) \geqslant 4$ by constructing an OA(6, 21). His method depended on a slight generalization of the following theorem.

**Theorem 4.3.1** *If a* PBD$(v, K, 1)$ *exists, then*

$$N(v) \geqslant \min_{k \in K} N(k) - 1.$$

**Example 4.3.1** Since a PBD$(22, \{4, 7\}, 1)$ exists (Example 1.4.3),

$$N(22) \geqslant \min\{N(4) - 1, N(7) - 1\} = 2.$$

(This was the first counterexample to the Euler conjecture.)

**Proof of Theorem 4.3.1** Let $q = \min_{k \in K} N(k)$. Then, for each $k \in K$, there exist $q$ MOLS of order $k$ and hence an OA$(q + 2, k)$ on $\{1, \ldots, k\}$. Without loss of generality we can suppose that the first row of the OA$(q + 2, k)$ is $11 \cdots 122 \cdots 2 \cdots kk \cdots k$, and that every other row begins

$12 \cdots k$. Delete the first row and the first $k$ columns from each array to obtain arrays $D_k$ with $q + 1$ rows and $k(k - 1)$ columns, in which the vertical pairs in any two rows are precisely all the ordered pairs of *distinct* elements from $\{1, \dots, k\}$.

Let the blocks of the PBD be $B_1, \dots, B_b$. For each such $B$, replace each entry $i$ in $D_{|B|}$ by the $i$th element of $B$, and denote the resulting array by $E_B$. String all such $E_B$s together in a row, and add the array

$$F = \begin{bmatrix} 1 & 2 & \cdots & v \\ 1 & 2 & \cdots & v \\ \vdots & & & \\ 1 & 2 & \cdots & v \end{bmatrix}_{(q + 1) \times v}.$$

Then $A = [E_{B_1} \cdots E_{B_b} F]$ has, by Lemma 1.4.2,

$$v + \sum_{k_i \in K} b_i k_i (k_i - 1) = v + v(v - 1) = v^2$$

columns. Further, it is an $OA(q + 1, v)$. For consider the $i$th and $j$th rows. To find where the pair $(a, b)'$ occurs in these rows, note that there is precisely one block $B$ containing $a$ and $b$; in $E_B$ the pair $(a, b)'$ will occur in the $i$th and $j$th rows exactly once, and clearly it will not occur outwith $E_B$. Thus an $OA(q + 1, v)$ exists, and so $N(v) \geqslant q - 1$.

**Example 4.3.2** Apply Theorem 4.3.1 to a $(7, 3, 1)$ design, i.e. to the seven-point plane of Example 1.1.3, and use the $OA(4, 3)$ exhibited as (4.3). The procedure described in the proof of the last theorem leads to the following $OA(3, 7)$:

```
1 2 4 1 2 4 2 3 5 2 3 5  ···  7 1 3 7 1 3 1 2 3 4 5 6 7
4 1 2 2 4 1 5 2 3 3 5 2  ···  3 7 1 1 3 7 1 2 3 4 5 6 7
2 4 1 4 1 2 3 5 2 5 2 3  ···  1 3 7 3 7 1 1 2 3 4 5 6 7
```
$$(4.4)$$

This is of course not very useful since it only yields $N(7) \geqslant 1$. However, it is of interest more generally that this $OA(3, 7)$ can be extended to an $OA(4, 7)$ by adding a further row:

```
1 2 4 1 2 4 2 3 5 2 3 5  ···  7 1 3 7 1 3 1 2 3 4 5 6 7
4 1 2 2 4 1 5 2 3 3 5 2  ···  3 7 1 1 3 7 1 2 3 4 5 6 7
2 4 1 4 1 2 3 5 2 5 2 3  ···  1 3 7 3 7 1 1 2 3 4 5 6 7
1 2 3 4 5 6 1 2 3 4 5 6  ···  1 2 3 4 5 6 7 7 7 7 7 7 7.
```

What makes it possible to add this extra row? In (4.4) the first columns of the seven groups of six columns are simply the blocks of the $(7, 3, 1)$ design. But these seven columns together form a Youden design

```
1 2 3 4 5 6 7
4 5 6 7 1 2 3
2 3 4 5 6 7 1
```

with each of $1,\ldots,7$ occurring once in each row. So if a 1 is placed under each of these seven columns then, in each row, each of $1,\ldots,7$ will be paired with 1 precisely once. The same reasoning shows that $2,\ldots,6$ can be placed under the remaining columns of these groups. Finally 7 is placed under each constant column at the end. Now it was shown in section 1.5 that, for any symmetric design, the blocks can be written in columns to form a Youden design; so the above procedure is applicable in each case of a symmetric design with $\lambda = 1$, establishing the following result which increases the estimate of Theorem 4.3.1 by 1.

**Theorem 4.3.2** *If a $(k^2 - k + 1, k, 1)$ design exists, then $N(k^2 - k + 1) \geqslant N(k)$.*

**Example 4.3.3** (a) Using a $(21, 5, 1)$ design, $N(21) \geqslant 4$.
(b) Using a $(57, 8, 1)$ design, $N(57) \geqslant 7$.
(Note that the Moore–MacNeish theorem yields only $N(21) \geqslant 2$, $N(57) \geqslant 2$.)

Theorem 4.3.1 will be a major tool in establishing that $N(n) \geqslant 2$ for all $n > 6$.

## 4.4 The collapse of the Euler conjecture

The theorems of the previous section are not quite strong enough to prove that $N(n) \geqslant 2$ for all $n > 6$. The breakthrough came with the following strengthening of Theorem 4.3.1, due to Bose and Shrikhande (1960).

**Theorem 4.4.1** *If there exists a* PBD$(v, k, 1)$ *with*

$$K = \{k_1, \ldots, k_r, k_{r+1}, \ldots, k_m\}$$

*where no two blocks of sizes in $\{k_1, \ldots, k_r\}$ intersect, then*

$$N(v) \geqslant \min\{N(k_1), \ldots, N(k_r), N(k_{r+1}) - 1, \ldots, N(k_m) - 1\}.$$

Note that this result gets rid of the $-1$ in Theorem 4.3.1 for some of the $k \in K$. Theorem 4.3.2 also did this, but in a much more limited way.

**Definition 4.4.1** If the blocks of sizes in $\{k_1, \ldots, k_r\}$ in a PBD$(v, K, 1)$ are all disjoint then they are said to form a *clear set* of blocks.

**Example 4.4.1** Take a plane of order 4, i.e. a $(21, 5, 1)$ BIBD. Choose three elements not all in the same block and delete them from the blocks in which they occur. This gives a PBD$(18, \{3, 4, 5\}, 1)$. There are three blocks of size 3, and they form a clear set since otherwise two of the blocks of the plane would intersect in at least two points. Thus

$$N(18) \geqslant \min\{N(3), N(4) - 1, N(5) - 1\} = 2.$$

**Proof of Theorem 4.4.1**  Let  $q = 1 + \min\{N(k_1), \ldots, N(k), N(k_{r+1}) - 1, \ldots, N(k_m) - 1\}$; then for each $i > r$ there are $q$ MOLS of order $k_i$. If we take them all to have first row $1\ 2\cdots k_i$ we can construct an OA$(q + 2, k_i)$, then delete the first row and the first $k_i$ columns to obtain an array $P_i$, with $q + 1$ rows and $k_i(k_i - 1)$ columns, in which the vertical pairs given by any two rows are precisely all the ordered pairs of distinct elements from $\{1, \ldots, k_i\}$. Next, for each $i \leqslant r$, simply define $P_i$ to be the OA$(q + 1, k_i)$ corresponding to $q - 1$ MOLS of order $k_i$ each with first row $1 \cdots k_i$. Then each $P_i$ has $q + 1$ rows, and $P_i$ has $k_i^2$ columns if $i \leqslant r$ but $k_i^2 - k_i$ columns if $i > r$.

Now take any block $B_j$ of the PBD. If it is of size $k_i$, replace each $h$ in $P_i$ by the $h$th element of $B_j$, so as to obtain an array $C_j$. Do this for each $j = 1, \ldots, b$. Finally let

$$C = \begin{bmatrix} C_1 & C_2 & \cdots & C_b & F \end{bmatrix},$$

where $F$ has one constant column for each element of the PBD not in any block of the clear set. The number of columns is

$$\sum_{i \leqslant r} b_i k_i^2 + \sum_{i > r} b_i k_i (k_i - 1) + v - \sum_{i \leqslant r} k_i b_i,$$

where $b_i$ is the number of blocks of size $k_i$. By Lemma 1.4.2 this is

$$\sum_{i=1}^{m} b_i k_i^2 - \sum_{i=1}^{m} b_i k_i + v = v + v^2 - v = v^2.$$

So the number of columns in $C$ is $v^2$, and $C$ is a $(q + 1) \times v^2$ array. We claim that $C$ is an OA$(q + 1, v)$, so that $N(v) \geqslant q - 1$ as required. Consider any two rows, and any two elements $x, y$ of the PBD. If $x = y$ the pair $(x, y)'$ occurs in $F$ or in a $C_j$ corresponding to a block of the clear set containing $x$. If $x \neq y$, $x$ and $y$ occur in a unique block $B_j$ and the pair $(x, y)'$ occurs in $C_j$.

**Example 4.4.2**  Take a $(41, 5, 1)$ design, as in Exercises 2, number 10, and remove three elements not all in the same block. This gives a PBD$(38, \{3, 4, 5\}, 1)$ in which the blocks of size 3 form a clear set. So

$$N(38) \geqslant \min\{N(3), N(4) - 1, N(5) - 1\} = 2.$$

Our next task is to construct a particularly useful PBD to which Theorem 4.4.1 can be applied.

**Theorem 4.4.2** *If* $N(m) \geqslant k - 1$ *then there exists a resolvable* PBD$(km, \{k, m\}, 1)$ *with* $m + 1$ *resolution classes in which the blocks of size* $m$ *form one of the resolution classes.*

**Proof** Since $N(m) \geqslant k - 1$, there exists an OA$(k + 1, m)$ with first row

$$11 \cdots 122 \cdots 2 \cdots mm \cdots m.$$

Consider the $m^2$ columns as being assigned to $m$ groups, those in group $i$ being those with $i$ in the first row position. Remove the first row of the OA and replace each $i$ in the resulting $j$th row by $i + (j - 1)m$; this gives an array on $km$ elements $1, \ldots, km$. Consider the columns of this array as the blocks of a design. We then have a design on $v = km$ elements with $m^2$ blocks each of size $k$. Further, by the properties of an OA, the blocks in each group contain in their union each of $1, \ldots, km$ exactly once; so the design is resolvable with $m$ resolution classes. It is not, however, balanced, since no two elements of the form $i + (j - 1)m$, $i' + (j - 1)m$ can lie in the same block. To obain a balanced design, add $k$ further blocks $B_1, \ldots, B_k$ defined by $B_j = \{i + (j - 1)m : 1 \leqslant i \leqslant m\}$. These blocks of size $m$ clearly form a further resolution class, and, together with the blocks of size $k$, form a PBD$(km, \{k, m\}, 1)$ with the required properties.

**Example 4.4.3** Take $m = 4$ and $k = 3$. The first two squares of Example 1.3.4 give the following OA$(4, 4)$:

| 1 | 1 | 1 | 1 | 2 | 2 | 2 | 2 | 3 | 3 | 3 | 3 | 4 | 4 | 4 | 4 |
|---|---|---|---|---|---|---|---|---|---|---|---|---|---|---|---|
| 1 | 2 | 3 | 4 | 1 | 2 | 3 | 4 | 1 | 2 | 3 | 4 | 1 | 2 | 3 | 4 |
| 1 | 2 | 3 | 4 | 2 | 1 | 4 | 3 | 3 | 4 | 1 | 2 | 4 | 3 | 2 | 1 |
| 1 | 2 | 3 | 4 | 4 | 3 | 2 | 1 | 2 | 1 | 4 | 3 | 3 | 4 | 1 | 2. |

Delete the first row, add 4 to each entry in the third row, add 8 to each entry in the fourth row to obtain

| 1 | 2 | 3 | 4 | 1 | 2 | 3 | 4 | 1 | 2 | 3 | 4 | 1 | 2 | 3 | 4 |
|---|---|---|---|---|---|---|---|---|---|---|---|---|---|---|---|
| 5 | 6 | 7 | 8 | 6 | 5 | 8 | 7 | 7 | 8 | 5 | 6 | 8 | 7 | 6 | 5 |
| 9 | 10 | 11 | 12 | 12 | 11 | 10 | 9 | 10 | 9 | 12 | 11 | 11 | 12 | 9 | 10. |

Take the columns as blocks and take the further blocks $\{1, 2, 3, 4\}$, $\{5, 6, 7, 8\}$, and $\{9, 10, 11, 12\}$ to obtain a resolvable PBD$(12, \{3, 4\}, 1)$.

**Corollary 4.4.3** *If* $N(m) \geqslant k - 1$ *and* $1 \leqslant x < m$ *then*

(a) *there exists a* PBD$(km + x, \{x, m, k, k + 1\}, 1)$, *and*
(b) $N(km + x) \geqslant \min\{N(m), N(x), N(k) - 1, N(k + 1) - 1\}$. $\qquad$ (4.5)

**Proof** By the above theorem, a resolvable PBD$(km, \{k, m\}, 1)$ exists with $m + 1$ resolution classes. Let $B_1, \ldots, B_k$ be, as before, blocks of size $m$

forming the $(m + 1)$th class. For each $i \leqslant x$, add a new element $\infty_i$ to each block of the $i$th class of blocks of size $k$, and introduce a new block $B_\infty = \{\infty_1, \ldots, \infty_x\}$. This gives a PBD$(km + x, \{x, m, k, k + 1\}, 1)$ in which the $k + 1$ blocks $B_1, \ldots, B_k, B_\infty$ partition the set of elements. If neither $m$ nor $x$ equals $k$ or $k + 1$ these $k + 1$ blocks form a clear set and (4.5) follows. In other cases, (4.5) still holds since the minimum function rejects $N(m)$ and/or $N(x)$ anyway. Note that if $x = 1$ the result still holds under the convention that $N(1) = \infty$.

**Example 4.4.4**

(a) $N(62) = \mathrm{N}(4 \cdot 13 + 10)$, so that $k = 4$, $m = 13$, $x = 10$. Since $N(13) = 12 \geqslant k - 1$ it follows that

$$N(62) \geqslant \min\{N(13), N(10), N(4) - 1, N(5) - 1\} = 2.$$

(b) $N(74) = N(4 \cdot 16 + 10) \geqslant \min\{N(16), N(10), N(4) - 1, N(5) - 1\} = 2.$
(c) $N(86) = N(4 \cdot 19 + 10) \geqslant \min\{N(19), N(10), N(4) - 1, N(5) - 1\} = 2.$
(d) $N(100) = N(7 \cdot 13 + 9) \geqslant \min\{N(13), N(9), N(7) - 1, N(8) - 1\} = 5.$

Several of these examples use part (b) of the corollary with $k = 4$. This value of $k$ yields the following general result.

**Theorem 4.4.4** *If $N(m) \geqslant 3$, $N(x) \geqslant 2$ and $1 \leqslant x < m$ then $N(4m + x) \geqslant 2$.*

This result, with various suitable choices of $m$, will be enough to prove the Bose–Parker–Shrikhande theorem. We shorten their original proof slightly by first showing that any multiple of 4 can be taken for $m$.

**Lemma 4.4.5** *$N(4t) \geqslant 3$ for all $t \geqslant 1$.*

**Proof** By the Moore–MacNeish theorem, $N(4t) \geqslant 3$ except, possibly, when $t$ is divisible by 3 but not by 9. To deal with these cases, write $4t = 2^{2a+b} \cdot 3u$, where $(u, 6) = 1$ and $b = 2$ or 3. If $u2^{2a} = 1$ then $N(4t) = N(3 \cdot 2^b)$. If $u2^{2a} \neq 1$, then $N(u2^{2a}) \geqslant 3$ so that $N(4t) \geqslant \min\{N(3 \cdot 2^b), 3\}$. So in either case we need only prove that $N(12) \geqslant 3$ and $N(24) \geqslant 3$. These are dealt with in Exercises 4, numbers 8 and 14.

At last we can prove the main theorem.

**Theorem 4.4.6** *$N(n) \geqslant 2$ for all positive integers $n \neq 2$ or 6.*

**Proof** In view of Theorem 4.1.5, only $n \equiv 2 \pmod 4$ need to be considered. Any such $n$ can be written as $n = 16k + y = 16(k - 1) + (16 + y)$

where $y = 2, 6\ 10$, or 14. But $N(16 + y) \geqslant 2$ for each such $y$: $N(18) \geqslant 2$ by Example 4.4.1, $N(22) \geqslant 2$ by Corollary 4.2.3, $N(26) \geqslant 2$ by Example 4.2.5, and $N(30) \geqslant \min\{N(3), N(10)\} = 2$. So, by Theorem 4.4.4 and Lemma 4.4.5

$$N(16k + y) = N(4 \cdot 4(k - 1) + (16 + y)) \geqslant 2$$

provided $k - 1 \geqslant 1$ and $16 + y < 4(k - 1)$, i.e. provided $k \geqslant 2$ and $4k - 4 > 30$, i.e. provided $k \geqslant 9$.

All that remains therefore is to verify that $N(n) \geqslant 2$ for all $n \equiv 2 \pmod 4$, $6 < n < 144$. All such $n \leqslant 30$ have already been discussed. Of the rest, Corollary 4.2.3 deals with 34, 46, 58, 70, 82, 94, 106, 118, 130, 142. The cases 38, 62, 74, 86 have been dealt with in Examples 4.4.2 and 4.4.4. Finally, $42 = 3 \times 14$, $50 = 5 \times 10$, $54 = 3 \times 18$, $66 = 3 \times 22$, $78 = 3 \times 26$, $90 = 3 \times 30$, $98 = 7 \times 14$, $102 = 3 \times 34$, $110 = 10 \times 11$, $114 = 3 \times 38$, $122 = 4(27) + 14$, $126 = 3 \times 42$, $134 = 4(27) + 26$, $138 = 3 \times 46$ deal with the rest.

## 4.5 Transversal designs

The proof of the existence of two MOLS of order $n$ whenever $n \neq 2$ or 6 depended on the use of pairwise balanced designs and orthogonal arrays. These orthogonal arrays are equivalent to sets of MOLS: they are not themselves PBDs, but they are equivalent to a special class of designs called *transversal designs*. It is by studying transversal designs that further information about MOLS can be obtained.

Transversal designs have in fact already been used, albeit in disguise, in the proof of Theorem 4.4.2. In that proof an $OA(k + 1, m)$ was taken and the first row removed, to obtain an $OA(k, n)$; then $(j - 1)m$ was added to each entry of the $j$th row of this array, $1 \leqslant j \leqslant k$. The columns were then taken as the blocks of a design. It was noted that this design was not balanced, since any two elements occurred together in a unique block only provided they did not both occur in $B_j = \{i + (j - 1)m : 1 \leqslant i > m\}$ for some $j$. Note that these sets $B_j$ are disjoint and form a partition of the whole set $\{1, \ldots, km\}$.

**Definition 4.5.1** A *transversal design* $TD(k, m)$ consists of a collection of $k$-subsets (*blocks*) of a $km$-set $X$, and a collection of $k$ disjoint $m$-subsets (*groups*) which partition $X$, such that (i) each block contains exactly one element from each group, and (ii) any pair of elements from different groups occurs in exactly one block.

In other words, a $TD(k, m)$ is just a $PBD(km, \{k, m\}, 1)$ in which the blocks of size $m$ partition the set of elements and are considered separately from the remaining blocks, being given the special name of 'groups'.

We can think of the base set $X$ as consisting of $k$ different 'types' of element, $m$ of each type, each block containing one element of each type.

**Example 4.5.1** $X = \{1, \ldots, 12\}$, $G_1 = \{1, 2, 3\}$, $G_2 = \{4, 5, 6\}$, $G_3 = \{7, 8, 9\}$, $G_4 = \{10, 11, 12\}$; take blocks $\{1, 4, 7, 10\}$, $\{1, 5, 8, 11\}$, $\{1, 6, 9, 12\}$, $\{2, 4, 8, 12\}$, $\{2, 5, 9, 10\}$, $\{2, 6, 7, 11\}$, $\{3, 4, 9, 11\}$, $\{3, 5, 7, 12\}$, $\{3, 6, 8, 10\}$. These form a $\mathrm{TD}(4, 3)$.

**Lemma 4.5.1** *In a* $\mathrm{TD}(k, m)$,

   (i) *each element is in m blocks*;
  (ii) *there are* $m^2$ *blocks*,
 (iii) *if* $k = m + 1$, *every two blocks intersect*;
 (iv) *if* $k < m + 1$, *each block has a block disjoint from it*;
  (v) *if* $k < m + 1$ *and* $k \geqslant 3$, *there are at least three mutually disjoint blocks*.

**Proof**

   (i) Suppose that the element $x$ occurs in $r$ blocks. Then there are $r(k - 1)$ other elements in these blocks, none repeated, and they must be precisely all the elements not in the same group as $x$. Thus $r(k - 1) = (k - 1)m$, i.e. $r = m$.
  (ii) If there are $b$ blocks, then, by (i), $bk = m(km)$, i.e. $b = m^2$.
 (iii) and (iv). If $x$ is in a block $B$, $x$ is in $m - 1$ other blocks; so $m - 1$ blocks intersect $B$ in $x$. Since $\lambda = 1$, no blocks can intersect $B$ in more than one point; so $k(m - 1)$ blocks intersect $B$. The remaining $m^2 - 1 - k(m - 1) = (m - 1)(m + 1 - k)$ blocks are disjoint from $B$. This number is 0 if $k = m + 1$ and is positive if $k < m + 1$.
  (v) By (iv), there are two disjoint blocks $A, B$. There are $k(k - 1)$ pairs $(a, b)$ with $a \in A$, $b \in B$, with $a, b$ in different groups, and these give $k^2 - k$ blocks which intersect both $A$ and $B$. But $k(m - 1)$ blocks intersect $A$ and the same number intersect $B$, so the number of blocks intersecting $A \cup B$ is $2k(m - 1) - k^2 + k$. Thus the number of blocks disjoint from both $A$ and $B$ is $m^2 - 2 - 2k(m - 1) + k^2 - k = (m - k)^2 + k - 2 > 0$.

**Lemma 4.5.2** *If a* $\mathrm{TD}(k, m)$ *exists then so does a* $\mathrm{TD}(h, m)$ *for each* $h < k$.

**Proof** Simply remove all of the elements of $k - h$ of the groups.

The concept of resolvability goes over to transversal designs.

**Definition 4.5.2** A $\mathrm{TD}(k, m)$ on a set $X$ is *resolvable* if the blocks can be put into $m$ classes each consisting of $m$ blocks which form a partition of $X$.

Of course the *groups* of a TD always partition the set. The next theorem imitates the step from affine to projective planes.

**Lemma 4.5.3** *A resolvable* TD$(k, m)$ *exists if and only if a* TD$(k + 1, m)$ *exists.*

**Proof** Suppose first that a resolvable TD$(k, m)$ exists, with groups $G_1, \ldots, G_k$ and with $m^2$ blocks $B_{ij}$, $1 \leqslant i \leqslant m$, $1 \leqslant j \leqslant m$, where $B_{i1}, \ldots, B_{im}$ form the $i$th resolution class. Introduce $m$ new elements $\infty_1, \ldots, \infty_m$, adjoin $\infty_i$ to each block of the $i$th class, and introduce a new group $G_\infty = \{\infty_1, \ldots, \infty_m\}$. This gives a TD$(k + 1, m)$.

Conversely, suppose that a TD$(k + 1, m)$ exists, with groups $G_1, \ldots, G_{k+1}$. Remove the elements of $G_{k+1}$ from the blocks. This leaves blocks of size $k$, and $k$ groups. Further, for each $x \in G_{k+1}$, the new blocks $B - \{x\}$, where $B$ was an old block containing $x$, form a partition of the set of $km$ remaining elements; so a resolvable TD$(k, m)$ is obtained.

**Example 4.5.2** Start with the TD$(4, 3)$ of Example 4.5.1, and remove all the elements of the fourth group. The resulting TD$(3, 3)$ is resolvable: its blocks are

$\{1, 4, 7\}, \{1, 5, 8\}, \{1, 6, 9\}, \{2, 4, 8\}, \{2, 5, 9\}, \{2, 6, 7\}, \{3, 4, 9\}, \{3, 5, 7\}, \{3, 6, 8\}$.

These form three resolution classes:

$$
\begin{array}{ccc}
\{1, 4, 7\} & \{2, 5, 9\} & \{3, 6, 8\} \\
\{1, 5, 8\} & \{2, 6, 7\} & \{3, 4, 9\} \\
\{1, 6, 9\} & \{2, 4, 8\} & \{3, 5, 7\}.
\end{array}
$$

Although transversal designs were not greatly used until around 1960, they had in fact already appeared in the work of Kirkman over 100 years previously!

**Example 4.5.3** (Kirkman, 1850) Take $X = \{1, 2, \ldots, p^\alpha\} \times \{1, \ldots, p\}$, so that $|X| = p^{\alpha+1}$. Let $k = p$, $m = p^\alpha$, and say that an element of $X$ is in the $i$th group if its second component is $i$. This gives $p = k$ groups of size $m$. Define $m^2$ blocks $B_{ij}$, $1 \leqslant i \leqslant p^\alpha$, $0 \leqslant j \leqslant p^\alpha - 1$ by

$$B_{ij} = \{(i, 1), (i + j, 2), (i + 2j, 3), \ldots, (i + (p - 1)j, p)\},$$

where the first components are all reduced mod $p^\alpha$ to lie in $\{1, 2, \ldots, p^\alpha\}$. Then each block is of size $k$. Further, any two elements from different groups occur together in exactly one block: for $(a, u), (b, v) \in B_{ij}(u < v)$ requires $i + (u - 1)j \equiv a$ and $i + (v - 1)j \equiv b$ (mod $p^\alpha$), whence $(v - u)j \equiv b - a$ (mod $p^\alpha$). This has a unique solution for $j$, since

$(v - u, p) = 1$, so that $j$ and hence $i$ are uniquely determined. Thus the blocks and groups form a TD($p, p^\alpha$) which is in fact resolvable since, for fixed $j$, the blocks $B_{ij}$ clearly partition $X$.

Kirkman's reason for constructing this transversal design was to enable him to construct BIBDs.

**Theorem 4.5.4** *For each prime $p$ and each $\alpha \geqslant 1$ there exists a resolvable $(p^\alpha, p, 1)$ design.*

**Proof** (Kirkman, 1850) Proceed by induction on $\alpha$. The case $\alpha = 1$ is trivial, consisting of just one block $\{1, \ldots, p\}$. Suppose then that the theorem is true for $\alpha = \beta$; then there exists a resolvable $(p^\beta, p, 1)$ design $D$. Form copies $D_1, \ldots, D_p$ of $D$, obtaining $D_i$ from $D$ by replacing each element $x$ of $D$ by $(x, i)$. Then the blocks of $D_1, \ldots, D_p$ are $p$-subsets of a set of $p^{\beta+1}$ elements such that any pair of elements with the *same* second component lie in exactly one of the blocks, and such that the resolution classes of $D_1, \ldots, D_p$ can be combined to give resolution classes for the whole set.

Now take the blocks of the TD($p, p^\beta$) constructed in Example 4.5.3; any two elements with *different* second components occur together in exactly one of these blocks, so that, with the blocks of the $D_i$, they give overall balance with $\lambda = 1$. Finally, the resolution classes of the TD($p, p^\beta$) give further resolution classes for the $(p^{\beta+1}, p, 1)$ design so formed.

As a corollary, Kirkman took $\alpha = 2$ to obtain resolvable $(p^2, p, 1)$ designs, and hence finite projective planes, for all primes $p$.

## 4.6 Transversal designs and orthogonal arrays

Transversal designs are really just orthogonal arrays in disguise; hence their importance in the study of MOLS.

**Theorem 4.6.1** *The following are equivalent:*

(i) *A TD($k, m$) exists.*
(ii) *An OA($k, m$) exists.*
(iii) *There exist $k - 2$ MOLS of order $m$.*

**Proof** Since the equivalence of (ii) and (iii) was established in Theorem 4.2.1, it is sufficient to show that (i) and (ii) are equivalent. Suppose first that a TD($k, m$) exists. Label its elements so that the groups are $G_1 = \{1, \ldots, m\}, \ldots, G_k = \{(k - 1)m + 1, \ldots, km\}$, and write down each block as a column, with elements in increasing order. This gives a $k \times m^2$ array in which each element in the $i$th row is in $G_i$. Now subtract $(i-1)m$ from each entry in the $i$th row. The resulting array is an $OA(k, m)$, for if the $i$th and $j$th rows are considered, the ordered pair $(a, b)$ will occur in these two rows in the column corresponding to the unique block of the TD which contains the elements $a + (i-1)m$ and $b - (j-1)m$.

Conversely, given an $OA(k, m)$ on elements $1, \ldots, m$, add $(i-1)m$ to each entry in the $i$th row to obtain a $TD(k, m)$ with groups $G_i$ as above.

**Example 4.6.1**  Start with three MOLS of order 4 as in Example 1.3.4. They are represented by the $OA(5, 4)$ of Example 4.2.1. This in turn is equivalent to the $TD(5, 4)$ with groups $G_1 = (1, \ldots, 4), \ldots, G_5 = \{17, \ldots, 20\}$, whose blocks are the columns of the following array:

| 1 | 1 | 1 | 1 | 2 | 2 | 2 | 2 | 3 | 3 | 3 | 3 | 4 | 4 | 4 | 4 |
|---|---|---|---|---|---|---|---|---|---|---|---|---|---|---|---|
| 5 | 6 | 7 | 8 | 5 | 6 | 7 | 8 | 5 | 6 | 7 | 8 | 5 | 6 | 7 | 8 |
| 9 | 10 | 11 | 12 | 10 | 9 | 12 | 11 | 11 | 12 | 9 | 10 | 12 | 11 | 10 | 9 |
| 13 | 14 | 15 | 16 | 16 | 15 | 14 | 13 | 14 | 13 | 16 | 15 | 15 | 16 | 13 | 14 |
| 17 | 18 | 19 | 20 | 19 | 20 | 17 | 18 | 20 | 19 | 18 | 17 | 18 | 17 | 20 | 19. |

We note that, combining the above theorem with Theorem 4.1.1, we obtain the following result.

**Theorem 4.6.2**

(i)  *A* $TD(q + 1, q)$ *exists whenever* $q$ *is a prime power.*
(ii)  *A resolvable* $TD(q, q)$ *exists whenever* $q$ *is a prime power.*

One use of TDs is to construct PBDs. Indeed, we can simply take the groups to be further blocks.

**Theorem 4.6.3**  *If a* $TD(k, m)$ *exists then so does a* $PBD(km, \{k, m\}, 1)$.

**Example 4.6.2**  From a $TD(5, 9)$ we obtain a $PBD(45, \{5, 9\}, 1)$.

Another method is to remove some of the elements.

**Theorem 4.6.4**  *Suppose that a* $TD(k + n, m)$ *exists. Let* $0 \leqslant u_i \leqslant m$ *for each* $i = 1, \ldots, n$. *Then a* $PBD(km + u_1 + \cdots + u_n, \{k, k + 1, \ldots, k + n, m, u_1, \ldots, u_n\}, 1)$ *exists.*

**Proof**  For each $i$, remove $m - u_i$ elements from the $(i + k)$th group. Each block can have at most $n$ elements removed, and so produces a block of size at least $k$. Also, the reduced groups are of size $u_i$. Taking the groups of size $> 1$ as further blocks gives the required PBD.

The choice $k = 3$, $n = 2$ gives the following corollary.

**Corollary 4.6.5** *If* $0 \leqslant u, v \leqslant m$ *and* $N(m) \geqslant 3$, *then a* PBD$(3m + u + v, \{3, 4, 5, u, v\}, 1)$ *exists.*

**Example 4.6.3** Take $m = 4$, $u = v = 1$ to obtain the existence of a PBD$(14, \{3, 4, 5\}, 1)$.

In the exercises we show how Corollary 4.6.5 leads to the existence of a PBD$(v, \{3, 4, 5\}, 1)$ for all $v \equiv 2$ (mod 3), $v \geqslant 11$. It then follows from Exercises 4, number 11, that a PBD$(v, \{3, 4, 5\}, 1)$ exists for all $v$ except 2, 6, and 8.

## 4.7 Group divisible designs

In a transversal design the groups partition the set of elements, and each block intersects each group in precisely one element. This of course requires that the size of each block is equal to the number of groups. To enable us to construct more PBDs we now need to consider a generalization to the situation where each block contains *at most* one element of each group. We recall the *group divisible designs* GDD$(v, k, m)$ introduced in Definition 1.4.2. Note that the definition requires that $m$ divides $v$ and $v \geqslant km$. We now make some elementary remarks about the existence of GDDs.

Suppose we take a $(v, k, 1)$BIBD and remove one of its elements. The resulting blocks of size $k - 1$ are all disjoint and form a partition of the set of $v - 1$ remaining elements. We can take these as the groups, and the remaining blocks as the blocks, of a GDD$(v - 1, k, k - 1)$. In particular, starting with a $(k^2, k, 1)$ design gives the following result.

**Theorem 4.7.1** *If $k$ is a prime power then a* GDD$(k^2 - 1, k, k - 1)$ *exists.*

**Example 4.7.1** Take the $(9, 3, 1)$ design of Example 1.1.9 and remove the element 9. This gives a GDD$(8, 3, 2)$; its blocks are $\{1, 2, 3\}$, $\{2, 4, 5\}$, $\{5, 6, 7\}$, $\{1, 5, 8\}$, $\{3, 7, 8\}$, $\{3, 4, 6\}$, $\{1, 4, 7\}$, $\{2, 6, 8\}$ and its groups are $\{2, 7\}$, $\{4, 8\}$, $\{3, 5\}$, $\{1, 6\}$.

The converse procedure of starting with a GDD$(v, k, k - 1)$ and adjoining a new element $\infty$ to each group, and then considering the augmented groups as further blocks, clearly gives rise to a $(v + 1, k, 1)$BIBD. So the following result is established.

**Theorem 4.7.2** *A* GDD$(v, k, k - 1)$ *exists if and only if a* $(v + 1, k, 1)$ BIBD *exists.*

Just as the idea of a BIBD can be generalized to that of a PBD by allowing blocks to have different sizes, so can the above idea of a GDD be generalized.

**Definition 4.7.1** A GDD($v, K, M$) on a $v$-set $X$ consists of a collection $\mathscr{G}$ of disjoint subsets of $X$ (called *groups*) which partition $X$, and with $|G| \in M$ for all $G \in \mathscr{G}$, and a collection $\mathscr{B}$ of subsets of $X$ (called *blocks*), with $|B| \in K$ for all $B \in \mathscr{B}$, such that:

(i) each block contains at most one element from each group;
(ii) each pair of elements from different groups occurs in exactly one block.

As in Theorem 4.6.3, a GDD yields a PBD if we consider the groups to be further blocks.

**Theorem 4.7.3** *If a GDD($v, K, M$) exists, then a PBD($v, K \cup M, 1$) exists.*

Similarly, as in Theorem 4.6.4, a TD yields a GDD: what was really done in that theorem was to construct a GDD and then call the groups blocks. We state the result in the case $n = 1$.

**Theorem 4.7.4** *If a TD($k + 1, m$) exists and $0 \leqslant u \leqslant m$, then a GDD($km + u$, $\{k, k + 1\}, \{m, u\}$) exists.*

**Example 4.7.2** (Bennett, 1987) We give an application of these ideas to the problem of finding lower bounds for $N(m)$, the maximum number of MOLS of order $m$. Suppose we have a TD($k + 1, m$) on $km + m$ elements. The blocks are of size $k + 1$ and there are $k + 1$ groups, each of size $m$. Take a block and delete $k + 1 - t$ of its elements, $0 \leqslant t \leqslant k + 1$, to obtain a GDD($mk + m + t - k - 1, \{t, k, k + 1\}, \{m - 1, m\}$). Note that there is just one block of size $t$. Add a new element $\infty$ to each group and take these augmented groups as further blocks. This gives a PBD($mk + m + t - k, \{k, k + 1, m, m + 1, t\}, 1$) in which there is just one block of size $t$ (provided $t \neq k, k + 1, m, m + 1$). Theorem 4.4.1 can now be applied. For example, take $k = 16$, $m = 31$, $t = 9$; this yields a PBD($520, \{16, 17, 31, 32, 9\}, 1$) with just one block of size 9, and Theorem 4.4.1 now yields

$$N(520) \geqslant \min\{N(16) - 1, N(17) - 1, N(31) - 1, N(32) - 1, N(9)\} = 8.$$

This improves the Moore–MacNeish bound of $N(520) = N(5 \cdot 8 \cdot 13) \geqslant 4$. For further examples, see Exercises 4, number 22.

Finally in this chapter, we note that an important result on TDs due to Wilson (1974$a$) leads to the result: $N(m) \geqslant 3$ for all $m \geqslant 4$, $m \neq 6$ or 10.

## 4.8   Exercises 4

1. Construct six MOLS of order 7.
2. What lower bounds does Theorem 4.1.2 give for $N(24)$, $N(69)$, $N(80)$, $N(95)$, $N(96)$, $N(105)$, $N(273)$, $N(993)$, $N(1023)$? These will be improved upon in later exercises.
3. Use the method of proof of Theorem 4.2.2 to construct an OA(4, 10). Hence obtain two MOLS of order 10.
4. *Three MOLS of order* 14. Construct an OA(5, 14) by imitating the proof of Theorem 4.2.2 and Example 4.2.5, taking

$$A_0 = \begin{bmatrix} \infty & 0 & 0 & 0 & 0 & 0 & 0 & 0 & 0 & 0 & 0 & 0 & 0 & 0 & 0 \\ 0 & \infty & 0 & 1 & 2 & 3 & 4 & 5 & 6 & 7 & 8 & 9 & 10 & 11 & 12 \\ 0 & 0 & \infty & 2 & 10 & 12 & 7 & 9 & 5 & 4 & 1 & 11 & 8 & 3 & 6 \\ 0 & 1 & 2 & \infty & 5 & 9 & 3 & 12 & 7 & 11 & 0 & 4 & 6 & 8 & 10 \\ 0 & 3 & 12 & 9 & 6 & \infty & 2 & 7 & 11 & 1 & 5 & 10 & 0 & 4 & 8 \end{bmatrix}$$

and verifying that $[E \; A_0 \; A_1 \; \cdots \; A_{12}]$ is an OA(5, 14), where $E' = (\infty, \infty, \infty, \infty, \infty)$. (Todorov, 1985)

5. Parker's proof that if $q > 3$ is a prime power congruent to 3 (mod 4) then there exist two MOLS of order $\frac{1}{2}(3q - 1)$. Let $\theta$ be a primitive element of GF($q$) and let

$$C_0 = \begin{bmatrix} x_1 & x_2 & \cdots & x_{(q-1)/2} \\ 0 & 0 & \cdots & 0 \\ \theta^2 & \theta^4 & \cdots & \theta^{q-1} \\ \theta^2(\theta+1) & \theta^4(\theta+1) & \cdots & \theta^{q-1}(\theta+1) \end{bmatrix}.$$

Let $C_1$, $C_2$, $C_3$ be obtained from $C_0$ by permuting the rows cyclically, and let $A_0 = C_0 C_1 C_2 C_3$. The $A_0$ has the property that the numerical differences between any two rows are the nonzero elements of GF($q$). For each $y \in$ GF($q$) let $A_y$ be obtained from $A_0$ by adding $y$ to each numerical entry in $A_0$; and define $B$ and $E$ as in Example 4.2.5. Then $[A_0 \; A_\theta \; \cdots \; A_{\theta^{q-1}} \; B \; E]$ is an OA(4, $\frac{1}{2}(3q - 1)$). (Parker, 1959$a$)

6. Construct a PBD(69, {7, 8, 9}, 1) from a plane of order 8, and use it to show that $N(69) \geqslant 5$.

7. Improve some of the estimates obtained in question 2 above, as follows: $N(95) \geqslant 6$, $N(96) \geqslant 6$, $N(105) \geqslant 4$, $N(273) \geqslant 16$, $N(993) \geqslant 31$. (Corollary 4.4.3 is particularly useful.)

8. Take a $(q^2, q, 1)$ design and delete just one element to obtain a PBD($q^2 - 1$, {$q, q - 1$}, 1) in which the ($q - 1$)-sets form a clear set. Deduce that $N(q^2 - 1) \geqslant N(q - 1)$ for all prime powers $q$. Does this help to improve estimates in question 2 above for $N(24)$, $N(80)$, $N(1023)$?

9. Derive from a finite projective plane of order $q$ a PBD($q^2 + q - x$, $\{q - x, q, q + 1\}, 1$) with just one block of size $q - x (x \leqslant q)$. Deduce that, for all prime powers $q$,

$$N(q^2 + q - x) \geqslant \min\{N(q - x), N(q) - 1, N(q + 1) - 1\}.$$

For example, $N(264) \geqslant 7$, $N(265) \geqslant 8$, $N(167) \geqslant 10$.

10. Suppose that a resolvable $(v, k, 1)$ design exists with $r$ resolution classes.
    (a) Show that $N(v + r) \geqslant \min\{N(r), N(k + 1) - 1\}$; deduce that $N(57) \geqslant 6$.
    (b) Show that, if $0 < x < r$, $N(v + x) \geqslant \min\{N(x), N(k) - 1, N(k + 1) - 1\}$; deduce that $N(54) \geqslant 4$, $N(50) \geqslant 5$.

11. *Existence of a PBD($v, \{3, 4\}, 1$) for all $v \equiv 0$, or 1 (mod 3), $v \neq 6$.*
    (i) Use projective or affine planes to deal with $v \leqslant 13$.
    (ii) Use a resolvable $(15, 3, 1)$ design (Example 1.2.1) to deal with $v = 15, 16, 18, 19$.
    (iii) Use $v = 3 \cdot 7 + h$ and Corollary 4.4.3(a) to deal with $v = 21, 22, 24, 25$, breaking down blocks of size 7 into blocks of size 3 by means of a $(7, 3, 1)$ design.
    (iv) Suppose it is known that a PBD($v, \{3, 4\}, 1$) exists for all $v \equiv 0$ or 1 (mod 3), $6 < v < 9m$ ($m \geqslant 3$). Consider $v = 3(3m) + h$, $h = 0, 1, 3, 4$, and $v = 3(3m + 1) + h$, $h = 3, 4$ to deal with all $v < 9(m + 1)$, breaking down blocks of sizes $3m$, $3m + 1$ into blocks of size 3, 4.
    (v) Existence now follows by induction.

12. Given a $(v, 3, 1)$ design $S$, where $N(v) \geqslant 3$, show how to construct a $(3v, 3, 1)$ design, as follows. Take copies of $S$ on $\{1, \ldots, v\}$, $\{v + 1, \ldots, 2v\}$ and $\{2v + 1, \ldots, 3v\}$: this gives $\frac{1}{2}v(v - 1)$ triples. Take three MOLS $A_0, A_1, A_2$ on $\{1, \ldots, v\}$, add $v$ to each entry of $A_1$ and $2v$ to each entry of $A_2$; and then obtain $v^2$ triples $T_{ij}$ where $T_{ij}$ consists of the $(i, j)$th entries of $A_0, A_1, A_2$. These $v^2 + \frac{1}{2}v(v - 1)$ triples form a $(3v, 3, 1)$ design.

13. *Regular symmetric Hadamard matrices.*
    (a) Suppose $N(2t) \geqslant t - 2$. We show how to construct a *regular* symmetric Hadamard matrix (i.e. one with constant row and column sums) of order $4t^2$. Take an OA($t, 4t^2$) and define $H = (h_{ij})$ by $h_{ii} = 1$ and

$$h_{ij} = \begin{cases} -1 & \text{if the } i\text{th and } j\text{th columns of the OA agree in some row;} \\ 1 & \text{otherwise.} \end{cases}$$

(Note that no two columns of the OA can agree in more than one row.)
    (i) $H$ is regular, with $2t^2 + t$ entries $+1$ and $2t^2 - t$ entries $-1$ in each row.

(ii) Any two rows of $H$ have $-1$ together in $t^2 - t$ places.

(iii) Any two rows of $H$ differ in $2t^2$ places, and hence are orthogonal. So $H$ is a regular symmetric Hadamard matrix of order $4t^2$.

(b) Construct a regular symmetric Hadamard matrix of order 36.

(c) Observe that a regular symmetric Hadamard matrix of order $2^{2u}$ exists, and hence so do the $(2^{2u}, 2^{2u-1} - 2^{u-1}, 2^{2u-2} - 2^{u-1})$ designs of Exercises 3, number 14. (Bush, 1973)

14. *Five MOLS of order* 12. It was shown by Dulmage, Johnson, and Mendelsohn in 1961 that $N(12) \geqslant 5$. Let $A$ be the reverse circulant matrix with first row 0 1 2 3 4 5, and let $B = A + 6J$. Define

$$L_1 = \begin{bmatrix} A & B \\ B & A \end{bmatrix},$$

and obtain $L_2, \ldots, L_5$ from $L_1$ by permuting the columns in such a way that

| | | | | | | | | | | | | |
|---|---|---|---|---|---|---|---|---|---|---|---|---|
| $L_2$ has first row | 0 | 6 | 8 | 2 | 7 | 1 | 9 | 11 | 4 | 10 | 5 | 3; |
| $L_3$ has first row | 0 | 3 | 6 | 1 | 9 | 11 | 2 | 8 | 5 | 4 | 7 | 10; |
| $L_4$ has first row | 0 | 8 | 1 | 11 | 5 | 9 | 3 | 10 | 2 | 7 | 6 | 4; |
| $L_5$ has first row | 0 | 4 | 11 | 10 | 2 | 7 | 8 | 6 | 9 | 1 | 3 | 5. |

Verify that $L_1, \ldots, L_5$ are mutually orthogonal.

15. *Four MOLS of order* 46.

(i) The non-zero fourth powers (or quartic residues) in $Z_{37}$ are $2^4 = 16$, $2^8 = 34$, $2^{12} = 26$, $2^{16} = 9$, $2^{20} = 33$, $2^{24} = 10$, $2^{28} = 12$, $2^{32} = 7$, $2^{36} = 1$. These form a $(37, 9, 2)$ difference set.

(ii) For each quartic residue $r$ in $Z_{37}$ let $M_r$ be the $6 \times 6$ circulant matrix whose first row is $(\infty_r, 34r, 14r, 21r, 13r, r)$. Verify that the numerical differences between rows 1 and 2 are $17r, 7r, 29r, 25r$, i.e. $2^{4(-)+3}$, $2^{4(-)}$, $2^{4(-)+1}$, $2^{4(-)+2}$, so that as $r$ varies, every non-zero residue (mod 37) occurs exactly once.

(iii) Let

$$A = \begin{bmatrix} M_{r_1} & (M_{r_1} + J) \cdots (M_{r_1} + 36J) & M_{r_2} \cdots (M_{r_9} + 36J) \end{bmatrix},$$

where $\infty_{r_i} + m = \infty_{r_i}$ for each $i$ and each $m \in Z_{37}$. Also let $B$ be an OA(6,9) on $\infty_{r_1}, \ldots, \infty_{r_9}$. Then [A B] is an OA(6, 46). (Wilson, 1974*b*]

16. $N(15) \geqslant 4$. Take the four squares $L_1, \ldots, L_4$ whose first rows are given by

| | | | | | | | | | | | | | | | |
|---|---|---|---|---|---|---|---|---|---|---|---|---|---|---|---|
| $L_1$: | 1 | 15 | 2 | 14 | 3 | 13 | 4 | 12 | 5 | 11 | 6 | 10 | 7 | 9 | 8 |
| $L_2$: | 1 | 14 | 3 | 11 | 6 | 9 | 8 | 7 | 10 | 4 | 13 | 12 | 5 | 15 | 2 |
| $L_3$: | 1 | 10 | 7 | 13 | 4 | 2 | 15 | 6 | 11 | 9 | 8 | 3 | 14 | 12 | 5 |
| $L_4$: | 1 | 6 | 11 | 10 | 7 | 15 | 2 | 5 | 12 | 14 | 3 | 9 | 8 | 4 | 13, |

in each case the $i$th row being obtained from the first by adding $i - 1$ (mod 15) to each entry. To see that this works, just note that the differences between the members of the vertical pairs in any two first rows are precisely the members of $Z_{15}$ once each. (Schellenberg, van Rees, Vanstone, 1978)

17. *Existence of a PBD($v, \{4, 5, 8, 9, 12\}, 1$) for all $v \equiv 0$ or 1 (mod 4), $v \geqslant 4$.* The following approach uses the fact that $N(4m) \geqslant 3$ for all $m$.
    (i) Use the finite planes to deal with each $v$ from 13 to 25.
    (ii) Use a resolvable $(28, 4, 1)$ design to deal with $v$ from 28 to 37.
    (iii) Use Corollary 4.4.3(a) to deal with all $v$ from 40 to 80, except for 60 (Theorem 4.6.3) and 64 (Theorem 4.5.4).
    (iv) Suppose that the required PBD has been established for all $v = 0$ or 1 (mod 4), $4 \leqslant v \leqslant 16m$ ($m \geqslant 5$). Consider $4(4m) + h$ to deal with $v$ from $16m + 1$ to $16m + 16$, using the induction hypothesis to break down blocks of sizes $13, 16, 4m$.

18. Construct a TD$(6, 5)$ and a resolvable TD$(5, 5)$.

19. By deleting eight elements from each of three groups of a TD$(10, 9)$ so that the three remaining elements of these groups are not all in the same block, construct a PBD($66, \{7, 8, 9\}, 1$). Deduce that $N(66) \geqslant 5$.

20. Show how to construct the following:
    (i) PBD($90, \{9, 10\}, 1$);
    (ii) GDD($55, \{7, 8\}, \{6, 7\}$);
    (iii) GDD($40, 5, 4$).

21. Existence of a PBD($v, \{3, 4, 5\}, 1$) for all $v \neq 1, 2, 6, 8$.
    (i) Show that no such design can exist for $v = 6, 8$.
    (ii) Observe that in view of question 11 we need only consider $v \equiv 2$ (mod 3), $v \geqslant 11$.
    (iii) Construct a PBD($11, \{3, 5\}, 1$) as in Exercises 1 number 27.
    (iv) Use Corollary 4.6.5 with $m = 4$ to deal with $v = 14, 17, 20$; with $m = 5$ for $v = 23$; with $m = 7$ for $v = 26, 29, 32, 35$, breaking down any blocks of size 7 so formed by using a $(7, 3, 1)$ design.
    (v) Suppose all $v < 18n$ have been dealt with. Use Corollary 4.6.5 with $m = 6n - 1$ to deal with all $v$ up to $18(n + 1)$.

22. Use the method of Example 4.7.2 to show that $N(522) \geqslant 10$ and $N(4060) \geqslant 26$.

23. Show that $N(58) \geqslant 5$ by using a PBD($58, \{7, 8, 9\}, 1$) which can be obtained from a $(64, 8, 1)$ design by adjoining $\infty$ to all blocks of one class and then removing seven elements from any block of another class.

24. Show that $N(62) \geqslant 3$ by adding $\infty_i$ to each block of the $i$th class ($1 \leqslant i \leqslant 5$) of a $(64, 8, 1)$ design and imitating the method of previous question to form a PBD($62, \{5, 7, 8, 9\}, 1$).

# Self-orthogonal Latin squares

## 5.1 SOLS and mixed doubles tournaments

In the previous chapter we presented a proof of the fact that, if $n \neq 2$ or 6, two MOLS of order $n$ exist. This result, the culmination of a great deal of effort spurred on by a conjecture almost 200 years old, is not, however, the end of the story. One development is to show that, for all $n \neq 2, 3, 6$, or (possibly) 10, there are in fact three MOLS of order $n$ and, in general, to show that, given any positive integer $m$, there exists an integer $n(m)$ such that $N(n) \geqslant m$ for all $n > n(m)$. However, another remarkable development will be dealt with in this chapter: it will be shown that, for every $n \neq 2, 3$, or 6, there exists a Latin square of order $n$ which is orthogonal to its transpose! Such Latin squares are useful in the construction of other combinatorial structures such as whist tournaments.

**Definition 5.1.1** A Latin square is *self-orthogonal* if it is orthogonal to its transpose. A self-orthogonal Latin square (SOLS) of order $n$ will be denoted by SOLS($n$).

**Example 5.1.1** The following is a SOLS(4)

$$
\begin{array}{cccc}
1 & 2 & 3 & 4 \\
4 & 3 & 2 & 1 \\
2 & 1 & 4 & 3 \\
3 & 4 & 1 & 2 \\
\end{array}
$$

Already SOLS of orders 10 and 14 have been exhibited (Examples 4.2.3 and 4.2.4). It is, in fact, quite easy to produce SOLS of infinitely many orders.

**Definition 5.1.2** A Latin square $A = (a_{ij})$ of order $n$ with entries in $Z_n$ is *cyclic* if $a_{i+1, j+1} = a_{i,j} + 1(\mathrm{mod}\ n)$ for all $i, j$.

**Theorem 5.1.1** (Mendelsohn, 1971)  *A cyclic* SOLS($n$) *exists whenever* $(n,6) = 1$.

**Proof**  Define the $n \times n$ square array $A = (a_{ij})$ by $a_{ij} \equiv 2i - j (\bmod n)$, $1 \leqslant a_{ij} \leqslant n$. First check that $A$ is a Latin square:

$$a_{ij} = a_{ik} \Rightarrow 2i - j \equiv 2i - k (\bmod n) \Rightarrow j \equiv k (\bmod n) \Rightarrow j = k;$$

$$a_{ij} = a_{kj} \Rightarrow 2i - j \equiv 2k - j (\bmod n) \Rightarrow 2i \equiv 2k (\bmod n)$$

$$\Rightarrow i \equiv k (\bmod n) \text{ since } n \text{ is odd}$$

$$\Rightarrow i = k.$$

Thus $A$ is a Latin square. Clearly, $A$ is cyclic. Now prove that $A$ and $A'$ are orthogonal, where $A' = (a'_{ij}) = (a_{ji})$. Suppose that $a_{ij} = a_{kl}$ and $a'_{ij} = a'_{kl}$, i.e. $a_{ij} = a_{kl}$ and $a_{ji} = a_{lk}$. Then $2i - j \equiv 2k - l$ and $2j - i \equiv 2l - k (\bmod n)$. Adding these congruences gives $i + j \equiv k + l (\bmod n)$, whence $2(k + l - j) - j \equiv 2k - l (\bmod n)$, i.e. $3l \equiv 3j (\bmod n)$. Since $(n,3) = 1$ this yields $l \equiv j (\bmod n)$, whence $l = j$ and $i = k$.

**Example 5.1.2**  The construction in the proof gives the following SOLS of orders 5 and 7.

```
                         1 7 6 5 4 3 2
          1 5 4 3 2      3 2 1 7 6 5 4
          3 2 1 5 4      5 4 3 2 1 7 6
          5 4 3 2 1      7 6 5 4 3 2 1
          2 1 5 4 3      2 1 7 6 5 4 3
          4 3 2 1 5      4 3 2 1 7 6 5
                         6 5 4 3 2 1 7
           SOLS(5)          SOLS(7)
```

A slight modification of the above proof yields the following.

**Theorem 5.1.2** (Mendelsohn, 1971)  *If $n$ is a prime power, $n \neq 2$ or 3, then a* SOLS($n$) *exists.*

**Proof**  A SOLS($n$) with the elements of GF($n$) as its entries will be constructed. Let GF($n$) = $\{c_1 = 0, c_2, \ldots, c_n\}$ and choose $\lambda \in$ GF($n$) such that $\lambda \neq 0$, $\lambda \neq 1$, $2\lambda \neq 1$. Then define the $n \times n$ array $A = (a_{ij})$ by

$$a_{ij} = \lambda c_i + (1 - \lambda) c_j.$$

As in the previous proof it is easily seen that $A$ is a Latin square. Suppose now that $a_{ij} = a_{kl}$ and $a'_{ij} = a'_{kl}$. Then $\lambda c_i + (1 - \lambda) c_j = \lambda c_k + (1 - \lambda) c_l$ and $\lambda c_j + (1 - \lambda) c_i = \lambda c_l + (1 - \lambda) c_k$; then adding these gives

$c_i + c_j = c_k + c_l$ so that $\lambda(c_k + c_l - c_j) + (1 - \lambda)c_j = \lambda c_k + (1 - \lambda)c_l$, i.e. $(1 - 2\lambda)c_j = (1 - 2\lambda c_l)$, whence $c_j = c_l$ since $2\lambda \neq 1$. Thus $j = l$ and $i = k$.

Note that this result guarantees the existence of a SOLS($n$) for $n = 4, 8, 16, \ldots$ and $9, 27, 81, \ldots$, which are not covered by Theorem 5.1.1. Our aim is to prove the following 1976 result of Brayton. Coppersmith, and Hoffman.

**Theorem 5.1.3** *If $n \neq 2, 3,$ or $6$, a SOLS($n$) exists.*

Before we set out on the proof of this, it is convenient to show how SOLS are related to a certain type of mixed doubles tournament.

**Definition 5.1.3** *A spouse-avoiding mixed doubles round robin tournament for $n$ couples (abbreviated to SAMDRR($n$)) is a tournament involving $n$ husband and wife couples; each match involves two players of opposite sex playing against two other players of opposite sex, and the matches are such that every two players of the same sex play against each other once, and each player plays with each member of the opposite sex (other than the spouse) exactly once as partner and once as opponent.*

Such tournaments were asked for by Dudeney (1917); he gave the following resolvable example for $n = 4$. where the couples are $(H_i, W_i)$, $1 \leqslant i \leqslant 4$.

**Example 5.1.3** A SAMDRR(4).

| | |
|---|---|
| $H_1W_3$ v $H_2W_4$ | $H_3W_1$ v $H_4W_2$ |
| $H_1W_4$ v $H_3W_2$ | $H_4W_1$ v $H_2W_3$ |
| $H_1W_2$ v $H_4W_3$ | $H_2W_1$ v $H_3W_4$ |

Note the resolvability: the games can be played in three rounds.

Before showing the connection between SOLS($n$) and SAMDRR($n$), we need a property of SOLS.

**Lemma 5.1.4** *If $A$ is a SOLS, its main diagonal is a transversal.*

**Proof** $A'$ has the same diagonal as $A$. If $a_{ii} = a_{jj} = k$ then the pair $(k, k)$ would occur twice in the join of $A$ and $A'$, contradicting orthogonality.

**Theorem 5.1.5** *A SAMDRR($n$) exists if and only if a SOLS($n$) exists.*

**Proof** Suppose a SAMDRR($n$) exists, and that the spouse pairs are $(H_i, W_i)$, $1 \leqslant i \leqslant n$. Define an $n \times n$ array $A = (a_{ij})$ by

$a_{ii} = i$;

$a_{ij} = l$ where $W_l$ is the partner of $H_i$ when $H_i$ plays $H_j$ $(i \neq j)$.

Since partnerships are never repeated, $A$ is clearly a Latin square. To prove $A$, $A'$ orthogonal, suppose that $a_{ij} = a_{IJ}$ and $a_{ji} = a_{JI}$. If $a_{ij} = l$ and $a_{ji} = m$ we have the match $H_iW_l$ v $H_jW_m$ and also the match $H_IW_l$ v $H_JW_m$. But $W_l$ and $W_m$ oppose each other only once, so $i = I$ and $j = J$. Thus $A$ is indeed a SOLS($n$).

Conversely, given a SOLS($n$), the entries can, by Lemma 5.1.4, be relabelled so that $a_{ii} = i$ for each $i$. If then $a_{ij} = l$ and $a_{ji} = m$, arrange the match $H_iW_l$ v $H_jW_m$.

**Example 5.1.4** Start with the SOLS(4) of Example 5.1.1, and relabel the entries to obtain a SOLS(4) with $(i, i)$th entry $i$:

$$
\begin{array}{cccc}
1 & 4 & 2 & 3 \\
3 & 2 & 4 & 1 \\
4 & 1 & 3 & 2 \\
2 & 3 & 1 & 4.
\end{array}
\tag{5.1}
$$

This yields a SAMDRR(4) with matches as follows:

$$
\begin{array}{lll}
H_1W_4 \text{ v } H_2W_3 & H_1W_2 \text{ v } H_3W_4 & H_1W_3 \text{ v } H_4W_2 \\
H_2W_4 \text{ v } H_3W_1 & H_2W_1 \text{ v } H_4W_3 & H_3W_2 \text{ v } H_4W_1.
\end{array}
$$

Although the SAMDRR obtained in this example is resolvable, it does not always happen that a SOLS yields a resolvable SAMDRR. The question of resolvability will be discussed in section 5.4.

## 5.2 New SOLS from old

In this section we show that constructions already met in the study of MOLS are also applicable to SOLS.

**Theorem 5.2.1** *If a SOLS($m$) and a SOLS($n$) exist, then a SOLS($mn$) also exists.*

**Proof** Imitate the proof of the Moore–MacNeish theorem (4.1.2). Let $A, B$ be SOLS of orders $m, n$ on $\{0, 1, \ldots, m - 1\}, \{0, 1, \ldots, n - 1\}$ respectively. Then with products as defined in that proof, $A \times B$ and $A' \times B'$ are MOLS of order $mn$. But it is easily seen that $(A \times B)' = A' \times B'$; so $A \times B$ is self-orthogonal.

**Corollary 5.2.2** *If $n = 2^\alpha 3^\beta 5^\gamma \cdots$ where $\alpha \neq 1$ and $\beta \neq 1$ then there exists a SOLS($n$).*

**Proof**  Follows immediately from Theorems 5.1.2 and 5.2.1.

**Example 5.2.1**  Since a SOLS(4) and a SOLS(5) exist, a SOLS(20) exists.

**Theorem 5.2.3**  *If a SOLS($m$) exists then a SOLS($3m + 1$) also exists.*

**Proof**  This is just Theorem 4.2.4.

**Example 5.2.2**  By Theorem 5.2.3 and Example 5.1.2 a SOLS(22) exists. Note that $n = 22$ is not covered by Corollary 5.2.2.

Next we present a rather nice result concerning PBDs.

**Theorem 5.2.4**  *Suppose that a PBD($v, K, 1$) exists and that, for each $k \in K$, a SOLS($k$) exists. Then a SOLS($v$) exists.*

**Proof**  For each $k$, replace each block of size $k$ by a SAMDRR($k$) on its elements. The union of these tournaments is a SAMDRR($v$); so by Theorem 5.1.5 a SOLS($v$) exists.

**Example 5.2.3**  From the affine plane of order 7 construct a PBD($50, \{7, 8\}, 1$). By Theorem 5.1.2 a SOLS(7) and a SOLS(8) exist; so by Theorem 5.2.4 a SOLS(50) exists.

**Example 5.2.4**  Remove an element from an affine plane of order 5 to get a PBD($24, \{4, 5\}, 1$). So a SOLS(24) exists.

Next recall Corollary 4.4.3 (a), one of the key results relating MOLS and PBDs. If we take $k = 4$ we obtain the fact that if $N(m) \geqslant 3$ and $1 \leqslant x < m$ then a PBD($4m + x, \{4, 5, m, x\}, 1$) exists. The following consequence of Theorem 5.2.4 now follows.

**Theorem 5.2.5**  *If $N(m) \geqslant 3$ and a SOLS($m$) and a SOLS($x$) exist, $1 \leqslant x < m$, then a SOLS($4m + x$) also exists.*

**Example 5.2.5**  (i) Take $m = 11$, $x = 7$: a SOLS(51) exists.

(ii) Take $m = 19$ $x = 10$: a SOLS(86) exists.

This theorem will be used in the proof of the Brayton–Coppersmith–Hoffman theorem; it will be applied in the case when $m$ is a multiple of 4. This application requires the facts that, for each $k \geqslant 1$, $N(4k) \geqslant 3$ and a SOLS($4k$) exists. The first of these was established in Lemma 4.4.5; the other is now confirmed.

**Theorem 5.2.6** *For each $k \geq 1$, a SOLS($4k$) exists.*

**Proof** Any $4k$ can be written as $4k = 2^\alpha 3^\beta u$ where $\alpha = 2$ or 3, $\beta = 0$ or 1, and where $u$ is a product of prime powers none of which is $2^1$ or $3^1$. By Corollary 5.2.2 a SOLS($u$) exists, so, by Theorem 5.2.1, we need only show the existence of a SOLS($n$) when $n$ is $2^\alpha$ or $3 \cdot 2^\alpha$, $\alpha = 2$ or 3, i.e. when $n = 4, 8, 12$ or 24. Of these, 4 and 8 are dealt with by Theorem 5.1.2 and 24 is in Example 5.2.4. Finally, a SOLS(12) is exhibited below.

**Example 5.2.6** A SOLS(12) due to Brayton, Coppersmith, and Hoffman.

| | | | | | | | | | | | |
|---|---|---|---|---|---|---|---|---|---|---|---|
| 0 | 8 | 3 | 6 | 2 | 9 | 11 | 1 | 10 | 5 | 7 | 4 |
| 10 | 1 | 9 | 4 | 7 | 3 | 5 | 6 | 2 | 11 | 0 | 8 |
| 4 | 11 | 2 | 10 | 5 | 8 | 9 | 0 | 7 | 3 | 6 | 1 |
| 9 | 5 | 6 | 3 | 11 | 0 | 2 | 10 | 1 | 8 | 4 | 7 |
| 1 | 10 | 0 | 7 | 4 | 6 | 8 | 3 | 11 | 2 | 9 | 5 |
| 7 | 2 | 11 | 1 | 8 | 5 | 0 | 9 | 4 | 6 | 3 | 10 |
| 5 | 7 | 4 | 11 | 1 | 10 | 6 | 2 | 9 | 0 | 8 | 3 |
| 11 | 0 | 8 | 5 | 6 | 2 | 4 | 7 | 3 | 10 | 1 | 9 |
| 3 | 6 | 1 | 9 | 0 | 7 | 10 | 5 | 8 | 4 | 11 | 2 |
| 8 | 4 | 7 | 2 | 10 | 1 | 3 | 11 | 0 | 9 | 5 | 6 |
| 2 | 9 | 5 | 8 | 3 | 11 | 7 | 4 | 6 | 1 | 10 | 0 |
| 6 | 3 | 10 | 0 | 9 | 4 | 1 | 8 | 5 | 7 | 2 | 11 |

Combining the last two theorems together we obtain the following key corollary.

**Corollary 5.2.7** *If $1 \leq x < 4k$ and a SOLS($x$) exists then a SOLS($16k + x$) also exists.*

## 5.3 The Brayton–Coppersmith–Hoffman theorem

Theorem 5.1.3 will be proved by making use of Corollary 5.2.7 and considering each residue class mod 16 separately. But before this is done a certain number of 'small' values of $n$ have to be disposed of separately, namely $n = 15, 18, 26, 30, 38$, and 42.

**Example 5.3.1** A SOLS(15) due to Franklin (1984). The cyclic Latin square with first row

| 15 | 2 | 11 | 6 | 10 | 13 | 5 | 14 | 4 | 7 | 12 | 1 | 9 | 8 | 3 |
|---|---|---|---|---|---|---|---|---|---|---|---|---|---|---|

is a SOLS(15).

**Example 5.3.2** A SOLS(18) due to Brayton *et al.* As in Example 4.2.3, cyclically develop

$$1 \quad 3 \quad 12 \quad 7 \quad 15 \quad 14 \quad 2 \quad 10 \quad 5 \quad 8 \quad 16 \quad 11 \quad 17 \quad 4 \quad 9 \quad 13 \quad \infty$$

mod 17, and then add the appropriate last row and column.

**Example 5.3.3** A SOLS(26). The OA(4, 26) constructed in Example 4.2.5 has the property that if $(a, b, c, d)'$ is a column then so is $(b, a, d, c)'$. As in the proof of Theorem 4.2.4, this guarantees that the OA yields a SOLS.

**Example 5.3.4** A SOLS(30) can be obtained by imitating the previous example, replacing the first quarter of $A_0$ by

| 0 | 0 | $x_1$ | $x_2$ | $x_3$ | $x_4$ | $x_5$ | $x_6$ | $x_7$ |
|---|---|---|---|---|---|---|---|---|
| 3 | 7 | 9 | 5 | 7 | 11 | 22 | 19 | 8 |
| 1 | 10 | 0 | 0 | 0 | 0 | 0 | 0 | 0 |
| 6 | 2 | 22 | 21 | 19 | 17 | 14 | 10 | 12 |

where numerical entries are in $Z_{23}$.

**Example 5.3.5** SOLS(38) This time start with

| 0 | 0 | 0 | 0 | $x_1$ | $x_2$ | $x_3$ | $x_4$ | $x_5$ | $x_6$ | $x_7$ |
|---|---|---|---|---|---|---|---|---|---|---|
| 8 | 3 | 6 | 1 | 11 | 27 | 9 | 19 | 6 | 18 | 23 |
| 3 | 1 | 12 | 4 | 0 | 0 | 0 | 0 | 0 | 0 | 0 |
| 15 | 11 | 16 | 18 | 2 | 26 | 7 | 9 | 20 | 13 | 16 |

where numerical entries are in $Z_{31}$.

**Example 5.3.6** SOLS(42) This time start with

| 0 | 0 | 0 | 0 | 0 | $x_1$ | $x_2$ | $x_3$ | $x_4$ | $x_5$ | $x_6$ | $x_7$ |
|---|---|---|---|---|---|---|---|---|---|---|---|
| 8 | 10 | 5 | 3 | 4 | 6 | 13 | 19 | 21 | 27 | 16 | 7 |
| 3 | 12 | 1 | 14 | 17 | 0 | 0 | 0 | 0 | 0 | 0 | 0 |
| 17 | 3 | 15 | 26 | 33 | 2 | 28 | 9 | 11 | 22 | 15 | 18 |

where numerical entries are in $Z_{35}$.

These constructions are all due to Brayton *et al.*

**Proof of Theorem 5.1.3** To prove the existence of a SOLS($n$) for all $n \neq 2, 3, 6$, we consider values of $n$ according to their residue class mod 16. The cases $n = 0, 4, 8, 12 \pmod{16}$ have already been covered by Theorem 5.2.6: for the other cases we use Corollary 5.2.7.

$n = 16k + 1$   Simply take $x = 1$ in Corollary 5.2.7.

$n = 16k + 2$   Here we write $16k + 2 = 16(k - 1) + 18$, so the result will follow whenever $18 < 4(k - 1)$, i.e. $k \geqslant 6$. But $k = 1$ is dealt with by Example 5.3.2; $k = 2$ and 5 yield to Theorem 5.2.3 since $34 = 3 \cdot 11 + 1$ and $82 = 3 \cdot 27 + 1$: $k = 3$ is dealt with by Example 5.2.3; and finally $k = 4$ yields to Theorem 5.2.4 since a PBD$(66, \{5, 14\}, 1)$ exists (Exercises 5, number 5) and a SOLS(14) exists (Example 4.2.4).

$n = 16k + 3$   Write $16k + 3 = 16(k - 1) + 19$, so that the result follows immediately provided $19 < 4(k - 1)$, i.e. $k \geqslant 6$. If $k = 1$, 4, or 5, $n$ is a prime and Theorem 5.1.2 can be applied; $k = 2$ yields to Theorem 5.2.1 since $35 = 5 \cdot 7$; and $k = 3$ yields to Theorem 5.2.4 since a PBD$(51, \{4, 5, 10, 11\}, 1)$ can be obtained from the points of five disjoint blocks of a resolvable $(121, 11, 1)$ design with four points of a block of another class removed.

$n = 16k + 5$   This yields to Corollary 5.2.7 provided $5 < 4k$, i.e. $k \geqslant 2$. But $k = 0$ gives $n$ prime, and $k = 1$ gives $n = 21$ which yields to Theorem 5.2.4 since a $(21.5.1)$ design exists.

$n = 16k + 6$   Here $16k + 6 = 16(k - 1) + 22$, and the result follows if $k \geqslant 7$. The cases $k = 1$ and 2 are dealt with by Examples 5.2.2 and 5.3.5; $k = 3$ is dealt with using a PBD$(54, \{4, 5, 10, 11\}, 1)$: $k = 4$ yields to Theorem 5.2.3 since $70 = 3 \cdot 23 + 1$: $k = 5$ uses a PBD$(86, \{5, 18\}, 1)$: and $k = 6$ yields to $102 = 4 \cdot 23 + 10$.

$n = 16k + 7$   This yields to Corollary 5.2.7 provided $k \geqslant 2$. But $k = 0, 1$ give $n$ prime.

$n = 16k + 9$   The only problem here is when $k \leqslant 2$. But $n$ is prime or a prime power in these cases.

$n = 16k + 10$   The only problem is if $k \leqslant 2$. The case $k = 0$ is dealt with in Example 4.2.3; $k = 1$ and 2 are dealt with in Examples 5.3.3 and 5.3.6.

$n = 16k + 11$   If $k = 0, 1$, or 2, $n$ is a prime power.

$n = 16k + 13$   If $k = 0, 1$, or 3, $n$ is a prime; $k = 2$ yields to $45 = 5 \cdot 9$.

$n = 16k + 14$   The cases $k = 0, 1$ are dealt with in Examples 4.2.4 and 5.3.4; $k = 2$ yields to $46 = 3 \cdot 15 + 1$, and $k = 3$ is dealt with using a PBD$(62, \{4, 5, 10, 13\}, 1)$.

$n = 16k + 15$   Here $k = 0$ is dealt with in Example 5.3.1; $k = 1, 2$ give $n$ prime, and $k = 3$ yields to $63 = 7 \cdot 9$.

## 5.4 Resolvable SAMDRR

The results of the previous sections establish the existence of a SAMDRR($n$) for all $n$ other than $1, 2, 3, 6$. Such tournaments could be played one game at a time, but it would be better if the games could be arranged in rounds so that every player plays in exactly one game in each round. This is of course impossible if $n$ is odd; in that case we might ask for one couple to sit out in each round.

**Definition 5.4.1**  A SAMDRR($n$) is *resolvable* if either

 (i) $n$ is even, and the games can be arranged in $n - 1$ rounds so that each player plays in exactly one game in each round; or
(ii) $n$ is odd, and the games can be arranged in $n$ rounds so that in round $i$ every player except the $i$th husband and wife plays in a game.

**Example 5.4.1**  The SAMDRR(4) of Example 5.1.4 is resolvable:

| Round 1: | $H_1W_4$ v $H_2W_3$ | $H_3W_2$ v $H_4W_1$ |
|---|---|---|
| Round 2: | $H_1W_2$ v $H_3W_4$ | $H_2W_1$ v $H_4W_3$ |
| Round 3: | $H_1W_3$ v $H_4W_2$ | $H_2W_4$ v $H_3W_1$ |

It was shown by Wallis (1979) that Mendelsohn's construction of SOLS($n$) in the case $(n, 6) = 1$ always yields a resolvable SAMDRR($n$).

**Theorem 5.4.1**  *If $(n, 6) = 1$, a cyclic resolvable SAMDRR($n$) exists.*

**Proof**  Since the $(i, j)$th entry in the SOLS($n$) constructed in the proof of Theorem 5.1.1 is $a_{ij} \equiv 2i - j \pmod{n}$, we obtain from the SOLS a SAMDRR($n$) with games

$$H_iW_{2i-j} \text{ v } H_jW_{2j-i} \quad (1 \leqslant i < j \leqslant n),$$

where $2i - j$ is interpreted mod $n$. On writing $i + j \equiv 2k \pmod{n}$ and putting $h \equiv k - i \equiv j - k \pmod{n}$, these games can be written as

$$H_{k-h}W_{k-3h} \text{ v } H_{k+h}W_{k+3h}.$$

Resolvability is shown by playing, in round $k$, those games given by that value of $k$, and $1 \leqslant h \leqslant \frac{1}{2}(n - 1)$.

**Example 5.4.2**  A resolvable SAMDRR(5).

| Round 1: | $H_5W_3$ v $H_2W_4$ | $H_4W_5$ v $H_3W_2$ |
|---|---|---|
| Round 2: | $H_1W_4$ v $H_3W_5$ | $H_5W_1$ v $H_4W_3$ |
| Round 3: | $H_2W_5$ v $H_4W_1$ | $H_1W_2$ v $H_5W_4$ |
| Round 4: | $H_3W_1$ v $H_5W_2$ | $H_2W_3$ v $H_1W_5$ |
| Round 5: | $H_4W_2$ v $H_1W_3$ | $H_3W_4$ v $H_2W_1$ |

We shall show how to construct a resolvable SAMDRR($n$) for many values of $n$. But before that we turn to the connection with SOLS. If a SAMDRR($n$) corresponds to a SOLS($n$), what *extra* property of the SOLS corresponds to the SAMDRR($n$) being *resolvable*?

**Theorem 5.4.2** *A resolvable SAMDRR($n$) exists if and only if there exists a SOLS($n$) $A$ and a symmetric Latin square $B$ of order $n$ which is orthogonal to $A$, having constant diagonal if $n$ is even and having the same diagonal as $A$ if $n$ is odd.*

**Proof** First suppose that a resolvable SAMDRR($n$) exists. Then by Theorem 5.1.5 a SOLS($n$) $A = (a_{ij})$ exists with $a_{ii} = i$ for each $i$. Define the symmetric matrix $B = (b_{ij})$ by

$$b_{ij} = k \text{ if } H_i \text{ plays against } H_j \text{ in the } k\text{th round } (i \neq j),$$

$$b_{ii} = \begin{cases} i & \text{if } n \text{ is odd} \\ n & \text{if } n \text{ is even.} \end{cases}$$

Then $B$ is clearly a symmetric Latin square. We now verify that $A$ and $B$ are orthogonal.

Suppose $a_{ij} = a_{IJ} = l$ and $b_{ij} = b_{IJ} = k$; it is required to prove that $i = I$ and $j = J$. If $i \neq j$ and $I \neq J$ then in the $k$th round there are games in which $H_i$ and $H_I$ both have $W_l$ as partner; but this is only possible if $i = I$, in which case $j = J$. If $i = j$ and $n$ is even then $b_{ij} = n = b_{IJ}$, so $I = J$; so $I = a_{II} = l = a_{ii} = i$ and similarly $j = J$. If $i = j$ and $n$ is odd, then $b_{ij} = i$ so $b_{IJ} = i$; also $a_{IJ} = a_{ij} = i$. So, if $I \neq J$, $H_I$ plays $H_J$ in round $i$ with $W_i$ as his partner, contradicting the fact that $W_i$ does not play in round $i$. So $I = J$ and $i = b_{IJ} = b_{II} = I$, whence $j = i = I = J$.

Conversely, given $A$ and $B$ as in the theorem, relabel entries so that $a_{ii} = i$ for each $i$. Then read off the games in round $k$ as follows: if $i \neq j$ and $b_{ij} = k$, play $H_i W_l$ v $H_j W_m$ where $a_{ij} = l$ and $a_{ji} = m$.

**Example 5.4.3** Take $A$ as the matrix given by (5.1). This SOLS(4) was used to obtain a SAMDRR(4) in Examples 5.1.4 and 5.4.1. The rounds displayed in Example 5.4.1 yield

$$B = \begin{bmatrix} 4 & 1 & 2 & 3 \\ 1 & 4 & 3 & 2 \\ 2 & 3 & 4 & 1 \\ 3 & 2 & 1 & 4 \end{bmatrix}$$

which is orthogonal to $A$.

**Example 5.4.4** For $A = (a_{ij})$, $a_{ij} \equiv 2i - j \pmod{n}$, where $(n, 6) = 1$, it was shown in Theorem 5.4.1 that the games in round $k$ of the corresponding resolvable SAMDRR are

$$H_{k-h} W_{k-3h} \text{ v } H_{k+h} W_{k+3h}.$$

The corresponding $B$ will have $b_{ii} = i$ for each $i \leqslant n$, and will have $b_{k-h,k+h} = k$ for each $h$. Thus $B$ will be a reverse circulant matrix. For example, if $n = 5$,

$$A = \begin{bmatrix} 1 & 5 & 4 & 3 & 2 \\ 3 & 2 & 1 & 5 & 4 \\ 5 & 4 & 3 & 2 & 1 \\ 2 & 1 & 5 & 4 & 3 \\ 4 & 3 & 2 & 1 & 5 \end{bmatrix} \quad \text{and} \quad B = \begin{bmatrix} 1 & 4 & 2 & 5 & 3 \\ 4 & 2 & 5 & 3 & 1 \\ 2 & 5 & 3 & 1 & 4 \\ 5 & 3 & 1 & 4 & 2 \\ 3 & 1 & 4 & 2 & 5 \end{bmatrix}.$$

In our search for constructions of resolvable SAMDRRs other than those given by Theorem 5.4.1 we first turn our attention to prime powers. Theorem 5.4.1 in fact guarantees the existence of such designs for all $n = p^\alpha$ where $p \neq 2$ or 3, but a different approach, using the method of differences, will give designs for all powers of 3 except 3 itself. The idea in this approach is to label the husbands and wives by the elements of $\text{GF}(p^\alpha)$. Suppose then that $\text{GF}(p^\alpha) = \{0, a_1, \ldots, a_{2k}\}$ where $n = p^\alpha = 2k + 1$; let $G$ be the additive group of the field, and try to construct a tournament in which the games in round $g$ are

$$\text{H}_{a_i+g}\text{W}_{a_{i+1}+g} \text{ v } \text{H}_{a_{k+i}+g}\text{W}_{a_{k+i+1}+g} \quad (1 \leqslant i \leqslant k)$$

where $a_{2k+1} = a_1$. To ensure that every player partners every player of the opposite sex (spouse excluded) exactly once, we need the differences $a_{i+1} - a_i$ $(1 \leqslant i \leqslant 2k)$ to be the non-zero elements of $\text{GF}(p^\alpha)$. To ensure that every player plays against every player of the opposite sex (spouse excluded) exactly once, we need the differences $a_{k+i+1} - a_i$ to be the non-zero elements of $\text{GF}(p^\alpha)$; and to ensure that every player plays against every other player of the same sex exactly once, the differences $a_{k+i} - a_i$ have to be the non-zero elements of $\text{GF}(p^\alpha)$ exactly once.

Now all of these requirements can be satisfied by the choice $a_i = \theta^i$ where $\theta$ is a primitive element of $\text{GF}(p^\alpha)$; for then $a_{i+1} - a_i = \theta^i(\theta - 1)$, $a_{k+i+1} - a_i = (\theta^{k+1} - 1)\theta^i$, $a_{k+i} - a_i = \theta^i(\theta^k - 1)$, where $\theta^k - 1 \neq 0$ since $\theta$ is a primitive element, and where $\theta^{k+1} - 1 \neq 0$ provided $k + 1 < 2k$, i.e. $p^\alpha > 3$. The following result due to Wallis (1979) has therefore been established.

**Theorem 5.4.3** *If* $n = p^\alpha$, *$p$ odd, $n > 3$, then a cyclic resolvable* SAMDRR($n$) *exists.*

**Corollary 5.4.4** *If* $n = p^\alpha$, *$p$ odd, $n > 3$, there exists a* SOLS($n$) *with idempotent symmetric mate.*

**Example 5.4.5** Take $n = 2 \cdot 3 + 1 = 7$, $\theta = 3$. Then $a_1 = 3$, $a_2 = 2$, $a_3 = 6$, $a_4 = 4$, $a_5 = 5$, $a_6 = 1$.

| | | | |
|---|---|---|---|
| Round 0 (or 7): | $H_3W_2 \vee H_4W_5$ | $H_2W_6 \vee H_5W_1$ | $H_6W_4 \vee H_1W_3$ |
| Round 1 | : | $H_4W_3 \vee H_5W_6$ | $H_3W_7 \vee H_6W_2$ | $H_7W_5 \vee H_2W_4$ |
| Round 2 | : | $H_5W_4 \vee H_6W_7$ | $H_4W_1 \vee H_7W_3$ | $H_1W_6 \vee H_3W_5$ |
| Round 3 | : | $H_6W_5 \vee H_7W_1$ | $H_5W_2 \vee H_1W_4$ | $H_2W_7 \vee H_4W_6$ |
| Round 4 | : | $H_7W_6 \vee H_1W_2$ | $H_6W_3 \vee H_2W_5$ | $H_3W_1 \vee H_5W_7$ |
| Round 5 | : | $H_1W_7 \vee H_2W_3$ | $H_7W_4 \vee H_3W_6$ | $H_4W_2 \vee H_6W_1$ |
| Round 6 | : | $H_2W_1 \vee H_3W_4$ | $H_1W_5 \vee H_4W_7$ | $H_5W_3 \vee H_7W_2$ |

The following is the corresponding SOLS and its symmetric mate.

$$
\begin{bmatrix}
1 & 7 & 6 & 5 & 4 & 3 & 2 \\
3 & 2 & 1 & 7 & 6 & 5 & 4 \\
5 & 4 & 3 & 2 & 1 & 7 & 6 \\
7 & 6 & 5 & 4 & 3 & 2 & 1 \\
2 & 1 & 7 & 6 & 5 & 4 & 3 \\
4 & 3 & 2 & 1 & 7 & 6 & 5 \\
6 & 5 & 4 & 3 & 2 & 1 & 7
\end{bmatrix}
\begin{bmatrix}
1 & 5 & 2 & 6 & 3 & 7 & 4 \\
5 & 2 & 6 & 3 & 7 & 4 & 1 \\
2 & 6 & 3 & 7 & 4 & 1 & 5 \\
6 & 3 & 7 & 4 & 1 & 5 & 2 \\
3 & 7 & 4 & 1 & 5 & 2 & 6 \\
7 & 4 & 1 & 5 & 2 & 6 & 3 \\
4 & 1 & 5 & 2 & 6 & 3 & 7
\end{bmatrix}
$$

We now turn to powers of 2. A resolvable SAMDRR(4) has already been exhibited in Example 5.4.1. and a resolvable SAMDRR(8) is now confirmed.

**Example 5.4.6** A resolvable SAMDRR(8) can be obtained from the following SOLS(8) $A$ and its symmetric mate $B$.

$$
A = \begin{bmatrix}
1 & 4 & 7 & 6 & 3 & 5 & 8 & 2 \\
7 & 2 & 1 & 3 & 6 & 8 & 5 & 4 \\
6 & 8 & 3 & 1 & 7 & 2 & 4 & 5 \\
8 & 6 & 5 & 4 & 2 & 7 & 1 & 3 \\
4 & 1 & 2 & 8 & 5 & 3 & 6 & 7 \\
2 & 7 & 4 & 5 & 8 & 6 & 3 & 1 \\
5 & 3 & 8 & 2 & 4 & 1 & 7 & 6 \\
3 & 5 & 6 & 7 & 1 & 4 & 2 & 8
\end{bmatrix}
$$

$$
B = \begin{bmatrix}
8 & 1 & 2 & 3 & 4 & 5 & 6 & 7 \\
1 & 8 & 5 & 6 & 7 & 4 & 2 & 3 \\
2 & 5 & 8 & 1 & 3 & 6 & 7 & 4 \\
3 & 6 & 1 & 8 & 2 & 7 & 4 & 5 \\
4 & 7 & 3 & 2 & 8 & 1 & 5 & 6 \\
5 & 4 & 6 & 7 & 1 & 8 & 3 & 2 \\
6 & 2 & 7 & 4 & 5 & 3 & 8 & 1 \\
7 & 3 & 4 & 5 & 6 & 2 & 1 & 8
\end{bmatrix}
$$

The existence of a resolvable SAMDRR($2^m$) for all $m \geqslant 2$ will follow from the following result of Wallis.

**Theorem 5.4.5** *If a resolvable SAMDRR($n$) exists, with $n$ even, then a resolvable SAMDRR($4n$) also exists.*

**Proof** Label the participants in the SAMDRR($4n$) to be constructed by $H_{ij}$, $W_{ij}$, $1 \leqslant i \leqslant n$, $1 \leqslant j \leqslant 4$. For fixed $j$, the $n$ couples $H_{ij}$, $W_{ij}$, $1 \leqslant i \leqslant n$, can play a resolvable SAMDRR($n$) $T_j$; so for the first $n - 1$ rounds of the required tournament play the $n - 1$ rounds of $T_1, \ldots, T_4$ in parallel. This gives $n - 1$ rounds in which all players with the same second suffix play with or against each other as required.

Next, for each $i \leqslant n - 1$, form three more rounds as follows. Take a resolvable SAMDRR($n$) $T$ and, for all games $H_a W_b$ v $H_c W_d$ in the $i$th round of $T$, play the games

$$H_{a1}W_{b2} \text{ v } H_{c3}W_{d4}, H_{a2}W_{b1} \text{ v } H_{c4}W_{d3}, H_{a3}W_{b4} \text{ v } H_{c1}W_{d2}, H_{a4}W_{b3} \text{ v } H_{c2}W_{d1}$$

as one round,

$$H_{a1}W_{b3} \text{ v } H_{c4}W_{d2}, H_{a3}W_{b1} \text{ v } H_{c2}W_{d4}, H_{a2}W_{b4} \text{ v } H_{c3}W_{d1}, H_{a4}W_{b2} \text{ v } H_{c1}W_{d3}$$

as another round, and

$$H_{a1}W_{b4} \text{ v } H_{c2}W_{d3}, H_{a4}W_{b1} \text{ v } H_{c3}W_{d2}, H_{a2}W_{b3} \text{ v } H_{c1}W_{d4}, H_{a3}W_{b2} \text{ v } H_{c4}W_{b1}$$

as a third round. This gives $3(n - 1)$ rounds, with games satisfying the requirements for pairs of players with different first and second suffixes.

Finally, we deal with pairs of players with the same first suffix. For one round take games

$$H_{i1}W_{i2} \text{ v } H_{i3}W_{14}, H_{i2}W_{i1} \text{ v } H_{i4}W_{i3} \quad (1 \leqslant n);$$

for another round take games

$$H_{i1}W_{i3} \text{ v } H_{i4}W_{i2}, H_{i2}W_{i4} \text{ v } H_{i3}W_{i1} \quad (1 \leqslant i \leqslant n);$$

and for the final round take games

$$H_{i1}W_{i4} \text{ v } H_{i2}W_{i3}, H_{i4}W_{i1} \text{ v } H_{i3}W_{i2} \quad (1 \leqslant i \leqslant n).$$

**Corollary 5.4.6** *A resolvable SAMDRR($2^n$) exists for all $n \geqslant 2$.*

**Proof** Every $2^n$ with $n \geqslant 2$ can be written as $4^k$ or $8 \cdot 4^k$; so apply the theorem repeatedly, starting with a tournament for four or eight couples.
Note that in terms of MOLS, this result has the following form.

**Corollary 5.4.7** *For each $n \geqslant 2$, there exists a SOLS($2^n$) with a constant-diagonal symmetric mate.*

As well as the existence results of Theorems 5.4.3 and 5.4.6 dealing with prime powers, we also have the following theorem which will lead to further SOLS with symmetric mates.

**Theorem 5.4.8** *Suppose that a resolvable SAMDRR($k$) exists for all $k \in K$ and that a PBD($v, K, 1$) exists, where $k$ is odd for all $k \in K$. Then a resolvable SAMDRR($v$) also exists.*

**Proof** For each $k \in K$ form a resolvable SAMDRR($k$) on each block $B$ of size $k$ in the PBD. For each $x \in B$, there will be one round in which the couple $H_x, W_x$ are not involved; let $C_B(x)$ denote the set of all games in that round. Then take the union of all $C_B(x)$ over all blocks containing $x$; these will form round $x$ of the required tournament.

**Example 5.4.7** Construct a PBD($57, \{7, 9\}, 1$) by starting with an OA($7, 8$) and proceeding as in the proof of Theorem 4.4.2. adding a new element $\infty$ to each block of size 8. Since a resolvable SAMDRR($k$) exists for $k = 7, 9$, a resolvable SAMDRR($57$) also exists. Thus there exists a SOLS($57$) with symmetric mate.

It is now known that a SOLS($n$) with symmetric mate exists for all but a finite number of values of $n$, the smallest unsolved case being, not unsurprisingly, $n = 10$. Such squares have applications to other topics in design theory; for example it will be shown later that they are useful in the construction of whist tournaments.

## 5.5 Some examples

In this section we construct a few important examples of SOLS with symmetric mate which are not covered by the results of the previous section. These designs will be needed in the final chapter. In accordance with Theorem 5.4.2, the SOLS and its mate will be idempotent if $n$ is odd, whereas, if $n$ is even, the SOLS will be idempotent but its mate will be *unipotent* (i.e. have constant diagonal).

**Example 5.5.1** Construction of a SOLS($33$) with idempotent symmetric mate. (This is based on the whist tournament for 33 players found by Baker (1975*b*); see section 11.6 for connections between whist tournaments and SOLS.) Consider the array $A_0$:

```
0  0  0  0  0  0  0  0  0  0  0  0  0  0  0  0  0  0  0  0  0  0  0  0  0  0  0  0  0  0  0  0  0
0  9 22 30 28  1  8 18 19 31  2  3 24 11 17  5 20 13 21 23  4 15 12 16  7 32 27 25 14 29  6 26 10
0 22  9 28 30  8  1 19 18  2 31 24  3 17 11 20  5 21 13  4 23 12 15  7 16 27 32 14 25  6 29 10 26
0 28 30  9 22 19 18  1  8 24  3 31  2 20  5 11 17  4 23 13 21  7 16 15 12 14 25 32 27 10 26 29  6
0 30 28 22  9 18 19  8  1  3 24  2 31  5 20 17 11 23  4 21 13 16  7 12 15 25 14 27 32 26 10  6 29.
```

Then, if $A_i$ is obtained from $A_0$ by adding $i$ (mod 33) to each entry, $[A_0 \, A_1 \cdots A_{32}]$ is an OA(5,33).

Use the last two rows of this array to specify rows and columns. Ignoring the first row, the resulting OA(4,33) gives a SOLS(33). For if $(i,j,k,l)'$ is one column then $(j,i,l,k)'$ is another. Further, the first row of the OA(5,33) gives a symmetric Latin square; for if $(i, \, , \, , k, l)'$ is one column, $(i, \, , \, , l, k)'$ is another. Finally all constructed squares are idempotent since the column $(i,i,i,i,i)'$ leads to the $(i,i)$th entry of each square being $i$.

The remaining examples in this section are due to Wang (1978). In each case a resolvable SAMDRR is constructed.

**Example 5.5.2** Construction of a resolvable SAMDRR(38). Label husbands and wives by $0, 1, \ldots, 29, \infty_1, \ldots, \infty_8$ and work mod 30. Consider the games $H_i W_j$ v $H_k W_l$ where suffixes are as in the table:

| $i$ | 29 | 7 | 3 | 1 | 4 | 9 | 19 | 10 | 20 | 22 | 28 | 26 | 17 | 11 | 27 | 18 | 15 | 14 | 5 |
| $j$ | 27 | 21 | 22 | 0 | 1 | 5 | 14 | 4 | 12 | 13 | 18 | 28 | 20 | 15 | 3 | 25 | 23 | 24 | 16 |
| --- | --- | --- | --- | --- | --- | --- | --- | --- | --- | --- | --- | --- | --- | --- | --- | --- | --- | --- | --- |
| $k$ | 13 | 24 | 21 | 2 | 6 | 12 | 23 | 16 | 25 | 0 | 8 | $\infty_1$ | $\infty_2$ | $\infty_3$ | $\infty_4$ | $\infty_5$ | $\infty_6$ | $\infty_7$ | $\infty_8$ |
| $l$ | 6 | 11 | 9 | $\infty_1$ | $\infty_2$ | $\infty_3$ | $\infty_4$ | $\infty_5$ | $\infty_6$ | $\infty_7$ | $\infty_8$ | 29 | 2 | 19 | 8 | 17 | 7 | 26 | 10 |

Note that each player plays precisely once in these games, so these games can be taken to form one round. We take round $R_x$ to consist of games

$$H_{i+x}W_{j+x} \text{ v } H_{k+x}W_{l+x} \quad (0 \leqslant x \leqslant 29),$$

(with the usual convention that $\infty_i + x = \infty_i$). Since the differences between opposing men in round $R_0$ are $14, 13, 12, 1, 2, 3, 4, 6, 5, 8, 10$ and their negatives, we get 30 rounds in which each $H_i$ opposes every $H_j$ except for $j = i \pm 7$, $i \pm 9$, $i \pm 11$, $i + 15$. Similarly we can check that each $W_i$ opposes every $W_j$ except for $j = i \pm 3$, $i \pm 7$, $i \pm 11$, $i + 15$; that each $H_i$ opposes every $W_j$ except for $j = i + 2$, $i + 9$, $i + 16$, $i + 19$, $i + 20$, $i + 24$, $i + 26$; and that each $H_i$ partners every $W_j$ except for $j = i + 1$, $i + 5$, $i + 9$, $i + 12$, $i + 13$, $i + 15$, $i + 16$. We use these missing combinations of players to make up the remaining rounds, as follows.

For round $R_{30}$ take the matches $H_i W_{i+1}$ v $H_{i+15}W_{i+16}$ and those of the first round of a resolvable SAMDRR(8) on the $\infty_i$.

For round $R_{31}$ take matches $H_i W_{i+5}$ v $H_{i+9}W_{i+24}$ ($i$ even) and the second round matches of the SAMDRR(8). Those with $i$ odd, along with the third round matches, will give $R_{32}$.

For rounds $R_{33}$ and $R_{34}$ take the matches $H_i W_{i+13}$ v $H_{i+11}W_{i+20}$ ($i$ even and odd respectively) along with the games of rounds 4 and 5 respectively of the SAMDRR(8).

Finally, for rounds $R_{35}$ and $R_{36}$, take the matches $H_i W_{i+12}$ v $H_{i+23}W_{i+9}$ ($i$ even and odd respectively) along with rounds 6 and 7 respectively of the SAMDRR(8).

Note that these 'extra rounds' can be specified as follows:

| $i$ | 0 | 0 | 0 | 0 |
|---|---|---|---|---|
| $j$ | 1 | 5 | 13 | 12 |
| $k$ | 15 | 9 | 11 | 23 |
| $l$ | 16 | 24 | 20 | 9 |

Similar constructions of resolvable SAMDRR were provided by Wang for several other values of $n \equiv 2 \pmod 4$. We finish this chapter by indicating briefly the constructions for $n = 18, 22,$ and $26$. It is left to the reader to check that each works.

**Example 5.5.3** Construction of a resolvable SAMDRR(18).

| $i$ | 10 | 1 | 8 | 5 | 7 | 0 | 4 | 6 | 13 |
|---|---|---|---|---|---|---|---|---|---|
| $j$ | 12 | 0 | 6 | 2 | 3 | 8 | 9 | 10 | 5 |
| $k$ | 2 | 3 | 11 | 9 | 12 | $\infty_1$ | $\infty_2$ | $\infty_3$ | $\infty_4$ |
| $l$ | 11 | $\infty_1$ | $\infty_2$ | $\infty_3$ | $\infty_4$ | 13 | 7 | 4 | 1 |

and

| 0 | 0 |
|---|---|
| 1 | 7 |
| 7 | 1 |
| 8 | 4 |

$\pmod{14}$.

**Example 5.5.4** SAMDRR(22)

| $i$ | 9 | 6 | 14 | 1 | 8 | 16 | 5 | 7 | 0 | 12 | 17 |
|---|---|---|---|---|---|---|---|---|---|---|---|
| $j$ | 17 | 12 | 7 | 0 | 6 | 13 | 1 | 9 | 3 | 4 | 11 |
| $k$ | 15 | 13 | 4 | 3 | 11 | 2 | 10 | $\infty_1$ | $\infty_2$ | $\infty_3$ | $\infty_4$ |
| $l$ | 10 | 2 | 8 | $\infty_1$ | $\infty_2$ | $\infty_3$ | $\infty_4$ | 14 | 5 | 16 | 15 |

and

| 0 | 0 |
|---|---|
| 1 | 6 |
| 9 | 0 |
| 10 | 9 |

$\pmod{18}$.

**Example 5.5.5** SAMDRR(26)

| $i$ | 15 | 5 | 0 | 11 | 19 | 1 | 13 | 6 | 9 | 2 | 4 | 17 | 18 |
|---|---|---|---|---|---|---|---|---|---|---|---|---|---|
| $j$ | 8 | 17 | 10 | 2 | 13 | 0 | 11 | 3 | 5 | 6 | 7 | 1 | 20 |
| $k$ | 21 | 12 | 8 | 20 | 7 | 3 | 16 | 10 | 14 | $\infty_1$ | $\infty_2$ | $\infty_3$ | $\infty_4$ |
| $l$ | 4 | 19 | 16 | 12 | 18 | $\infty_1$ | $\infty_2$ | $\infty_3$ | $\infty_4$ | 9 | 14 | 15 | 21 |

and

| 0 | 0 |
|---|---|
| 1 | 9 |
| 11 | 1 |
| 12 | 18 |

$\pmod{22}$.

## 5.6 Exercises 5

1. Write down a SOLS(13) using Theorem 5.1.1, and exhibit a resolvable SAMDRR(13) by using Theorem 5.4.1.
2. Use Theorem 5.2.3 to give another construction of a SOLS(13).
3. Use Theorem 5.4.3 to construct a resolvable SAMDRR(13).
4. Show that the SOLS($n$) of Theorem 5.1.1 has symmetric mate $B$ where $b_{ij} = \frac{1}{2}(i + j)$. Observe how this fits in with Theorems 5.4.1 and 5.4.2.

5. The proof of the Brayton–Coppersmith–Hoffman theorem requires the existence of certain PBDs. Construct them as follows:
    (i) PBD(66,{5,14},1): take a FPP of order 13, and retain only the elements of five of the blocks which contain a chosen element $x$.
    (ii) PBD(54,{4,5,10,11},1): take the elements of five blocks from the same resolution class of a (121,11,1) design, and then delete one element.
    (iii) PBD(86,{5,18},1): imitate (i).
    (iv) PBD(62,{4,5,10,13},1): imitate (ii), this time deleting three elements.
6. Check that the constructions of SOLSs in Examples 5.3.1 to 5.3.6 actually work.
7. Show that if we take $n = 5$ and $\theta = 2$ in Theorem 5.4.3 (so that $a_1 = 2$, $a_2 = 4$, $a_3 = 3$, $a_4 = 1$), we obtain the tournament of Example 5.4.2 with H and W interchanged.
8. Check that the constructions of Examples 5.5.3 to 5.5.5 work.
9. Can any of the PBDs constructed in question 5 above be used in Theorem 5.4.8?

# 6

# Steiner systems

## 6.1 Existence of Steiner triple systems

A Steiner triple system is simply a block design with $k = 3$ and $\lambda = 1$. This choice of parameters is one of the simplest which could be made, and the corresponding designs form one of the first families of designs for which existence was established.

**Definition 6.1.1** A *Steiner triple system* STS($v$) of order $v$ is a $(v, 3, 1)$ design.

If an STS($v$) exists it will have $b = \frac{1}{6}v(v - 1)$ blocks, and $r = \frac{1}{2}(v - 1)$. It follows that $v$ must be odd, $v = 2r + 1$, and $b = \frac{1}{3}r(2r + 1)$, so that $r \equiv 0$ or $1 \pmod 3$. Putting $r = 3n$ or $r = 3n + 1$ gives $v = 6n + 1$ or $6n + 3$, so the following result is established.

**Lemma 6.1.1** *An* STS($v$) *can exist only if* $v \equiv 1$ *or* $3 \pmod 6$.

It was Kirkman in 1847 who first established the existence of an STS($v$) for all $v \equiv 1$ or $3 \pmod 6$, $v \geqslant 3$. Without knowing of Kirkman's work. Steiner came across the problem of the existence of such designs in a geometrical context in 1853, and publicized the problem. The designs were therefore to become known as Steiner triple systems. This is particularly unfair on Kirkman, since not only did Kirkman provide an existence proof, but Steiner did not. Reiss published a proof shortly afterwards which is essentially identical to Kirkman's, and later on many other mathematicians, beginning with Moore in 1893, gave a variety of different proofs. Kirkman's own proof has been undeservedly neglected. It is a step-by-step construction, as is the more famous proof due to Moore which is often presented in combinatorial texts. We have described Kirkman's approach in Anderson (1990), and we outline Moore's construction, as adapted by Hilton, in the exercises.

We have already met several constructions of STS; Corollary 2.4.3 dealt completely with the case $v = 3$ (mod 6), and Corollary 2.2.4 dealt with the case $v = q$ where $q \equiv 1$ (mod 6) is a prime power. In what follows we give three complete existence proofs for all $v \equiv 1$ or 3 (mod 6). The first gives a direct construction for each $v$; the second uses pairwise balanced designs, and the third, in section 6.2, uses Skolem triples to produce *cyclic* designs.

But before proceeding to these, we give a useful construction first used by Kirkman.

**Lemma 6.1.2** *If an STS($n$) exists then an STS($2n + 1$) also exists.*

**Proof** By Lemma 6.1.1, $n$ must be odd. Take a league schedule for $n + 1$ teams, labelled $1, \ldots, n + 1$, and an STS($n$) on $x_1, \ldots, x_n$. For each $i = 1, \ldots, n$ adjoin $x_i$ to each pair occurring on day $i$ of the league schedule. The resulting set of triples, along with those of the STS($n$) on $x_1, \ldots, x_n$, clearly form an STS($2n + 1$).

**Example 6.1.1** Take $n = 7$ and the league schedule of Example 1.2.3, with $\infty$ written as 8. Also take the STS(7) with triples $x_1 x_2 x_4, x_2 x_3 x_5, \ldots, x_7 x_1 x_3$. The construction of the lemma yields an STS(15) with the following triples:

| | | | |
|---|---|---|---|
| $\{1, 8, x_1\}$ | $\{2, 7, x_1\}$ | $\{3, 6, x_1\}$ | $\{4, 5, x_1\}$ |
| $\{2, 8, x_2\}$ | $\{3, 1, x_2\}$ | $\{4, 7, x_2\}$ | $\{5, 6, x_2\}$ |
| $\{3, 8, x_3\}$ | $\{4, 2, x_3\}$ | $\{5, 1, x_3\}$ | $\{6, 7, x_3\}$ |
| $\{4, 8, x_4\}$ | $\{5, 3, x_4\}$ | $\{6, 2, x_4\}$ | $\{7, 1, x_4\}$ |
| $\{5, 8, x_5\}$ | $\{6, 4, x_5\}$ | $\{7, 3, x_5\}$ | $\{1, 2, x_5\}$ |
| $\{6, 8, x_6\}$ | $\{7, 5, x_6\}$ | $\{1, 4, x_6\}$ | $\{2, 3, x_6\}$ |
| $\{7, 8, x_7\}$ | $\{1, 6, x_7\}$ | $\{2, 5, x_7\}$ | $\{3, 4, x_7\}$ |

$\{x_1, x_2, x_4\}$ $\{x_2, x_3, x_5\}$ $\{x_3, x_4, x_6\}$ $\{x_4, x_5, x_7\}$ $\{x_5, x_6, x_1\}$ $\{x_6, x_7, x_2\}$ $\{x_7, x_1, x_2\}$.

***Skolem's method* (Skolem, 1958)**

First we deal with the case $v = 6m + 3$. We are looking for $b = \frac{1}{6}v(v - 1) = (2m + 1)(3m + 1)$ triples. Write the numbers $0, 1, \ldots, 6m + 2$ in three rows of $2m + 1$ numbers as follows:

| 0 | 1 | 2 | 3 | $\cdots$ | $2m - 1$ | $2m$ |
|---|---|---|---|---|---|---|
| $2m + 1$ | $2m + 2$ | $2m + 3$ | $2m + 4$ | $\cdots$ | $4m$ | $4m + 1$ |
| $4m + 2$ | $4m + 3$ | $4m + 4$ | $4m + 5$ | $\cdots$ | $6m + 1$ | $6m + 2$. |

Now take the $2m + 1$ triples given by the columns:
$$A_i = \{i, i + 2m + 1, i + 4m + 2\} \quad 0 \leqslant i \leqslant 2m.$$

Further, for each pair $\{a, b\}$ in any row, define $c$ to be the entry in the next row such that $2c \equiv a + b$ (mod $2m + 1$), or, equivalently, such that $c \equiv (m + 1)(a + b)$ (mod $2m + 1$). Here, the next row after the bottom row is taken to be the top row. Note that $c$ cannot occur in the same column as $a$

or $b$; for if $c \equiv a \pmod{2m + 1}$ then $2a = a + b$, i.e. $a \equiv b \pmod{2m + 1}$, i.e. $a$ and $b$ are in the same column as well as the same row. Now there are $\binom{2m+1}{2} = m(2m + 1)$ pairs of numbers in each row, so we obtain $3m(2m + 1)$ triples $\{a, b, c\}$. We assert that these triples, together with the triples $A_i$, form an STS($6m + 3$).

Certainly there are $2m + 1 + 3m(2m + 1) = (2m + 1)(3m + 1)$ triples, as required. We have to show that each pair of numbers from $0, \ldots, 6m + 2$ occurs in exactly one triple. Certainly any two numbers in the same column lie in exactly one $A_i$ and in no other triple; and any two in the same row also clearly lie in just one triple. So now consider two numbers in different rows and columns: $a$ in one row, and $c$ in the next row, $a \neq c \pmod{2m + 1}$. These two numbers occur together in a triple if and only if there is a number $b$ in the same row as $a$ for which $a + b \equiv 2c \pmod{2m + 1}$, i.e. $b \equiv 2c - a \pmod{2m + 1}$. But clearly there is precisely one $b$ in the same row as $a$ satisfying this congruence.

**Example 6.1.2** Take $v = 9$, i.e. $m = 1$. Consider the array

$$
\begin{array}{ccc}
0 & 1 & 2 \\
3 & 4 & 5 \\
6 & 7 & 8.
\end{array}
$$

The columns give three triples. Next, given $a$ and $b$ in the same row, we have to find $c$ in the next row such that $c \equiv 2(a + b) \pmod 3$; but since $c$ cannot be in the same column as $a$ or $b$, $c$ is simply found by taking the entry in the next row and in the *remaining* column. The triples of the STS(9) so constructed are:

$$036, 147, 258, 015, 024, 123, 348, 357, 456, 672, 681, 780.$$

The case $v = 6m + 1$ is deals with by a similar but more complicated method. Here we have to construct $b = \frac{1}{6}v(v - 1) = m(6m + 1)$ triples. This time consider the array

$$
\begin{array}{cccc|cccc}
0 & 1 & \cdots & m - 1 & m & m + 1 & \cdots & 2m - 1 \\
2m & 2m + 1 & \cdots & 3m - 1 & 3m & 3m + 1 & \cdots & 4m - 1 \\
4m & 4m + 1 & \cdots & 5m - 1 & 5m & 5m + 1 & \cdots & 6m - 1
\end{array}
$$

Note that the array is divided into two halves, and that the $(6m + 1)$th element $6m$ is missing. As the triples of the desired STS($6m + 1$) we take

(i) the triples given by the first $m$ columns:

$$B_i = \{i, 2m + i, 4m + i\} \qquad 0 \leqslant i \leqslant m - 1;$$

(ii) the triples $C_i = \{m + i, 2m + i, 6m\}$ $\qquad 0 \leqslant i \leqslant m - 1;$

$$D_i = \{3m + i, 4m + i, 6m\} \qquad 0 \leqslant i \leqslant m - 1;$$

$$E_i = \{5m + i, i, 6m\} \qquad 0 \leqslant i \leqslant m - 1;$$

(iii) the triples $\{a, b, c\}$ where $a$ and $b$ are in the same row of the array and $c$ is in the next row, such that

$$\text{if } a + b \text{ is even, } 2c \equiv a + b \pmod{2m}, \ c \text{ in left half of row,}$$

$$\text{if } a + b \text{ is odd, } 2c \equiv a + b - 1 \pmod{2m}, \ c \text{ in right half of row.}$$

This gives $4m + 3\binom{2m}{2} = m(6m + 1)$ triples as required: it remains to show that any two numbers occur together in exactly one triple.

Certainly, $6m$ occurs with each other number exactly once, and any two numbers in the same row occur together in exactly one triple, namely one of type (iii). If two numbers occur in a column in the left half of the array, they occur in exactly one triple of type (i); further, they cannot occur in a triple of type (iii), for if $a \equiv c \pmod{2m}$ with $a$ and $c$ in the left half, then $2c \equiv a + b \pmod{2m}$ gives $a \equiv b \pmod{2m}$, contradicting the fact that $a$ and $b$ are in the same row. So two numbers in a left column are in a unique triple. Next consider two numbers in a column on the right: $a$ in one row, $c$ in the next, $a \equiv c \pmod{2m}$. They will occur together in the unique triple $\{a, b, c\}$ of type (iii) where $b$ is given by $2a \equiv a + b - 1 \pmod{2m}$, i.e., $b \equiv a + 1 \pmod{2m}$, and clearly they cannot occur together in any other type of triple.

Finally consider any two numbers in different rows and columns: $a$ in one row, $c$ in the next. If $c$ is in the right half, determine $b$ in the same row as $a$ by $b \equiv 2c - a + 1 \pmod{2m}$; the unique $b$ gives the unique triple $\{a, b, c\}$ containing $a$ and $c$. If $c$ is in the left half and $c \equiv a \pmod{m}$, then $a$ and $c$ are in a unique triple of type (ii); further, they cannot also occur together in a triple of type (iii) since then $2c \equiv a + b \pmod{2m}$, where $2c \equiv 2a \pmod{2m}$, so that $a \equiv b \pmod{2m}$, whereas $a$ and $b$ are in the same row. Finally, if $c$ is in the left half and $c \not\equiv a \pmod{m}$, then $a$ and $c$ lie in the unique triple $\{a, b, c\}$ of type (iii) where $b$ is given by $2c \equiv a + b \pmod{2m}$: this gives a triple as long as $b \not\equiv a \pmod{2m}$; but this would give $a \equiv c \pmod{m}$.

**Example 6.1.3**   Take $v = 13$, i.e. $m = 2$. Consider the array

$$
\begin{array}{cccc}
0 & 1 & 2 & 3 \\
4 & 5 & 6 & 7 \\
8 & 9 & 10 & 11.
\end{array}
$$

The triples of the STS(13) are:

(i) $\{0, 4, 8\}$, $\{1, 5, 9\}$

(ii) $\{2, 4, 12\}$, $\{3, 5, 12\}$, $\{6, 8, 12\}$, $\{7, 9, 12\}$, $\{10, 0, 12\}$, $\{11, 1, 12\}$

(iii) $\{0, 1, 6\}$, $\{0, 2, 5\}$, $\{0, 3, 7\}$, $\{1, 2, 7\}$, $\{1, 3, 4\}$, $\{2, 3, 6\}$;
$\{4, 5, 10\}$, $\{4, 6, 9\}$, $\{4, 7, 11\}$, $\{5, 6, 11\}$, $\{5, 7, 8\}$, $\{6, 7, 10\}$;
$\{8, 9, 2\}$, $\{8, 10, 1\}$, $\{8, 11, 3\}$, $\{9, 10, 3\}$, $\{9, 11, 0\}$, $\{10, 11, 2\}$.

Note how the triples obtained by this method 'move down' the array, e.g. the triple $\{0, 1, 6\}$ moves down one row to become $\{4, 5, 10\}$.

## PBD approach

We close this section by giving another method of constructing Steiner triple systems, this time making use of pairwise balanced designs. It was shown in Exercises 4, number 11, that a PBD$(u, \{3, 4\}, 1)$ exists for all $u \equiv 0$ or 1 (mod 3), $u \neq 6$. The proof made use of the existence of two MOLS of order $n$ for $n \neq 2$ or 6. We can use these PBDs to construct an STS$(v)$ for all $v \equiv 1$ or 3 (mod 6) other than 13, which has to be dealt with separately.

If $v \equiv 1$ or 3 (mod 6) then $v = 2u + 1$ where $u \equiv 0$ or 1 (mod 3). Let $S$ be a set of $2u + 1$ elements, namely $\infty$ and two copies $x_1, x_2$ of each $x \in \{1, \ldots, u\}$. Corresponding to each block $B = \{x, y, z\}$ of a PBD$(u, \{3, 4\}, 1)$ $D$ on $\{1, \ldots, u\}$, construct an STS$(7)$ on $\{\infty, x_1, x_2, y_1, y_2, z_1, z_2\}$ in such a way that the blocks of the STS containing $\infty$ are $\{\infty, x_1, x_2\}$, $\{\infty, y_1, y_2\}$, $\{\infty, z_1, z_2\}$. Similarly construct an STS$(9)$ corresponding to each block of size 4 in $D$. The set of all triples so formed constitutes an STS$(v)$ on $S$: $\infty$ clearly occurs with each element once; $x_1, x_2$ occur together once, and if $x \neq y$, $x_i$ and $y_j$ occur together in precisely one triple of the STS corresponding to the one block $B$ containing both $x$ and $y$.

## 6.2 Cyclic Steiner triple systems

A Steiner triple system STS$(v)$ on the elements of $Z_v$ is said to be *cyclic* if, whenever $\{a, b, c\}$ is a triple, so also is $\{a + 1, b + c, c + 1\}$.

**Example 6.2.1** (a) The triples $\{1 + i, 2 + i, 4 + i\}$, $0 \leqslant i \leqslant 6$, form a cyclic STS$(7)$. (b) The triples $\{i, 1 + i, 4 + i\}$, $\{i, 2 + i, 8 + i\}$, $0 \leqslant i \leqslant 12$, form a cyclic STS$(13)$ in $Z_{13}$.

Clearly a cyclic STS$(v)$ exists whenever a $(v, 3, 1)$ difference system in $Z_v$ exists, the triples being the translates of the triples of the difference system. More generally, if a $(v, 3, 1)$ difference system $D$ exists in a finite field GF$(q)$ then the translates form an STS$(q)$ which we also describe as cyclic.

**Example 6.2.2** By Corollary 2.2.4, a cyclic STS$(q)$ exists whenever $q \equiv 1$ (mod 6) is a prime power.

Note that a STS$(v)$ has $\frac{1}{6}v(v - 1)$ triples, so that the construction of a cyclic STS from a difference system can only be achieved if $\frac{1}{6}(v - 1)$ is an integer, for each triple of the difference system will give rise to $v$ triples. The problem of constructing a cyclic STS($v$), $v \equiv 1$ (mod 6) was studied by Netto and Heffter in the 1890s. Realizing that they were looking for $(v, 3, 1)$ difference systems, i.e. collections of triples which give rise to differences which are precisely all the non-zero elements of $Z_v$ exactly once, they observed that some systems can be constructed in a particularly attractive way.

**Theorem 6.2.1** *Suppose that* $\{1, \ldots, 3m\}$ *can be partitioned into* $m$ *triples* $\{a, b, c\}$ *such that* $a + b = c$ *or* $a + b + c \equiv 0$ (mod $6m + 1$). *Then the triples* $\{0, a, a + b\}$ *form a* $(6m + 1, 3, 1)$ *difference system and so lead to the construction of a cyclic* STS($6m + 1$).

**Proof** The differences arising from the triple $\{0, a, a + b\}$ are $\pm a$, $\pm b$, $\pm (a + b)$, i.e. $\pm a$, $\pm b$, $\pm c$. So altogether the differences are just the non-zero elements $\pm 1, \ldots, \pm 3m$ of $Z_{6m+1}$.

**Example 6.2.3** Take $m = 4$. The triples $\{1, 5, 6\}, \{2, 8, 10\}, \{4, 7, 11\}, \{3, 9, 12\}$ form a partition of $\{1, \ldots, 12\}$ and hence yield a cyclic STS(25).

Note that in this example we had $a + b = c$ in each triple, i.e. we did not use the alternative condition $a + b + c \equiv 0$ at all; note further that $1, \ldots, m$ occurred in different triples of the partition.

**Definition 6.2.1** A *Skolem triple system* of order $m$ is a partition of $\{1, \ldots, 3m\}$ into $m$ triples $\{i, a_i, i + a_i\}$, $1 \leqslant i \leqslant m$.

When can such a system of triples be found?

**Lemma 6.2.2** *A Skolem triple system of order* $m$ *can exist only if* $m \equiv 0$ *or* 1 (mod 4).

**Proof** Suppose that the $m$ triples $\{i, a_i, i + a_i\}$, $1 \leqslant i \leqslant m$, form such a system. Then

$$\sum_{i=1}^{m} i + \sum_{i=1}^{m} a_i + \sum_{i=1}^{m} (i + a_i) = 1 + \cdots + 3m = \tfrac{1}{2} 3m(3m + 1),$$

i.e.

$$\sum_{i=1}^{m} i + \sum_{i=1}^{m} a_i = \tfrac{3}{4} m(3m + 1).$$

Thus $m(3m + 1)$ is divisible by 4, so that $m \equiv 0$ or 1 (mod 4).

Skolem (any many other mathematicians since) showed how to construct such triple systems for all $m \equiv 0$ or 1 (mod 4). (Skolem, 1957)

**Example 6.2.4** The triples

$$\{1, 14, 15\}, \{2, 7, 9\}, \{3, 10, 13\}, \{4, 8, 12\}, \{5, 6, 11\}$$

form a Skolem triple system of order 5.

**Theorem 6.2.3** *If $m \equiv 0$ or 1 (mod 4) then a Skolem triple system of order $m$ exists.*

**Proof** If $m = 4k$ then the triples

$$
\begin{array}{ll}
1, 12k - 1, 12k & \\
2r + 1, 10k - r - 1, 10k + r & 1 \leqslant r \leqslant k - 1 \\
2k + 2r - 1, 5k - r, 7k + r - 1 & 1 \leqslant r \leqslant k - 1 \\
4k - 1, 5k, 9k - 1 & \\
2r, 6k - r, 6k + r & 1 \leqslant r \leqslant k - 1 \\
2k, 8k - 1, 10k - 1 & \\
2k + 2r, 9k - 1 - r, 11k + r - 1 & 1 \leqslant r \leqslant k - 1 \\
4k, 6k, 10k &
\end{array}
$$

form a Skolem triple system. Note that, if $k = 1$, we omit those triples defined for $1 \leqslant r \leqslant k - 1$; the remaining triples are

$$\{1, 11, 12\}, \{3, 5, 8\}, \{2, 7, 9\}, \{4, 6, 10\}$$

which form a Skolem triple system of order 4 different from the one exhibited in Example 6.2.3.

If $m = 4k + 1$ then the triples

$$
\begin{array}{ll}
1, 12k + 2, 12k + 3 & \\
2r + 1, 10k - r, 10k + r + 1 & 1 \leqslant r \leqslant k - 1 \\
2k + 2r - 1, 5k - r + 1, 7k + r & 1 \leqslant r \leqslant k - 1 \\
4k - 1, 5k + 1, 9k & \\
4k + 1, 8k, 12k + 1 & \\
2r, 6k + 1 - r, 6k + 1 + r & 1 \leqslant r \leqslant k - 1 \\
2k, 10k, 12k & \\
2k + 2r, 9k - r, 11k + r & 1 \leqslant r \leqslant k - 1 \\
4k, 6k + 1, 10k + 1 &
\end{array}
$$

work.

Note in passing that the triples of a Skolem system yield cyclic league schedules. For example, in Example 6.2.4 replace each triple $\{i, a_i, i + a_i\}$ by the pair $\{a_i - 4, i + a_i - 4\}$ to obtain pairs

$$\{10, 11\}, \{3, 5\}, \{6, 9\}, \{4, 8\}, \{2, 7\}$$

which, with $\{1,\infty\}$, give a possible choice of fixtures for the first day of a cyclic league schedule; for their differences are $1, 2, 3, 4, 5$.

Skolem suggested that, although Skolem triple systems cannot exist when $m \equiv 2$ or 3 (mod 4), a similar approach might be possible if triples partitioning $\{1, 2, \ldots, 3m - 1; 3m + 1\}$ instead of $\{1, \ldots, 3m\}$ were found. The idea here is to replace $3m$ by its negative $-3m = 3m + 1$ (mod $6m + 1$); this corresponds to using the alternative condition $a + b + c \equiv 0$ of Theorem 6.2.1 just once. Such triples were duly constructed by O'Keefe (1961). It should be mentioned, however, that Davies (1959) had already solved an equivalent problem (see Langford's problem, Exercises 6, number 11).

**Definition 6.2.2** An *O'Keefe triple system* of order $m$ is a partition of $\{1, \ldots, 3m - 1; 3m + 1\}$ into $m$ triples $\{i, a_i, i + a_i\}$, $1 \leqslant i \leqslant m$.

It is easy to establish the following condition for existence.

**Lemma 6.2.4** *An O'Keefe triple system of order $m$ can exist only if $m \equiv 2$ or 3 (mod 4).*

**Example 6.2.5**
(a)  $m = 2$. The triples $\{1, 3, 4\}$, $\{2, 5, 7\}$ form an O'Keefe system of order 2.
(b)  $\{1, 4, 5\}$, $\{2, 6, 8\}$, $\{3, 7, 10\}$ form an O'Keefe system of order 3.
(c)  $\{1, 16, 17\}$, $\{2, 8, 10\}$, $\{3, 12, 15\}$, $\{4, 7, 11\}$, $\{5, 9, 14\}$, $\{6, 13, 19\}$ form an O'Keefe system of order 6.
(d)  $\{1, 9, 10\}$, $\{2, 17, 19\}$, $\{3, 11, 14\}$, $\{4, 16, 20\}$, $\{5, 8, 13\}$, $\{6, 12, 18\}$, $\{7, 15, 22\}$ form an O'Keefe system of order 7.

**Theorem 6.2.5** *If $m \equiv 2$ or 3 (mod 4) then an O'Keefe triple system of order $m$ exists.*

**Proof** We need only deal with $m \geqslant 10$ since smaller values are dealt with in the example above.
  If $m = 4k + 2$ the triples

$$1, 11k + 6, 11k + 7$$
$$2r + 1, 10k + 4 - r, 10k + r + 5 \qquad\qquad 1 \leqslant r \leqslant k$$
$$2k + 3, 10k + 4, 12k + 7$$
$$2k + 3 + 2r, 9k + 4 - r, 11k + r + 7 \qquad\qquad 1 \leqslant r \leqslant k - 2$$
$$4k + 1, 6k + 4, 10k + 5$$
$$2r, 6k + 4 - r, 6k + 4 + r \qquad\qquad 1 \leqslant r \leqslant 2k + 1$$

form an O'Keefe system provided $k \geqslant 2$, i.e. $m \geqslant 10$.

If $m = 4k + 3$ the triples

$$1, 9k + 7, 9k + 8$$
$$2r + 1, 10k + 8 - r, 10k + 9 + r \qquad\qquad 1 \leqslant r \leqslant k - 1$$
$$2k + 1, 10k + 9, 12k + 10$$
$$2k + 1 + 2r, 9k + 7 - r, 11k + 8 + r \qquad 1 \leqslant r \leqslant k$$
$$4k + 3, 6k + 5, 10k + 8$$
$$2r, 6k + 5 - r, 6k + 5 + r \qquad\qquad 1 \leqslant r \leqslant 2k + 1$$

work provided $k \geqslant 1$, i.e. $m \geqslant 7$.

**Corollary 6.2.6**  *A cyclic* STS($6m + 1$) *exists for all* $m \geqslant 1$.

**Proof**  If $m \equiv 0$ or $1 \pmod 4$, the triples of a Skolem system of order $m$ satisfy the conditions of Theorem 6.2.1. If $m \equiv 2$ or $3 \pmod 4$, the triples of an O'Keefe system of order $m$ satisfy the conditions if the triple $\{a, b, 3m + 1\}$ is replaced by the triple $\{a, b, 3m\}$; for, since $a + b = 3m + 1$, $a + b + 3m = 6m + 1 \equiv 0 \pmod{6m + 1}$.

**Example 6.2.6**  To obtain a cyclic STS(19), start with an O'Keefe system of order 3 as in Example 6.2.5(b), and replace 10 by 9 to obtain the following triple system: $\{1, 4, 5\}$, $\{2, 6, 8\}$, $\{3, 7, 9\}$. Then, as in Theorem 6.2.1, the triples $\{0, 1, 5\}$, $\{0, 2, 8\}$, $\{0, 3, 10\}$ form a difference system (mod 19) which yields a cyclic STS(19) with triples $\{i, 1 + i, 5 + i\}$, $\{i, 2 + i, 8 + i\}$, $\{i, 3 + i, 10 + i\}$, $0 \leqslant i \leqslant 18$.

If $v \equiv 3 \pmod 6$, say $v = 6m + 3$, the number of triples in an STS($v$) is $v(m + \frac{1}{3})$ which is *not* a multiple of $v$, so we cannot hope to obtain a cyclic design from a difference system as in the case $v \equiv 1 \pmod 6$. However, the method can be adapted. As an example, consider the following STS(15) exhibited by Netto.

**Example 6.2.7**  A cyclic STS(15).

| | | |
|---|---|---|
| $\{0, 2, 8\}$ | $\{0, 1, 4\}$ | $\{0, 5, 10\}$ |
| $\{1, 3, 9\}$ | $\{1, 2, 5\}$ | $\{1, 6, 11\}$ |
| $\{2, 4, 10\}$ | $\{2, 3, 6\}$ | $\{2, 7, 12\}$ |
| $\vdots$ | $\vdots$ | $\{3, 8, 13\}$ |
| $\{14, 1, 7\}$ | $\{14, 0, 3\}$ | $\{4, 9, 14\}$ |

Note that the sets $\{0, 2, 8\}$, $\{0, 1, 4\}$ give every non-zero difference (mod 15) except 5 and 10, and that these differences are obtained from $\{0, 5, 10\}$, which is developed cyclically, not through all its translates, but through a third of them. This reflects the term $\frac{1}{3}$ in the expression $b = v(m + \frac{1}{3})$

above. In general, to construct a cyclic STS($6m + 3$), we look for a difference system which gives all non-zero differences (mod $6m + 3$) except $2m + 1$ and $4m + 2$.

**Definition 6.2.3** A *modified difference system* of triples (mod $6m + 3$) is a set of $m$ triples such that the differences obtained from them are precisely all the non-zero elements of $Z_{6m+3}$ other than $2m + 1$ and $4m + 2$.

**Theorem 6.2.7** *A cyclic STS($6m + 3$) exists if a modified difference system of triples (mod $6m + 3$) exists.*

**Proof** Take all the translates of the triples of the system; these will contain every pair except those differing by $2m + 1$ (or $4m + 2$). Take also the first $2m + 1$ translates of $\{0, 2m + 1, 4m + 2\}$.

The construction of modified difference systems can be accomplished by using triple systems similar to Skolem triple systems. Such systems were independently discussed by Rosa (1966) and Hilton (1969).

**Definition 6.2.4** A *Rosa triple system* of order $m$ is a partition of $\{1, 2, \ldots, 2m; \ 2m + 2, \ldots, 3m + 1\}$ or $\{1, \ldots, 2m; \ 2m + 2, \ldots, 3m; \ 3m + 2\}$ into $m$ triples $\{i, a_i, i + a_i\}$, $1 \leqslant i \leqslant m$.

As in Lemmas 6.2.2 and 6.2.4 it is easy to check that the first set is appropriate if $m \equiv 0$ or 3 (mod 4), whereas the second set is appropriate if $m \equiv 1$ or 2 (mod 4).

**Example 6.2.8** Rosa triple systems of orders (a) 3, (b) 5.

(a) $\{1, 9, 10\}$, $\{2, 4, 6\}$, $\{3, 5, 8\}$.
(b) $\{1, 8, 9\}$, $\{2, 13, 15\}$, $\{3, 14, 17\}$, $\{4, 6, 10\}$, $\{5, 7, 12\}$.

**Theorem 6.2.8** *A Rosa triple system of order $m$ exists for all $m \geqslant 2$.*

**Proof** First consider the case $m = 4k$, and partition the set $\{1, \ldots, 8k; \ 8k + 2, \ldots, 12k + 1\}$ into triples as follows.

$$
\begin{aligned}
&1, 6k - 1, 6k \\
&2r + 1, 6k - r - 1, 6k + r && 1 \leqslant r \leqslant k - 1 \\
&2k + 1, 7k, 9k + 1 \\
&2k + 2r + 1, 5k - r, 7k + r + 1 && 1 \leqslant r \leqslant k - 1 \\
&2r, 10k + 1 - r, 10k + 1 + r && 1 \leqslant r \leqslant k - 1 \\
&2k, 10k + 1, 12k + 1 \\
&2k + 2r, 9k + 1 - r, 11k + r + 1 && 1 \leqslant r \leqslant k - 1 \\
&4k, 7k + 1, 11k + 1.
\end{aligned}
$$

For $m = 4k + 3$, $k \geqslant 1$, take triples

$$1, 11k + 9, 11k + 10$$
$$2r + 1, 10k - r + 8, 10k + r + 9 \qquad\qquad 1 \leqslant r \leqslant k - 1$$
$$2k + 1, 8k + 8, 10k + 9$$
$$2k + 3, 9k + 8, 11k + 11$$
$$2k + 2r + 3, 9k + 8 - r, 11k + r + 11 \qquad\qquad 1 \leqslant r \leqslant k - 1$$
$$4k + 3, 6k + 5, 10k + 8$$
$$2r, 6k + 5 - r, 6k + 5 + r \qquad\qquad 1 \leqslant r \leqslant 2k + 1.$$

The case $m = 3$ is deals with separately as in Example 6.2.8.
For $m = 4k + 1$, $k \geqslant 2$, take the triples

$$1, 11k + 4, 11k + 5$$
$$2r + 1, 10k + 2 - r, 10k + r + 3 \qquad\qquad 1 \leqslant r \leqslant k$$
$$2k + 3, 10k + 2, 12k + 5$$
$$2k + 2r + 3, 9k - r + 2, 11k + r + 5 \qquad\qquad 1 \leqslant r \leqslant k - 2$$
$$4k + 1, 6k + 2, 10k + 3$$
$$2r, 6k + 2 - r, 6k + 2 + r \qquad\qquad 1 \leqslant r \leqslant 2k.$$

The missing case $k = 1$ is dealt with in Example 6.2.8.
Finally, for $m = 4k + 2$, $k \geqslant 2$, take triples

$$1, 11k + 5, 11k + 6$$
$$2r + 1, 6k + 3 - r, 6k + 4 + r \qquad\qquad 1 \leqslant r \leqslant 2k$$
$$2, 12k + 6, 12k + 8$$
$$2r, 10k + 5 - r, 10k + 5 + r \qquad\qquad 2 \leqslant r \leqslant k - 1$$
$$2k, 8k + 6, 10k + 6$$
$$2k + 2r, 9k + 6 - r, 11k + 6 + r \qquad\qquad 1 \leqslant r \leqslant k - 1$$
$$4k, 6k + 4, 10k + 4$$
$$4k + 2, 6k + 3, 10k + 5.$$

For $m = 2$ take the triples $\{1, 3, 4\}$, $\{2, 6, 8\}$; for $m = 6$ take $\{1, 16, 17\}$, $\{2, 18, 20\}$, $\{3, 8, 11\}$, $\{4, 10, 14\}$, $\{5, 7, 12\}$, $\{6, 9, 15\}$.

Having constructed Rosa triple systems in all cases, we can now establish the existence of cyclic STS($6m + 3$) for all $m \geqslant 2$.

**Theorem 6.2.9**  *A cyclic* STS($6m + 3$) *exists for all* $m \geqslant 2$.

**Proof**  By Theorem 6.2.7 it suffices to construct a modified difference system of triples (mod $6m + 3$). If $m \equiv 0$ or $3$ (mod 4), the triples $\{i, a_i, i + a_i\}$ of a Rosa system give rise to the triples $\{0, i, i + a_i\}$, which give the differences $\pm i$, $\pm a_i$, $\pm (i + a_i)$, i.e. $1, \ldots, 2m$; $2m + 2, \ldots, 3m + 1$ and their negatives, i.e. every member of $Z_{6m+3}$ apart from $2m + 1$ and

$4m + 2$. If $m \equiv 1$ or $2 \pmod 4$, again consider the triples of a Rosa system; they yield the triples $\{0, i, i + a_i\}$ which give the differences $\pm$, $\pm a_i$, $\pm(i + a_i)$, i.e. $1, \ldots, 2m$; $2m + 2, \ldots, 3m$; $3m + 2$ and their negatives, i.e. every non-zero member of $Z_{6m+3}$ apart from $2m + 1$ and $4m + 2$. Thus in all cases the triples $\{0, i, i + a_i\}$ form a modified difference system.

**Example 6.2.9** Construction of a cyclic STS(27). We have $27 = 6m + 3$ where $m = 4$. A Rosa triple system of order 4 is:

$$\{1, 5, 6\}, \{2, 11, 13\}, \{4, 8, 12\}, \{3, 7, 10\}.$$

This yields the modified difference system

$$\{0, 1, 6\}, \{0, 2, 13\}, \{0, 4, 12\}, \{0, 3, 10\}$$

which in turn gives rise to a cyclic STS(27) whose triples are the following:

$$
\begin{array}{ll}
\{i, 1 + i, 6 + i\} & 0 \leqslant i \leqslant 26 \\
\{i, 2 + i, 13 + i\} & 0 \leqslant i \leqslant 26 \\
\{i, 4 + i, 12 + i\} & 0 \leqslant i \leqslant 26 \\
\{i, 3 + i, 10 + i\} & 0 \leqslant i \leqslant 26 \\
\{i, 9 + i, 18 + i\} & 0 \leqslant i \leqslant 8.
\end{array}
$$

The above method, using Skolem, O'Keefe or Rosa triple systems, was not the first complete construction of cyclic STS. It was Peltesohn (1939) who first constructed a cyclic STS($v$) for all $v \equiv 3 \pmod 6$, $v > 9$. But we have presented the method above because Skolem triples have turned out to be of use in other combinatorial constructions. Further, it should be remarked that no cyclic STS(9) exists.

The next major problem concerning Steiner triple systems is that of resolvability. This is of course the Kirkman schoolgirls problem, and it is dealt with in the next chapter.

## 6.3  Introduction to *t*-designs

As was mentioned in the first section of this book, Kirkman's study of Steiner triple systems grew out of a prize problem in the *Lady's and Gentleman's Diary* of 1844. The problem was concerned with a system of $k$-subsets of a $v$-set such that no $t$-subset occurs more than once in any of the $k$-subsets. In the case of $t = 2$, with each 2-subset in exactly one $k$-subset, we get a BIBD with $\lambda = 1$; we now have a look at the case where $t > 2$.

**Definition 6.3.1** Let $t, v, k, \lambda$ be integers with $v > k \geqslant t$ and $\lambda > 0$. A $t$-$(v, k, \lambda)$ *t-design* is a collection $\mathscr{D}$ of $k$-subsets (blocks) of a $v$-set $S$ such that every $t$-subset of $S$ is contained in exactly $\lambda$ blocks.

We observe that a 2-design is just a BIBD. In fact, every $t$-design with $t > 2$ is also a BIBD, in view of the following result.

**Lemma 6.3.1**   *If $s < t$ then every t-design is also an s-design.*

**Proof**   Let $\lambda_s$ denote the number of blocks containing a given $s$-subset $A$. Counting in two ways the number of pairs $(D, B)$, where $D$ is a $t$-subset containing $A$, and $B$ is a block containing $D$, we get

$$\lambda \binom{v-s}{t-s} = \lambda_s \binom{k-s}{t-s}. \tag{6.1}$$

This number $\lambda_s$ is therefore independent of the choice of $A$; so the design is also an $s$-$(v, k, \lambda_s)$ design.

**Notes**

(i) Putting $\lambda = \lambda_t$, we obtain

$$\lambda_{t-1} = \lambda_t \left( \frac{v-t+1}{k-t+1} \right). \tag{6.2}$$

(ii) The number of blocks in the $t$-design is

$$b = \lambda_0 = \lambda \binom{v}{t} \bigg/ \binom{k}{t}. \tag{6.3}$$

(iii) Each element occurs in $r$ blocks where

$$r = \lambda_1 = \lambda \binom{v-1}{t-1} \bigg/ \binom{k-1}{t-1}. \tag{6.4}$$

**Example 6.3.1**   The following subsets of $S = \{1, \ldots, 8\}$ are the blocks of a 3-$(8, 4, 1)$ design:

$\{1,2,4,8\}$ $\{2,3,5,8\}$ $\{3,4,6,8\}$ $\{4,5,7,8\}$ $\{5,6,1,8\}$ $\{6,7,2,8\}$ $\{7,1,3,8\}$
$\{3,5,6,7\}$ $\{4,6,7,1\}$ $\{5,7,1,2\}$ $\{6,1,2,3\}$ $\{7,2,3,4\}$ $\{1,3,4,5\}$ $\{2,4,5,6\}$.

It is straightforward to check that each 3-subset of $S$ occurs in exactly one block: we now study its structure.

Given a $t$-design $\mathscr{D}$, there is a natural way of exhibiting a $(t-1)$-design embedded in it. Take any $t$-design; let $\mathscr{B}$ denote the collection of its blocks, and take any element $P$. Then let $\mathscr{B}_P$ denote the collection of all sets $B - \{P\}$, where $B$ is a block containing $P$.

**Lemma 6.3.2**   *The sets in $\mathscr{B}_P$ are the blocks of a $(t-1)$-$(v-1, k-1, \lambda)$ design $\mathscr{D}_P$.*

**Proof**   Any $(t-1)$-subset $A$ occurs in a set $B - \{P\}$ as often as $A \cup \{P\}$ occurs in a block in $\mathscr{B}$, i.e. $\lambda$ times.

**Definition 6.3.2** The design $\mathscr{D}_P$ is called a *derived design*. Conversely, if a *t*-design is isomorphic to $\mathscr{D}_P$ for some $(t + 1)$-design $\mathscr{D}$ and some $P$, then $\mathscr{D}$ is called an *extension*.

**Example 6.3.2** The 3-design exhibited in Example 6.3.1 is an extension of the seven-point plane; take $P$ as the point 8 to see this. Note that the blocks not containing 8 are just the complements of the lines of the plane.

When can a *t*-design be extended to a $(t + 1)$-design?

**Lemma 6.3.3** *If a t-$(v, k, \lambda)$ design with b blocks is extendable, then $k + 1$ divides $b(v + 1)$.*

**Proof** By eqn (6.3) we have $b = \lambda \dfrac{v(v - 1) \cdots (v - t + 1)}{k(k - 1) \cdots (k - t + 1)}$, and the number of blocks in the extension is

$$\lambda \frac{(v + 1)v \cdots (v - t + 1)}{(k + 1)k \cdots (k - t + 1)} = \frac{(v + 1)b}{k + 1}.$$

**Example 6.3.3** If a finite projective plane of order $n$ is extendable to a 3-design then $n + 2$ must divide $(n^2 + n + 2)(n^2 + n + 1)$, and so must divide $n^2(n^2 - 1)$. The remainder obtained by dividing $n^4 - n^2$ by $n + 2$ is 12; so for extendability we must have $n + 2$ a factor of 12. Thus $n = 2, 4$, or 10.

(a) $n = 2$. Example 6.3.1 shows the plane *can* be extended in this case.
(b) $n = 4$. It is known that the 2-$(21, 5, 1)$ design can be extended three times to yield a 5-$(24, 5, 1)$ design; see the remarks at the end of the next section.
(c) $n = 10$. It is now known that no finite projective plane of order 10 exists.

Although extending finite projective planes is not very fruitful, we have more success when we try to extend more general symmetric designs. The key observation lies in the following lemma.

**Lemma 6.3.4** *If $a, b, c$ are distinct elements of a $(v, k, \lambda)$ design, if $n_0$ denotes the number of blocks containing none of $a, b, c$ and if $n_{abc}$ denotes the number of blocks containing all of $a, b, c$, then*

$$n_0 + n_{abc} = b + 3\lambda - 3r \tag{6.5}$$

*(which is independent of the choice of $a, b, c$).*

**Proof** Let $n_a$ denote the number of blocks containing $a$, and let $n_{ab}$ denote the number of blocks containing $a$ and $b$. By the inclusion–exclusion principle,

$$n_0 = b - (n_a + n_b + n_c) + (n_{ab} + n_{bc} + n_{ca}) - n_{abc}$$
$$= b - 3r + 3\lambda - n_{abc}.$$

Note that for a symmetric design we obtain $n_0 + n_{abc} = v - 3k + 3\lambda$.

**Corollary 6.3.5** *Every Hadamard design can be extended to a 3-design.*

**Proof** Let $\mathcal{H}$ be a Hadamard $(4m - 1, 2m - 1, m - 1)$ design. Take the complements of the blocks of $\mathcal{H}$, and also all the blocks of $\mathcal{H}$ with a new element $\infty$ adjoined to each. This gives $8m - 2$ sets all of size $2m$; we show that each 3-subset is contained in precisely $m - 1$ of them. Certainly the set $\{\infty, a, b\}$ occurs in $B \cup \{\infty\}$ precisely when $\{a, b\}$ occurs in $B$, i.e. $m - 1$ times. Further, if none of $a, b, c$ is $\infty$, the number of sets $\bar{B}$ or $B \cup \{\infty\}$ containing $a, b, c$ is just $n_0 + n_{abc}$, which, by eqn (6.5), is just $4m - 1 - 3(2m - 1) + 3(m - 1) = m - 1$.

**Definition 6.3.3** Any $3\text{-}(4m, 2m, m - 1)$ design is called a *Hadamard 3-design*.

**Corollary 6.3.6** *There exist infinitely many 3-designs.*

The Hadamard 3-designs were not the first family of 3-designs to be constructed. It will probably by now come as no surprise to learn that Kirkman, back in 1853, showed that a $3\text{-}(2^n, 4, 1)$ design exists for all $n \geqslant 2$. We shall construct such designs in the next section.

In view of Corollary 6.3.5, it might be hoped that other families of symmetric 2-designs can be extended to 3-designs. However, it was proved by Cameron (1973) that, apart from the Hadamard designs, the only possibilities are those with $\lambda = 3$ and $k = 39$, and those with $k = \lambda^2 + 3\lambda + 1$. Of these, the only one known to be extendable is the FPP of order 4.

If we try to obtain 3-designs by extending non-symmetric 2-designs, the most obvious candidates to start with are the affine plane $(q^2, q, 1)$ designs. The condition of Lemma 6.3.3 is automatically satisfied since $k + 1 = q + 1$ and $b = q^2 + q = q(q + 1)$, and it turns out that all such designs can in fact be extended to $3\text{-}(q^2 + 1, q + 1, 1)$ designs. These 3-designs are known as *inversive* or *Möbius* planes. (See Hughes and Piper, 1985)

**Example 6.3.4** A $3\text{-}(10, 4, 1)$ design can be obtained by taking the translates (mod 10) of $\{1, 2, 4, 5\}$, $\{1, 2, 3, 7\}$, $\{1, 3, 5, 8\}$ as blocks. If we take $P = 10$

and form the derived design $\mathcal{D}_P$, we obtain the blocks

$$\{1,3,4\}, \{1,7,8\}, \{2,3,9\}, \{6,7,9\}, \{1,2,6\}, \{1,5,9\},$$
$$\{4,8,9\}, \{4,5,6\}, \{2,4,7\}, \{2,5,8\}, \{3,6,8\}, \{3,5,7\},$$

which form (the unique) $(9,3,1)$ affine plane.

For values of $t > 3$, $t$-designs have been hard to find. For example, it was not until 1972 that an infinite family of 5-designs was known. Alltop showed that a $5\text{-}(2^n + 2, 2^{n-1} + 1, (2^{n-1} - 3)(2^{n-2} - 1))$ design exists for all $n \geqslant 4$. Then it was not until 1984 that any 6-designs were known, when Magliveras and Leavitt published a $6\text{-}(33,8,36)$ design. But then in 1987 Tierlinck published the remarkable result that $t$-designs exist for all values of $t$.

These developments are well beyond the scope of this book. Instead, we now turn to look at $t$-designs for which $\lambda = 1$.

## 6.4   Steiner systems

A Steiner system is simply a $t$-design with $\lambda = 1$.

**Definition 6.4.1**   A *Steiner system* $S(t,k,v)$ is a collection of $k$-subsets (blocks) of a $v$-set $S$ such that every $t$-subset of $S$ is contained in exactly one of the blocks.

Note that we use $S(t,k,v)$ as an equivalent of $t\text{-}(v,k,1)$. Note too that Steiner systems $S(2,k,v)$ are just $(v,k,1)$ BIBDs.

In 1853 Steiner posed the problem of the existence of designs $S(t,k,v)$ in the special case $k = t + 1$. The case $t = 2, k = 3$ corresponds to what are now known as Steiner triple systems, and was of course already solved by Kirkman. The next case, that of $S(3,4,v)$, was not fully dealt with until Hanani constructed them in all possible cases in 1960. Kirkman showed that they existed whenever $v$ is a power of 2, and we shall obtain this result shortly. But first we establish a few basic lemmas.

**Lemma 6.4.1**   *A Steiner system* $S(t,k,v)$ *has* $\binom{v}{t}\Big/\binom{k}{t}$ *blocks.*

**Proof**   This is eqn (6.3) in the case $\lambda = 1$.

**Lemma 6.4.2**   *If an* $S(t,k,v)$ *exists, then an* $S(t-1,k-1,v-1)$ *also exists.*

**Proof**   This is just Lemma 6.3.2 in the case $\lambda = 1$.

**Lemma 6.4.3** *If an S(t, k, v) exists, then all of*

$$\binom{v}{t}\bigg/\binom{k}{t}, \ \binom{v-1}{t-1}\bigg/\binom{k-1}{t-1}, \ldots, \binom{v-t+1}{1}\bigg/\binom{k-t+1}{1}$$

*must be integers.*

**Proof** These are just the numbers of blocks in $S(t, k, v)$, $S(t-1, k-1, v-1)$, etc.

Lemma 6.4.3 can be used to obtain necessary conditions for the existence of any $S(t, k, v)$. For example, if an $S(2, k, v)$ exists, then $\binom{v}{2}\big/\binom{k}{2}$ and $\binom{v-1}{1}\big/\binom{k-1}{1}$ must be integers, i.e. $\dfrac{v(v-1)}{k(k-1)}$ and $\dfrac{v-1}{k-1}$ must be integers. So for an $S(2, 3, v)$ to exist we require $v(v-1) \equiv 0 \pmod 6$ and $v - 1 \equiv 0 \pmod 2$, whence $v \equiv 1$ or $3 \pmod 6$, as we saw in Lemma 6.1.1. Similarly we can obtain conditions for an $S(3, 4, v)$.

**Lemma 6.4.4** *If an S(3, 4, v) exists, then $v \equiv 2$ or $4 \pmod 6$.*

**Proof** We could use the conditions of Lemma 6.4.3, which demand that $\frac{1}{4}\binom{v}{3}$, $\frac{1}{3}\binom{v-1}{2}$, and $\frac{1}{2}\binom{v-2}{1}$ should all be integers. But it is easier to note that if an $S(3, 4, v)$ exists then an $S(2, 3, v-1)$ exists, so we must have $v - 1 \equiv 1$ or $3 \pmod 6$, i.e. $v \equiv 2$ or $4 \pmod 6$.

**Example 6.4.1** The design of Example 6.3.4 is an $S(3, 4, 10)$.

**Definition 6.4.2** An $S(3, 4, v)$ is called a *Steiner quadruple system* and is denoted by SQS($v$).

The 3-$(2^n, 4, 1)$ designs constructed by Kirkman in 1853 are thus examples of Steiner quadruple systems. We can establish the existence of such designs very easily by making use of the following doubling construction which goes at least as far back as Witt (1938).

**Theorem 6.4.5** *If an SQS(v) exists then an SQS(2v) also exists.*

**Proof** Let $S_1$ be an SQS($v$) on $X$ and let $S_2$ be an SQS($v$) on $Y$ where $X$ and $Y$ are disjoint $v$-sets. Since $v$ is even (by Lemma 6.4.4), we can construct a resolvable $(v, 2, 1)$ design on $X$ with resolution classes $F_1, \ldots, F_{v-1}$, and another on $Y$ with resolution classes $G_1, \ldots, G_{v-1}$. Now form a quadruple system on $X \cup Y$ by taking as quadruples

(a) all the quadruples in $S_1$;
(b) all the quadruples in $S_2$;
(c) all quadruples $\{x_1, x_2, y_1, y_2\}$ where, for some $i$, $\{x_1, x_2\} \in F_i$ and $\{y_1, y_2\} \in G_i$. It is clear that any three elements of $X$ will occur in just one quadruple, of type (a), and similarly any three elements of $Y$ will occur in just one quadruple of type (b). Next consider $x_1, x_2, y$. There is just one $i$ for which $\{x_1, x_2\} \in F_i$; form a quadruple of type (c) by taking $\{y_1, y_2\}$ to be the unique block in $G_i$ containing $y$.

**Corollary 6.4.6**    *An SQS($2^n$) exists for all $n \geq 2$.*

During the early years of this century a few small SQSs were constructed: an SQS(10) was found in 1908 but, surprisingly, it was not until 1935 that Bays and de Weck found an SQS(14). It was not until 1960 that it was finally shown by Hanani that the necessary conditions of Lemma 6.4.4 are in fact sufficient for the existence of an SQS($v$).

**Theorem 6.4.7** (Hanani, 1960)    *An SQS($v$) exists if and only if $v \equiv 2$ or $4$ (mod 6), $v \geqslant 4$.*

What about the existence of $S(t, k, v)$ with $t > 3$? So far, only a handful of such designs are known to exist.

(a) Systems $S(5, 6, 12)$ and $S(5, 8, 24)$ were known to Witt (1938). Their existence of course implies that of $S(4, 5, 11)$ and $S(4, 7, 23)$. The (essentially unique) $S(5, 8, 24)$ is a truly remarkable design, closely related to Leech's 24-dimensional lattice and Golay's perfect binary error-correcting code. (See Anderson (1974, 1989) and Cameron and van Lint (1991).)
(b) Denniston (1976) proved the existence of $S(5, 7, 28)$ and $S(5, 6, v)$ with $v = 24, 48,$ and $84$.
(c) Mills (1978) proved the existence of $S(5, 6, 72)$.
(d) In 1991, Grannell and Griggs constructed $S(5, 6, 108)$ and, more recently, with Mathon, $S(5, 6, 132)$.

So far, no Steiner system with $t > 5$ has been discovered.

## 6.5  Exercises 6

1. Construct an STS(15) using (i) Skolem's, (ii) PBD method.
2. Construct a cyclic STS(25) using Skolem triples.
3. Let $n$ be odd, $n$ not a multiple of 3. Construct an STS($3n$) on $\{1, \ldots, 3n\}$ as follows. There are $\frac{1}{2}(n-1)$ pairs $\{s, t\}$ with $s + t \equiv 0$

(mod $n$), $s \not\equiv 0$, $t \not\equiv 0$ (mod $n$), $s \equiv t \equiv 1$ (mod 3); form the triples $\{r, r+s, r+t\}$, $0 \leqslant r \leqslant 3n-1$ and $\{r, r+n, r+2n\}$, $0 \leqslant r \leqslant n-1$. In particular obtain an STS(15).

4. *Another proof of Lemma 6.1.2.* Take an STS($n$) on $x_1, \ldots, x_n$, and take $n+1$ other elements $y_1, \ldots, y_{n+1}$. Replace each triple $\{x_i, x_j, x_k\}$ of the STS($n$) by four triples: $\{x_i, x_j, x_k\}$, $\{y_i, y_j, x_k\}$, $\{y_i, x_j, y_k\}$, $\{x_i, y_j, y_k\}$. Also take the $n$ triples $\{x_i, y_i, y_{n+1}\}$. (Netto, 1893)

5. *A product theorem.* Show that if a STS($m$) and a STS($n$) both exist then an STS($mn$) also exists, as follows. Let $S_1$ be an STS($m_1$) on $\{x_1, \ldots, x_m\}$ and let $S_2$ be an STS($n$) on $\{y_1, \ldots, y_n\}$. Define $mn$ elements $z_{ij}$, $1 \leqslant i \leqslant m$, $1 \leqslant j \leqslant n$.

   (i) For each triple $\{x_i, x_j, x_k\}$ of $S_1$ form $n$ triples $\{z_{iu}, z_{ju}, z_{ku}\}$, $1 \leqslant u \leqslant n$.

   (ii) For each triple $\{y_u, y_v, y_w\}$ of $S_2$, form $m$ triples $\{z_{iu}, z_{iv}, z_{iw}\}$, $1 \leqslant i \leqslant m$.

   (iii) For each pair of triples, $\{x_i, x_j, x_k\}$ from $S_1$, $\{y_u, y_v, y_w\}$ form $S_2$, form six triples: $\{z_{iu}, z_{jv}, z_{kw}\}$, $\{z_{iu}, z_{jw}, z_{kv}\}$, $\{z_{iw}, z_{ju}, z_{kv}\}$, $\{z_{iw}, z_{jv}, z_{ku}\}$, $\{z_{iv}, z_{ju}, z_{kw}\}$, $\{z_{iv}, z_{jw}, z_{kw}\}$. (Moore, 1893)

6. Verify that if $\theta$ is a primitive root of the prime $p = 6 \cdot 2^s n + 1$, where $s \geqslant 0$ and $n$ is odd, then

$$\{\theta^{2^{s+1}i+j}, \theta^{2^{s+1}(i+n)+j}, \theta^{2^{s+1}(i+2n)+j}\},$$

$0 \leqslant i \leqslant n-1$, $0 \leqslant j \leqslant 2^{s-1}$, form a $(p, 3, 1)$ difference system and hence yields a STS($p$). (Anstice, 1853)

7. Show that if $\theta$ is a primitive root of the prime $p = 6t+1$ such that $\theta + 1 = \theta^u$ for some $u \equiv 2$ (mod 3), then $\{\theta^{3i}, \theta^{3i+1}, \theta^{3i+2}\}$, $0 \leqslant i \leqslant t-1$, form a $(p, 3, 1)$ difference system which yields a STS($p$). Find such a $\theta$ for $p = 19$.

8. Suppose an STS($n$) $S$ exists and three MOLS of order $n$ exist. Construct an STS($3n$) as follows. Take copies of $S$ on $\{1, \ldots, n\}$, $\{n+1, \ldots, 2n\}$, $\{2n+1, \ldots, 3n\}$ respectively. Take three MOLS on $\{1, \ldots, n\}$; add $n$ to each entry of the second one and $2n$ to each entry of the third. Form $n^2$ triples, one corresponding to each position in the squares, consisting of the three entries in the MOLS. Along with the triples of the three copies of $S$, these form an STS($3n$).

9. Let $T$ be a collection of triples on $\{1, \ldots, n\}$ such that no pair of elements occurs in more than one of the triples of $T$. Show that the number of triples in $T$ is $\leqslant f(n) = \left[ \dfrac{n}{3} \left[ \dfrac{n-1}{2} \right] \right]$, with equality when $n \equiv 1$ or 3 (mod 6). Schonheim (1966) showed that max $|T|$ is $f(n)$ provided $n \not\equiv 5$ (mod 6), and max $|T| = f(n) - 1$ if $n \equiv 5$ (mod 6).

   (i) If $n = 6k$, $f(n) = 6k^2 - 2k$; show how to attain this bound by starting with a STS($6k+1$).

(ii) If $n = 6k + 2$, $f(n) = 6k^2 + 2k$; show how to attain this bound.
(iii) If $n = 6k + 4$, show that the following triples attain the bound:

$$\{6k + 4, x, x + 3k + 1\}, \qquad 1 \leqslant x \leqslant 3k + 1$$
$$\{y, y + 3i + 1, y + 6i + 3\}, \qquad 1 \leqslant y \leqslant 6k + 3, 0 \leqslant i \leqslant k - 1$$

considered mod $6k + 3$.

(iv) If $n = 6k + 5$, show that $|T| = f(n)$ is impossible. Show that the bound $f(n) - 1$ is attained by the following triples:

$$\{6k + 4, 1 + 2j(3k + 1), 1 + (2j + 1)(3k + 1)\} \quad 1 \leqslant j \leqslant 3k + 1$$
$$\{6k + 5, 1 + (2j + 1)(3k + 1), 1 + (2j + 2)(3k + 1)\} \quad 1 \leqslant j \leqslant 3k + 1$$
$$\{y, y + 3i + 1, y + 6i + 3\} \quad 1 \leqslant y \leqslant 6k + 3, 0 \leqslant i \leqslant k - 1.$$

(Schonheim, 1966; Hilton, 1969)

10. *Existence of a PBD$(v, \{3, 5\}, 1)$ for all odd $v > 1$.*
    (i) Show that a PBD$(v, \{3, 5\}, 1)$ cannot exist if $v$ is even.
    (ii) Note that an STS deals with $v \equiv 1$ or $3 \pmod{6}$; so only $v \equiv 5$ or 11 $\pmod{12}$ remain to be considered.
    (iii) Case $v = 12m + 5$. Take an STS$(6m + 3)$ on $X = \{0, \pm 1, \ldots, \pm 3m\}$, including the blocks $\{0, i, -i\}$, $1 \leqslant i \leqslant 3m$. Remove 0 to obtain a PBD$(6m + 2, \{2, 3\}, 1)$ in which the blocks of size 2 are $\{i, -i\}$. Corresponding to each triple $\{x, y, z\}$ form four blocks:

    $$\{(x, 1), (y, 1), (z, 1)\}, \ \{(x, 1), (y, -1), (z, -1)\},$$
    $$\{(x, -1), (y, 1), (z, -1)\}, \ \{(x, -1), (y, -1), (z, 1)\}.$$

    These blocks, together with the sets $\{(i, 1), (i, -1), (-i, 1), (-i, -1), \infty\}$ form a PBD$(12m + 5, \{3, 5\}, 1)$ on the set $S = \{\infty\} \cup (X \times \{-1, 1\})$.

    (iv) The case $v = 12m + 11$. From a resolvable $(6m + 6, 2, 1)$ design construct a PBD$(12m + 11, \{3, 6m + 5\}, 1)$. Replace each block of size $6m + 5$ by the blocks of a PBD$(6m + 5, \{3, 5\}, 1)$ on its elements.

11. *Langford's problem.* Given two bricks of each of $n$ different colours, can we arrange them in a line so that the two bricks of the $i$th colour are separated by $i$ other bricks? For example, 2 3 1 2 1 3 is a solution for three colours, and 4 1 3 1 2 4 3 2 is a solution for four colours. Note that these solutions can be converted into Skolem triple systems as follows: add 0 0 at the end, and then write the numbers 1 to $2n + 2$ underneath:

| 2 | 3 | 1 | 2 | 1 | 3 | 0 | 0 |
|---|---|---|---|---|---|---|---|
| 1 | 2 | 3 | 4 | 5 | 6 | 7 | 8 |

Then pair together numbers in the second row if they lie under the same number: $\{1,4\}, \{2,6\}, \{3,5\}, \{7,8\}$; these yield a Skolem system if each pair $\{a, b\}$ is replaced by the triple $\{b - a, a + n + 1, b + n + 1\}$: $\{3,5,8\}, \{4,6,10\}, \{2,7,9\}, \{1,11,12\}$.

(i) Show this procedure always yields a Skolem triple system.

(ii) Deduce that the Langford problem can have a solution only if $n \equiv 3$ or $4 \pmod 4$.

(iii) Show how to obtain a solution to the Langford problem in each case by working backwards from a Skolem system of order $n + 1$ in which one of the triples is $\{1, 3n + 2, 3n + 3\}$. (Note that the systems exhibited in the text have this triple.) (Langford, 1958; Davies, 1959)

12. *Moore's 1893 construction of an* STS($v$) *in all possible cases, as simplified by Hilton (1972).*

(i) Note the product theorem of Exercises 6, number 5. Observe that the STS($mn$) constructed there contains an STS($m$) and an STS($n$) as subsystems.

(ii) Suppose that there exist systems STS($n_i$), $1 \leqslant i \leqslant 3$, on sets $A_1, A_2, A_3$, where $n_3 < n_2$, $A_3 \subset A_2$, $A_2 - A_3 = \{1, \ldots, n_2 - n_3\}$, $n_1 \geqslant 3$. Then there exists an STS($n_3 + n_1(n_2 - n_3)$) whose triples are:

(a) $\{x_i, x_j, x_k\} \in$ STS($n_3$);

(b) $\{x_i, (x_j, y_r), (x_k, y_r)\}$ for all $\{x_i, x_j, x_k\} \in$ STS($n_2$) with $x_i \in A_3$, $x_j, x_k \in A_2 - A_3$, $y_r \in A_1$;

(c) $\{(x_i, y_r), (x_j, y_r), (x_k, y_r)\}$ for all $(x_i, x_j, x_k) \in$ STS($n_2$) with $x_i, x_j, x_k \in A_2 - A_3$, $y_r \in A_1$;

(d) $\{(x_i, y_r), (x_j, y_s), (x_k, y_t)\}$ for all $\{x_i, x_j, x_k\} \in A_2 - A_3$ with $x_i + x_j + x_k \equiv 0 \pmod{n_2 - n_3}$ and all $\{y_r, y_s, y_t\} \in$ STS($n_1$).

(iii) Construct all STS($v$) up to $v = 33$. The cases $v = 1, 3, 7, 9, 13$, are easy. Then use $15 = 1 + 7(3 - 1)$, $19 = 1 + 9(3 - 1)$, $21 = 3 \times 7$, $25 = 1 + 3(9 - 1)$, $27 = 3 \times 9$, $31 = 1 + 15(3 - 1)$, $33 = 3 + 3(13 - 3)$.

(iv) Assuming that all STS($v$) have been constructed for $v < 36t$, obtain those for $36t < v < 36(t + 1)$ as follows.

| | |
|---|---|
| $36t + 1 = 1 + 3((12t + 1) - 1)$ | $36t + 19 = 1 + (16t + 3)(7 - 1)$ |
| $36t + 3 = 1 + (18t + 1)(3 - 1)$ | $36t + 21 = 3 + (6t + 3)(9 - 3)$ |
| $36t + 7 = 1 + (6t + 1)(7 - 1)$ | $36t + 25 = 1 + 3((12t + 9) - 1)$ |
| $36t + 9 = 3 + (6t + 1)(9 - 3)$ | $36t + 27 = 1 + (18t + 13)(3 - 1)$ |
| $36t + 13$ below | $36t + 31 = 1 + (18t + 15)(3 - 1)$ |
| $36t + 15 = 1 + (18t + 7)(3 - 1)$ | $36t + 33 = 3 + 3((12t + 13) - 3)$. |

(v) Deal with $36t + 13$ as follows. Write as

$$36t + 13 = 1 + (3t + 1) \cdot 3 \cdot 2^2.$$

If $t \equiv 1 \pmod 4$ then $t = 4u + 1$ and $3t + 1 = 4(3u + 1)$; so we can write

$$36t + 13 = 1 + (3x + 1) \cdot 3 \cdot 2^{2r+2}$$

where $x \equiv 0$, 2, or 3 (mod 4). Deduce that

$$36t + 13 = 1 + (6n + 1)((3 \cdot 2^{2r+2} + 1) - 1) \qquad (x \text{ even})$$
$$\text{or} \quad 1 + (18n + 15)((2^{2r+3} + 1) - 1) \qquad (x \text{ odd}).$$

These provide a construction, except when $x = 0$.

(vi) Finally consider $36t + 13 = 1 + 3 \cdot 2^{2r+2}$. In this case

$$t = \frac{4^r - 1}{4 - 1} = 1 + 2^2 + \cdots + 2^{2r-2}.$$

Then, if $r = 2s + 2, s \geqslant 0$, we have $36t + 13 = 3 + (18u + 1)(13 - 3)$ for some $u$, and if $r = 2s + 3, s \geqslant 0$, we have $36t + 13 = 9 + (18u + 1)(49 - 9)$ for some $u$. Finally, if $r = 1$, $t = 1$ so $36t + 13 = 49 = 7 \times 7$. (Moore, 1893; Hilton, 1972)

13. *Proof of the existence of $(v, 4, 1)$ designs* whenever $v \equiv 1$ or 4 (mod 12). We shall imitate the PBD approach to triple systems, using Exercises 4, number 17.

   (i) Use Theorem 1.1.1 to show that a $(v, 4, 1)$ design can exist only if $v \equiv 1$ or 4 (mod 12).

   (ii) Use finite planes to deal with $v = 13, 16$; and use Exercises 2, numbers 8 and 23, to deal with $v = 25, 28, 37$.

   (iii) Note that if $v \equiv 1$ or 4 (mod 12) then $v = 3u + 1$ where $u \equiv 0$ or 1 (mod 4), and recall from Exercises 4, number 17, that a PBD$(u, \{4, 5, 8, 9, 12\}, 1)$ on $\{1, \ldots, u\}$ exists for all such $u$. Let $S$ be a set of $v = 3u + 1$ elements, namely $\infty$ and three copies $x_1, x_2, x_3$ of each $x \in \{1, \ldots, u\}$. Corresponding to each block $B$ of the PBD construct a $(3|B| + 1, 4, 1)$ design on the set consisting of $\infty$ and the copies of the elements of $B$, ensuring that the sets $\{\infty, x_1, x_2, x_3\}$ are among the blocks. The blocks obtained in this way from all the blocks $B$ form a $(v, 4, 1)$ design. (Hanani, 1961)

14. Use eqn (6.1) and the inclusion–exclusion principle to show that the complement of a $t$-design (i.e. the set of complements of its blocks) is also a $t$-design.

15. Show that every 3-$(4m, 2m, m - 1)$ design $\mathscr{D}$ is an extension of a Hadamard $(4m - 1, 2m - 1, m - 1)$ design, as follows.

   (i) Any two blocks of $\mathscr{D}$ meet in 0 or $m$ points; each point is in $4m - 1$ blocks.

   (ii) Let $B$ be any fixed block in $\mathscr{D}$, and let $t$ denote the number of blocks $C \neq B$ which intersect $B$. By counting in two different ways the pairs $(b, C)$ with $b \in B$, $b \in C$, $C \neq B$, show that $2m(4m - 2) = tm$, so that $t = 8m - 4$.

(iii) $\mathscr{D}$ has $8m - 2$ blocks.

(iv) Thus if $B$ is a block, so is its complement in $\mathscr{D}$.

16. Show that if $\mathscr{H}$ is a twice extendable symmetric $(v, k, \lambda)$ design, then $\mathscr{H}$ must be a finite projective plane of order 4, as follows.

    (i) By considering the number of blocks in the second extension, show that $(k + 1)(k + 2)$ must divide $v(v + 1)(v + 2)$.

    (ii) Apply this to each of the three cases given by Cameron's theorem. All are impossible except those with $\lambda = 1$. (Cameron, 1973)

17. Show that if an $S(2, k, v)$ exists then $v - 1 \geqslant k(k - 1)$. Show further that if strict inequality occurs then $v \geqslant k^2$.

18. Deduce from the previous problem that if an $S(t, k, v)$ exists then $v - t + 1 \geqslant (k - t + 1)(k - t + 2)$. Hence show that if an $S(5, 8, v)$ exists then $v \geqslant 24$.

19. Show that a cyclic SQS(20) can be obtained by taking the translates (mod 20) of $\{0, 1, 3, 4\}$, $\{0, 1, 2, 11\}$, $\{0, 1, 5, 16\}$, $\{0, 2, 6, 8\}$, $\{0, 2, 4, 12\}$, $\{0, 3, 6, 13\}$, $\{0, 3, 9, 14\}$, $\{0, 1, 6, 7\}$, $\{0, 1, 9, 12\}$, $\{0, 1, 8, 13\}$, $\{0, 2, 7, 9\}$, $\{0, 2, 5, 17\}$, $\{0, 3, 7, 16\}$, $\{0, 4, 8, 14\}$, and the first five translates of $\{0, 5, 10, 15\}$. (R. K. Jain)

20. The existence of an $S(5, 8, 24)$ suggests that an $S(6, 9, 25)$ might exist. Show that no such design can in fact exist.

# 7

# Kirkman triple systems

## 7.1 Early history

Query 6 on page 48 of the *Lady's and Gentleman's Diary* for 1850 was posed by Kirkman as follows.

Fifteen young ladies in a school walk out three abreast for seven days in succession: it is required to arrange them daily, so that no two shall walk twice abreast.

Writing later that year, Kirkman said of the problem that it was 'hardly worthy of mention if it were not that it has excited, as I hear, some attention among a far higher class of readers than those for whom the first 48 pages of the Diary are intended. I hit upon this pretty problem four years ago...'. It had certainly aroused interest. In modern terminology, Kirkman was asking for a resolvable $(15, 3, 1)$ design, and the more general problem of constructing a resolvable $(v, 3, 1)$ design for other values of $v$ soon followed. Indeed, by March 1850, Kirkman was being informed by Rev. James Mease of Kilkenny that resolvable $(3^n, 3, 1)$ designs always exist.

Since a $(v, 3, 1)$ design is of course just a Steiner triple system, the general problem being considered here is that of constructing a resolvable STS$(v)$ for as many values of $v$ as possible. Kirkman himself produced a resolvable STS(15), solving his 15 schoolgirls problem, and one of his solutions is particularly worthy of mention since it makes use of the seven-point plane and a league schedule for eight teams, slightly altering the method of Example 6.1.1. The following is essentially Kirkman's solution. Label the girls $1, \ldots, 8, x_1, \ldots, x_7$. We require seven sets of five triples, each set partitioning the set of girls. Imitating the usual cyclic construction of an STS(7), use $x_1, x_2, x_4$ on the first day, $x_2, x_3, x_5$ on the second day, and so on. On each day we then have to find four further triples. Since each pair $x_i, x_j$ already occurs in a triple, the remaining four triples on each day will each contain one $x$; one way to do this would be to take the four remaining $x_i$ with the four pairs in one round of a league schedule on $\{1, \ldots, 8\}$. A little trial and error produces the following cyclic solution using the schedule of Example 1.2.3.

**Example 7.1.1** Resolvable STS(15).

| Day 1: | $x_1, x_2, x_4$ | $x_7, 8, 1$ | $x_5, 2, 7$ | $x_3, 3, 6$ | $x_6, 4, 5$ |
|---|---|---|---|---|---|
| Day 2: | $x_2, x_3, x_5$ | $x_1, 8, 2$ | $x_6, 3, 1$ | $x_4, 4, 7$ | $x_7, 5, 6$ |
| Day 3: | $x_3, x_4, x_6$ | $x_2, 8, 3$ | $x_7, 4, 2$ | $x_5, 5, 1$ | $x_1, 6, 7$ |
| Day 4: | $x_4, x_5, x_7$ | $x_3, 8, 4$ | $x_1, 5, 3$ | $x_6, 6, 2$ | $x_2, 7, 1$ |
| Day 5: | $x_5, x_6, x_1$ | $x_4, 8, 5$ | $x_2, 6, 4$ | $x_7, 7, 3$ | $x_3, 1, 2$ |
| Day 6: | $x_6, x_7, x_2$ | $x_5, 8, 6$ | $x_3, 7, 5$ | $x_1, 1, 4$ | $x_4, 2, 3$ |
| Day 7: | $x_7, x_1, x_3$ | $x_6, 8, 7$ | $x_4, 1, 6$ | $x_2, 2, 5$ | $x_5, 3, 4$ |

Kirkman in fact used a non-cyclic league schedule, but his method was essentially the one here. We shall reconsider his solution when discussing Room squares in a later chapter.

Among the early writers on the general schoolgirls problem were R.R. Anstice (1852–3), who described a method for dealing with the case $v = 2p + 1$ where $p$ is a prime of the form $6m + 1$ (see Theorem 7.1.5) and A.H. Frost (1871) who studied $v = 4^m - 1$. All of these have $v \equiv 3$ (mod 6).

**Lemma 7.1.1** *A resolvable* STS($v$) *can exist only if* $v \equiv 3$ (mod 6).

**Proof** An STS($v$) can exist only if $v \equiv 1$ or 3 (mod 6). For resolvability we also need $v$ divisible by 3; so $v \equiv 1$ is eliminated.

We shall call a resolvable STS($v$) a *Kirkman triple system* KTS($v$) of order $v$. A KTS($v$) certainly exists for $v = 3$ (trivial), $v = 9$ (affine plane) and $v = 15$ (original problem). The case $v = 27$ was solved by Mease but is a special case of $v = 3^n$, which in turn is a special case of the following result proved by Kirkman in 1850 and later given a simple group-theoretic proof by E.H. Moore. (For Kirkman's proof, see Theorem 4.5.4.)

**Theorem 7.1.2** *For each prime $p$ and each positive integer $n$, a resolvable* $(p^n, p, 1)$ *design exists.*

**Proof** (Moore, 1896) Let $G = Z_p \oplus Z_p \oplus \cdots \oplus Z_p$, where there are $n$ summands; so $|G| = p^n$. For each subgroup $H$ of order $p$, the cosets of $H$ partition the set of elements of $G$ into $p$-sets, i.e. the cosets of $H$ form a resolution class. So construct a resolution class for each subgroup $H$ of order $p$. We claim that there classes yield a resolvable $(p^n, p, 1)$ design. All we have to do is prove that each pair of distinct elements of $G$ occurs in precisely one coset of one such $H$.

First suppose that $x, y \in G$ occur together in more than one coset; then $x - y \in H$ and $x - y \in K$ for subgroups $H, K$ of $G$ of order $p$, $H \neq K$. Then $H \cap K = \{0\}$, where 0 is the identity element of $G$; so $x - y = 0$, i.e. $x = y$, contradiction. So $x, y$ can occur together in *at most* one coset. Finally, note that every element of $G$ other than 0 has order $p$; so $x - y$ generates a subgroup $L$ of order $p$, and $x, y$ are in the same coset of $L$.

**Corollary 7.1.3** *A* KTS($3^n$) *exists for all positive integers n.*

Another family of Kirkman triple systems is given by the following result.

**Theorem 7.1.4** *If* $q = 6m + 1$ *is a prime power, a* KTS($3q$) *exists.*

**Proof** This is just Theorem 2.4.4.

**Example 7.1.2** Construct a KTS(21) by taking $m = 1$, $q = 7$ in the proof of Theorem 2.4.4. The generating blocks are listed in Example 2.4.5, and the 10 resolution classes are:

$G_1$: $\quad 0_1 0_2 0_3 \quad 3_1 6_1 5_1 \quad 3_2 6_2 5_2 \quad 3_3 6_3 5_3 \quad 2_1 4_2 1_3 \quad 2_2 4_3 1_1 \quad 2_3 4_1 1_2$

$G_2$: $\quad 1_1 1_2 1_3 \quad 4_1 0_1 6_1 \quad 4_2 0_2 6_2 \quad 4_3 0_3 6_3 \quad 3_1 5_2 2_3 \quad 3_2 5_3 2_1 \quad 3_3 5_1 2_2$

$\quad \vdots$

$G_7$: $\quad 6_1 6_2 6_2 \quad 2_1 5_1 4_1 \quad 2_2 5_2 4_2 \quad 2_3 5_3 4_3 \quad 1_1 3_2 0_3 \quad 1_2 3_3 0_1 \quad 1_3 3_1 0_2$

$G_8$: $\quad 3_1 6_2 5_3 \quad 4_1 0_2 6_3 \quad 5_1 1_2 0_3 \quad 6_1 2_2 1_3 \quad 0_1 3_2 2_3 \quad 1_1 4_2 3_3 \quad 2_1 5_2 4_3$

$G_9$: $\quad 3_2 6_3 5_1 \quad 4_2 0_3 6_1 \quad 5_2 1_3 0_1 \quad 6_2 2_3 1_1 \quad 0_2 3_3 2_1 \quad 1_2 4_3 3_1 \quad 2_2 5_3 4_1$

$G_{10}$: $\quad 3_3 6_1 5_2 \quad 4_3 0_1 6_2 \quad 5_3 1_1 0_2 \quad 6_3 2_1 1_2 \quad 0_3 3_1 2_2 \quad 1_3 4_1 3_2 \quad 2_3 5_1 4_2$

Another family of Kirkman triple systems is given in the following theorem which generalizes Anstice's result. Note that the sets $B_i$ yield a cyclic STS($q$).

**Theorem 7.1.5** *If* $q = 6m + 1$ *is a prime power, a* KTS($2q + 1$) *exists.*

**Proof** Let $\theta$ be a primitive element of GF($q$), so that $\theta^{6m} = 1$ and $\theta^{3m} = -1$; choose $u$ so that $\theta^m + 1 = 2\theta^u$. We shall use the method of mixed differences, essentially Theorem 2.4.1 with $t = 2$, with an extra element $\infty$. Consider the sets

$$A = \{0_1, 0_2, \infty\}$$
$$B_i = \{\theta_2^{i+u+m}, \theta_2^{i+u+3m}, \theta_2^{i+u+5m}\} \qquad 0 \leqslant i \leqslant m-1$$
$$C_i = \{\theta_1^i, \theta_1^{i+m}, \theta_2^{i+u}\} \qquad 0 \leqslant i \leqslant m-1$$
$$D_i = \{\theta_2^{i+2m+u}, \theta_1^{i+2m}, \theta_1^{1+3m}\} \qquad 0 \leqslant i \leqslant m-1$$
$$E_i = \{\theta_2^{i+4m+u}, \theta_1^{i+4m}, \theta_1^{i+5m}\} \qquad 0 \leqslant i \leqslant m-1$$

We show that these sets form a mixed difference system. The pure $(1,1)$ differences are $\pm\theta^i(\theta^m - 1)$, $\pm\theta^{i+2m}(\theta^m - 1)$, $\pm\theta^{i+4m}(\theta^m - 1)$, i.e. $(\theta^m - 1)$ times all the non-zero elements of $GF(q)$, i.e. all the non-zero elements of $GF(q)$. The $(2,2)$ differences are $\pm\theta^{i+u+m}(\theta^{2m} - 1)$, $\pm\theta^{i+u+3m}(\theta^{2m} - 1)$, $\pm\theta^{i+u+5m}(\theta^{2m} - 1)$, i.e. $\theta^{u+m}(\theta^{2m} - 1)$ times $\pm\theta^i$, $\pm\theta^{i+2m}$, $\pm\theta^{i+4m}$, i.e. each non-zero element of $GF(q)$ once. The mixed $(2,1)$ differences are $0 - 0$ and $\theta^i(\theta^u - 1)$, $\theta^i(\theta^u - \theta^m)$, $\theta^{i+2m}(\theta^u - 1)$, $\theta^{i+2m}(\theta^u - \theta^m)$, $\theta^{i+4m}(\theta^u - 1)$, $\theta^{i+4m}(\theta^u - \theta^m)$. Since $\theta^u - \theta^m = 1 - \theta^u = \theta^{3m}(\theta^u - 1)$, these are 0, and $\theta^u - 1$ times $\theta^i$, $\theta^{i+3m}$, $\theta^{i+2m}$, $\theta^{i+5m}$, $\theta^{i+4m}$, $\theta^{i+m}$, i.e. all elements of $GF(q)$ exactly once. Finally, the $(1,2)$ differences are just the negatives of the $(2,1)$ differences. So the sets do indeed form a mixed difference system, and the translates therefore form a $(2q + 1, 3, 1)$ design.

Further, the design is resolvable. For the sets of the difference system themselves form a partition of the set of $2q + 1$ elements, and hence the translates give $q - 1$ other resolution classes.

**Example 7.1.3** We can construct a KTS(15) by taking $q = 7$, $m = 1$, $\theta = 3$, $u = 2$. The five sets $A_1, \ldots, E_1$ are

$$\{0_1, 0_2, \infty\}, \{6_2, 5_2, 3_2\}, \{1_1, 3_1, 2_2\}, \{2_1, 6_1, 4_2\}, \{4_1, 5_1, 1_2\}.$$

By taking the translates of these blocks (mod 7) we obtain the design of Example 2.4.4.

Having obtained various families of Kirkman triple systems, it is natural to ask if there is any method of combining systems together to obtain new, larger systems. An early method of this type is due to G.L. Harison (alias A. Bray) whose result is described in the book by Rouse-Ball (1940). It is convenient to describe this method in terms of transversal designs.

**Theorem 7.1.6** *If a KTS(3m) and a KTS(3n) both exist, then a KTS(3mn) also exists.*

**Proof** Let $S = \{1, \ldots, 3mn\}$, and split $S$ up into $3m$ sets $A_i$, $0 \leqslant i < 3m$, where $A_i = \{ni + 1, \ldots, ni + n\}$. Take the sets $A_i$ as the elements of a KTS($3m$), with first resolution class

$$\{A_0, A_1, A_2\}, \{A_3, A_4, A_5\}, \ldots, \{A_{3m-3}, A_{3m-2}, A_{3m-1}\}. \quad (7.1)$$

The three sets in each triple are disjoint and have $3n$ elements in their union, so we can form a KTS($3n$) on these elements with $\frac{1}{2}(3n - 1)$ resolution classes. If we do this for each of the $m$ sets of $3n$ elements in a triple, and put the resolution classes together, we obtain $\frac{1}{2}(3n - 1)$ resolution classes of the elements of $S$ into triples; we take these as the first $\frac{1}{2}(3n - 1)$ classes of the KTS($3mn$) to be constructed. Note that a pair of elements of $S$ will occur in one of the blocks of one of these classes if and only if the elements belong to the same set $A_i$ or are members of two different sets $A_i$ which belong to the same triple in (7.1).

Consider now any other resolution class of the KTS($3m$), say

$$\{A_0, A_3, A_6\}, \ldots . \quad (7.2)$$

On the set $A_0 \cup A_3 \cup A_6$ form a resolvable TD($3, n$), taking $A_0, A_3, A_6$ as groups. Such a design exists by Theorem 4.6.1 and Lemma 4.5.3 since $N(n) \geqslant 2$ for all odd $n$; further, by Lemma 4.5.1 it has $n$ resolution classes. The reason for taking a TD here is of course that we do not want pairs of elements of $S$ in the same group, i.e. in an $A_i$, to occur together again in a triple. If we form such a TD($3n$) for each triple in (7.2), and combine the resulting resolution classes together, we obtain $n$ further resolution classes for the required KTS($3mn$). Repeating for each of the other classes of the KTS($3m$) we obtain altogether

$$\tfrac{1}{2}(3n - 1) + \tfrac{1}{2}(3m - 3)n = \tfrac{1}{2}(3mn - 1)$$

resolution classes as required. Each pair of elements in sets $A_i, A_j$ of different triples in (7.1) will occur together in exactly one triple arising in the TD($3, n$) constructed on the elements of the unique triple of sets containing $A_i$ and $A_j$.

The results obtained so far establish the existence of a KTS($v$) for the following $v < 100$: $v = 3, 9, 15, 21, 27 = 3^3$, $39 = 3 \cdot 13$, $45 = 3 \cdot 3 \cdot 5$, $51 = 2 \cdot 25 + 1$, $57 = 3 \cdot 19$, $63 = 3 \cdot 3 \cdot 7$, $75 = 3 \cdot 25$, $81 = 3^4$, $87 = 2 \cdot 43 + 1$, $93 = 3 \cdot 31$, $99 = 2 \cdot 49 + 1$. The only missing values are 33 and 69.

We now show the existence of a KTS(33). The idea behind the construction is suggested by the following elegant KTS(9):

$$
\begin{array}{lll}
\{\infty, 1, 5\} & \{2, 3, 8\} & \{6, 7, 4\} \\
\{\infty, 2, 6\} & \{3, 4, 1\} & \{7, 8, 5\} \\
\{\infty, 3, 7\} & \{4, 5, 2\} & \{8, 1, 6\} \\
\{\infty, 4, 8\} & \{5, 6, 3\} & \{1, 2, 7\}
\end{array}
$$

This has cyclic structure: $\infty$ remains unaltered, and the other elements move cyclically (mod 8). The triple $\{2,3,8\}$ gives differences $\pm1, \pm2, \pm5$, and the triple $\{\infty,1,5\}$ gives the difference 4; so the eight translates of $\{2,3,8\}$ and the first four translates of $\{\infty,1,5\}$ yield an STS(9). *Resolvability* depends on the fact that the translates $\{2+i,3+i,8+i\}$ and $\{2+j,3+j,8+j\}$, $j=i+4$, can be taken together in the same class.

**Example 7.1.4** A KTS(33). Work mod 32, and take the following as the first day's triples: $\{\infty,1,17\}$ and

$$\{2,11,16\},\{4,6,10\},\{5,13,30\},\{7,8,19\},\{9,28,31\},$$
$$\{18,27,32\},\{20,22,26\},\{21,29,14\},\{23,24,3\},\{25,12,15\}.$$

Another approach to a KTS(33) will be given in the next section; that approach will also yield a KTS(69). A different construction of a KTS(69) is given in Exercises 7, number 1.

In the next section we present a proof of the existence of KTS($6w+3$) for all $w \geqslant 0$. This result is usually referred to as the theorem of Ray-Chaudhuri and Wilson, since they were the first to publish a proof (in 1971). However, it has recently become known that the Chinese mathematician Lu Jiaxi had constructed a proof at least six years before. He was unable to publish his work due to the Cultural Revolution, but it has now, at last, been published, in English, in (Lu Jiaxi 1990).

### 7.2 The theorem of Ray-Chaudhuri, Wilson, and Lu Jiaxi

In this section we describe the use made of pairwise balanced designs in the proof of the existence of a KTS($v$) for all $v \equiv 3$ (mod 6). Pairwise balanced designs PBD($v, K, 1$) have already been seen to have the ability to pass on the existence of MOLS from the cases of size $k \in K$ to the case of size $v$; similarly they pass on the existence of Kirkman triple systems.

**Theorem 7.2.1** *Suppose that a* PBD($v, K, 1$) *exists and that a* KTS($2k+1$) *exists for each* $k \in K$. *Then a* KTS($2v+1$) *also exists.*

**Proof** For each block $B$ of size $k$ take a KTS($2k+1$) on the set consisting of $\infty$ and elements $x_1, x_2$ corresponding to each $x \in B$, in such a way that, for each $x \in B$, $\{\infty, x_1, x_2\}$ is a triple of the KTS. Take the collection $\mathscr{C}$ of all triples formed in this way from all the blocks $B$, including each triple $\{\infty, x_1, x_2\}$ just once. Clearly an STS($2v+1$) is formed: the elements $x_i, y_j$ ($x \neq y$) occur together in one of the triples of the KTS formed from the unique block $B$ containing both $x$ and $y$. We now show the STS is resolvable. For each element $z$ of the PBD, let $R_z$ denote the set of all triples in the resolution classes of the KTSs containing $\{\infty, z_1, z_2\}$. Then each $R_z$ will form a resolution class of the STS$(2v + 1)$, giving $v$ classes altogether.

**Example 7.2.1**  Since a $(16, 4, 1)$ design and a KTS$(9)$ both exist, if follows immediately that a KTS$(33)$ exists.

**Example 7.2.2**  Existence of a KTS$(69)$.
Since a KTS$(9)$ and a KTS$(15)$ both exist, it suffices to establish the existence of a PBD$(34, \{4, 7\}, 1)$. Take three copies of each member of $Z_9$ and consider the following sets:

$$A_1 = \{0_0, 1_1, 2_0, 4_1\}, \qquad A_2 = \{0_1, 1_2, 2_1, 4_2\}, \qquad A_3 = \{0_2, 1_0, 2_2, 4_0\},$$
$$B_1 = \{0_0, 1_0, 4_2\}, \qquad B_2 = \{0_1, 1_1, 4_0\}, \qquad B_3 = \{0_2, 1_2, 4_1\},$$
$$C_1 = \{0_0, 2_2, 5_0\}, \qquad C_2 = \{0_1, 2_0, 5_1\}, \qquad C_3 = \{0_2, 2_1, 5_2\},$$
$$D = \{0_0, 0_1, 0_2\}.$$

These form a mixed difference system which yields a PBD$(27, \{3, 4\}, 1)$. This PBD has seven parallel classes. $D$ and its translates form one class; for each $i = 1, 2, 3$, the sets $B_i$, $B_i + 3$, $B_i + 6$, $B_{i+1} + 1$, $B_{i+1} + 4$, $B_{i+1} + 7$, $B_{i+2} + 2$, $B_{i+2} + 5$, $B_{i+2} + 8$ form another class, where the suffixes are interpreted mod 3; for each $i$, the sets $C_i$, $C_i + 3$, $C_i + 6$, $C_{i-1} + 1$, $C_{i-1} + 4$, $C_{i-1} + 7$, $C_{i-2} + 2$, $C_{i-2} + 5$, $C_{i-2} + 8$ form another class. Add a new element $\infty_i$ to the sets of the $i$th of these seven classes, $1 \leqslant i \leqslant 7$, and introduce a further block consisting of these seven new elements. This gives a PBD$(34, \{4, 7\}, 1)$ as required. We recall in this construction that, for example $B_1 + 3$ denotes the set $\{3_0, 4_0, 7_2\}$.

Theorem 7.2.1 will be applied to a particular PBD which will enable the following result to be proved.

**Theorem 7.2.2**  *Suppose that a KTS$(6t + 3)$ and a KTS$(6h + 3)$ both exist, where $0 \leqslant h \leqslant t$ and $N(t) \geqslant 3$. Then a KTS$(6w + 3)$ exists where $w = 4t + h$.*

**Proof**  Suppose for the moment that a PBD$(3w + 1, \{4, 3t + 1, 3h + 1\}, 1)$ exists. Then, since a KTS$(9)$ exists, it will follow immediately from Theorem 7.2.1 that a KTS$(6w + 3)$ also exists. It therefore remains to show that this particular PBD can be constructed.

**Theorem 7.2.3**  *Suppose that $N(t) \geqslant 3$ and $0 \leqslant h \leqslant t$. Then a PBD$(12t + 3h + 1, \{4, 3t + 1, 3h + 1\}, 1)$ exists.*

**Proof**  Since $N(t) \geqslant 3$, a TD$(5, t)$ exists. Remove $t - h$ elements from one of its groups; the resulting design has four groups of size $t$ and one group of size $h$, and blocks of sizes 4 or 5; so a GDD$(4t + h, \{4, 5\}, \{t, h\})$ is obtained. Take such a GDD on a set $X$ of $v = 4t + h$ elements, and let $X^*$ denote the set obtained from $X$ by replacing each $x \in X$ by three copies, $x_1, x_2, x_3$. If in each group we replace each element by its three copies, we obtain groups of sizes $3t$ or $3h$. If we replace each element of each block by its three copies, we do *not* obtain a GDD since each block will have no or three members of a given group in it. We overcome this problem as follows. First note that a GDD(12, 4, 3) exists (namely a TD(4, 3)), and a GDD(15, 4, 3) also exists (by Theorem 4.7.1); so if, for each block $B$, we take the set of 12 or 15 elements arising from it and form a GDD($3|B|, 4, 3$) in which each group consists of the three copies of an element of $B$, then we can replace each $B$ by the blocks of this GDD and thus obtain a GDD($3v, 4, \{3t, 3h\}$) on $X^*$ which we shall denote by $D^*$. If we then adjoin a new element $\infty$ to each group of $D^*$ and take these augmented groups as further blocks, we finally obtain a PBD($3v + 1$, $\{4, 3t + 1, 3h + 1\}, 1$) design, $D$, say. In it, $\infty$ occurs with each other element exactly once since the groups of $D^*$ partition $X^*$; any two copies $x_i, x_j$ of the same element $x$ of $X$ occur together in precisely one of the groups of $D^*$ and in none of its blocks, and so occur together in precisely one of the blocks of $D$; and finally if $x \neq y$, then $x_i$ and $y_j$ occur in precisely one of the blocks of the GDD($3|B|, 4, 3$) on the unique block $B$ containing $x$ and $y$, and hence in a unique block of $D$.

Thus Theorem 7.2.2 is now established, and we are in position to prove the Ray-Chaudhuri–Wilson–Lu theorem.

**Theorem 7.2.4**   *A KTS($6w + 3$) exists for all $w \geqslant 0$.*

**Proof**   In the previous section we checked that a KTS($6w + 3$) exists for all $6w + 3 < 100$, except for $w = 11$; but this case has been dealt with in Example 7.2.2. So we can now suppose that $w \geqslant 17$.

The cases $w = 17$ to $20$ are dealt with by taking $t = 4$ and $h = 1, \ldots, 4$ in Theorem 7.2.2. The cases $w = 21$ to $25$ are then dealt with by taking $t = 5$ and $h = 1, \ldots, 5$. We cannot choose $t = 6$ since $N(6) = 1$; the case $w = 26$, i.e. $6w + 3 = 159$, can be dealt with by using Theorem 7.1.5, since $159 = 2 \cdot 79 + 1$, and $w = 27$ yields to Harison's theorem (Theorem 7.1.6) since $6 \cdot 27 + 3 = 165 = 3 \cdot 5 \cdot 11$.

For $w = 28$ to $35$, take $t = 7$ and $h = 0, \ldots, 7$; for $w = 36$ to $45$, take $t = 9$ and $h = 0, \ldots, 9$; for $w = 46$ to $55$, take $t = 11$ and $h = 2, \ldots, 11$; for $w = 56$ to $65$, take $t = 13$ and $h = 4, \ldots, 13$; and for $w = 66$ to $71$, take $t = 16$ and $h = 2, \ldots, 7$.

Thereafter, $w \geqslant 72$, so put $w = 24x + y$, $x \geqslant 3$, $0 \leqslant y \leqslant 23$, and write $w = 4(6x - 1) + (4 + y)$ or $w = 4(6x + 1) + (y - 4)$ according as $y \leqslant 4$ or $y > 4$. The Moore–MacNeish theorem gives $N(6x \pm 1) \geqslant 3$; and $4 + y < 6x - 1$ if $y \leqslant 4$ and $x \geqslant 3$ whereas, if $y > 4$, $y - 4 \leqslant 19 \leqslant 6x + 1$. Theorem 7.2.2 therefore yields the existence of a KTS($6w + 3$).

## 7.3 Sylvester's problem

Early on in the study of the Kirkman schoolgirls problem, an interesting extension was suggested by Sylvester. There are $\binom{15}{3} = 455$ different triples of elements from a set of 15 elements, and 455 is just $13 \times 35$. Since every KTS(15) contains 35 triples, the following problem naturally suggests itself.

### Sylvester's problem

Can the 455 triples of elements of a 15-set be grouped into 13 different KTS(15) designs? In other words, can a different solution to the schoolgirls problem be used during each of the 13 weeks of a school term so that no triple is ever repeated?

This generalizes to the case of $6n + 3$ elements.

### General Sylvester problem

Can the $\binom{6n+3}{3} = (6n + 1)(3n + 1)(2n + 1)$ 3-subsets of a $(6n + 3)$-set be grouped into $6n + 1$ disjoint systems KTS($6n + 3$)?

**Example 7.3.1** The case of $n = 1$, i.e. of nine schoolgirls, was solved by Kirkman in 1850. There are 84 triples arising from a 9-set: they are to be grouped so as to form seven KTS(9). Label the elements $1, \ldots, 9$ and consider the following seven squares.

| 123 | 124 | 457 | 781 | 145 | 478 | 712 |
|-----|-----|-----|-----|-----|-----|-----|
| 456 | 567 | 891 | 234 | 678 | 912 | 345 |
| 789 | 893 | 236 | 569 | 923 | 356 | 689 |

From each square we can obtain a KTS(9) as follows. The rows give one class, the columns give another class, and the six triples consisting of three numbers from different rows and columns can be grouped to form two other classes. For example, for the first square we obtain the following KTS(9), where the rows of triples are the resolution classes:

$$\{1,2,3\} \quad \{4,5,6\} \quad \{7,8,9\}$$
$$\{1,4,7\} \quad \{2,5,8\} \quad \{3,6,9\}$$
$$\{1,5,9\} \quad \{2,6,7\} \quad \{3,4,8\}$$
$$\{3,5,7\} \quad \{1,6,8\} \quad \{2,4,9\}.$$

Kirkman failed to solve the case $n = 15$, although he did manage to arrange the 455 triples in 91 resolution classes, thus constructing a resolvable $(15, 455, 91, 3, 13)$ design. Unfortunately the 91 classes cannot be grouped to form 13 Kirkman triple systems. It was in fact not until 1974 that Sylvester's problem was finally solved: a solution was published by R.H.F. Denniston, who later also showed that the general Sylvester problem can be solved whenever $6n + 3$ is a power of 3.

**Example 7.3.2** Denniston's solution to Sylvester's problem. Take as the 15 objects: $0, 1, \ldots, 12, A, B$. The following is an arrangement for the first week, since it is a KTS(15):

| | | | |
|---|---|---|---|
| $0, 1, 9$ | $2, 4, 12$ | $5, 10, 11$ | $7, 8, A$ | $3, 6, B$ |
| $0, 2, 7$ | $3, 4, 8$ | $5, 6, 12$ | $9, 11, A$ | $1, 10, B$ |
| $0, 3, 11$ | $1, 7, 12$ | $6, 8, 10$ | $2, 5, A$ | $4, 9, B$ |
| $0, 4, 6$ | $1, 8, 11$ | $2, 9, 10$ | $3, 12, A$ | $5, 7, B$ |
| $0, 5, 8$ | $1, 2, 3$ | $6, 7, 9$ | $4, 10, A$ | $11, 12, B$ |
| $0, 10, 12$ | $3, 5, 9$ | $4, 7, 11$ | $1, 6, A$ | $2, 8, B$ |
| $1, 4, 5$ | $2, 6, 11$ | $3, 7, 10$ | $8, 9, 12$ | $0, A, B.$ |

Although triples are by definition unordered, we have taken care to write down the elements of each numerical triple in order (mod 13). We now claim that if we derive from this an arrangement for week $i$, $2 \leqslant i \leqslant 13$, by adding $i - 1$ to each numerical entry (mod 13), then the 13 KTS(15)s so formed will provide a solution to Sylvester's problem. Certainly, each triple consisting of two numerical entries and A will occur precisely once since the pairs $\{7, 8\}, \{9, 11\}, \{2, 5\}, \{3, 12\}, \{4, 10\}, \{1, 6\}$ form a different system. Similarly each triple consisting of two numbers and B will occur once; and clearly the pair $\{A, B\}$ will occur in a triple with each number exactly once. So it remains to prove that each numerical triple will occur exactly once.

Note that each triple $\{a, b, c\}$ gives rise to a cyclic difference triple $(b - a, c - b, a - c)$; for example, $\{1, 7, 12\}$ gives rise to different triple $(6, 5, 2)$ which is *cyclic* in the sense that it is considered the same as $(5, 2, 6)$, the difference triple obtained if the original triple had been rewritten as $\{7, 12, 1\}$. Now if we write down the cyclic difference triples for each triple in the KTS(15) above, we find that they are all different. On the other hand, if a triple occurred twice, as translates of two different triples of the KTS(15), then these two triples would have to have the same cyclic difference triple. So indeed no triple occurs twice.

## 7.4 Exercises 7

1. *A direct construction of a* KTS(69). Take three copies of each element of $Z_{23}$. The following sets form a mixed difference system which yields a $(69, 3, 1)$ design. $\{0_1, 0_2, 0_3\}$, $\{18_1, 19_1, 15_1\}$, $\{18_i, 19_i, 22_i\}$ $(i = 2, 3)$, $\{2_3, 15_3, 17_3\}$, $\{2_i, 4_i, 17_i\}$ $(i = 1, 2)$, $\{22_1, 15_2, 4_3\}$, $\{0_1, 18_2, 11_3\}$, $\{0_2, 18_3, 11_1\}$, and the following each for $i = 1, 2, 3$: $\{1_i, 8_i, 13_i\}$, $\{3_i, 9_i, 14_{i+1}\}$, $\{7_i, 21_i, 6_{i+1}\}$, $\{5_i, 20_{i+1}, 11_{i+2}\}$, $\{10_i, 16_{i+1}, 12_{i+2}\}$, $\{0_i, 13_{i+1}, 16_{i+2}\}$, $\{0_i, 10_{i+1}, 14_{i+2}\}$, $\{0_i, 2_{i+1}, 22_{i+2}\}$. Each of the last

nine of these sets, along with its translates, forms a resolution class. Similarly the two sets $\{0_1, 18_2, 11_3\}$ and $\{0_2, 18_3, 11_1\}$ with their translates form two more classes. Finally, the remainder of the above sets form a resolution class and their translates give 22 more classes. (Ray-Chaudhuri and Wilson, 1971)

2. The existence of a KTS($6m + 3$) implies the existence of a PBD($9m + 4, \{4, 3m + 1\}, 1$); thus $N(9m + 4) \geqslant \min(2, N(3m + 1) - 1)$. Use this to give alternative proofs of $N(58) \geqslant 2$, $N(94) \geqslant 2$, $N(130) \geqslant 2$.

3. The existence of a KTS($6m + 3$) implies the existence of a PBD($9m + 3, \{3, 4, 3m\}, 1$) in which the sets of size 3 form a clear set. Deduce that $N(30) \geqslant 2$, $N(66) \geqslant 2$.

4. Use the existence of Kirkman triple systems to show that a PBD($v, \{3, 4\}, 1$) exists whenever $v \equiv 0$ or $1 \pmod 3$ $v \neq 6$. (See also Exercises 4, number 11.)

5. Use Kirkman triple systems to show the existence of the following designs: PBD($31, \{4, 10\}, 1$), PBD($76, \{4, 25\}, 1$).

6. *Resolvable* ($12u + 4, 4, 1$) *designs.* The proof of the existence of such resolvable designs for all $u \geqslant 1$ is outlined. Roughly speaking, it runs parallel to the proof of the existence of Kirkman triple systems.

   (i) Resolvable ($12u + 4, 4, 1$) designs exist whenever $4u + 1$ is a prime power (Exercises 2, number 23).

   (ii) If $N(n) \geqslant 3$ and if resolvable ($4m, 4, 1$) and ($4n, 4, 1$) designs exist, then a resolvable ($4mn, 4, 1$) design also exists. (Imitate the proof of Theorem 7.1.6; you need a resolvable TD($4, n$).)

   (iii) Construction of a resolvable ($100, 4, 1$) design. Take four copies of each element of GF(25), which has primitive element $\theta$ satisfying $\theta^2 = 2\theta + 2$ over $Z_5$. Take the following sets:

   $$\{0_1, 0_2, 0_3, 0_4\}, \{1_1\,\theta_1^2, \theta_1^5, \theta_1^{19}\}, \{\theta_1^6, \theta_1^8, \theta_1^{11}, \theta_1\},$$

   $$\{\theta_2^6, \theta_2^8, \theta_2^{11}, \theta_2\}, \{\theta_2^{12}, \theta_2^{14}, \theta_2^{17}, \theta_2^7\}, \{\theta_3^{12}, \theta_3^{14}, \theta_3^{17}, \theta_3^7\},$$

   $$\{\theta_3^{18}, \theta_3^{20}, \theta_3^{23}, \theta_3^{13}\}, \{\theta_4^{18}, \theta_4^{20}, \theta_4^{23}, \theta_4^{13}\}, \{1_4, \theta_4^2, \theta_4^5, \theta_4^{19}\},$$

   $$\{\theta_1^v, \theta_2^{v+6}, \theta_3^{v+12}, \theta_4^{v+18}\} \quad (v = 3, 4, 7, 9, 10, 12\text{-}18, 20\text{-}23),$$

   $$\{\theta_1^u, \theta_2^{u+6}, \theta_3^{u+12}, \theta_4^{u+18}\} \quad (u = 0, 1, 2, 5, 6, 8, 11, 19).$$

   Each of the last eight, with its translates, forms a resolution class. The remaining sets form a class; their translates give 24 other classes.

   (iv) We can now deal with all $12u + 4 \leqslant 124$.

   (v) Suppose that a PBD($v, K, 1$) exists and that a resolvable ($3k + 1, 4, 1$) design exists for all $k \in K$. Then a resolvable ($3v + 1, 4, 1$) design also exists.

(vi) If $N(t) \geqslant 4$ and $0 \leqslant h \leqslant t$, then a PBD$(20t + 4h + 1, \{5, 4t + 1, 4h + 1\}, 1)$ exists. (Imitate the proof of Theorem 7.2.3. For full details see Theorem 11.5.4.)

(vii) It follows that if $N(t) \geqslant 4$, $0 \leqslant h \leqslant t$, and resolvable $(12t + 4, 4, 1)$ and $(12h + 4, 4, 1)$ designs exist, then a resolvable $(60t + 12h + 4, 4, 1)$ design exists.

(viii) Deal with all $12u + 4$, $u < 90$. For $u = 11$ use (v) with a TD$(5, 9)$. All other $u < 90$ can be dealt with by above methods, except for $12u + 4 = 172, 232, 388$. For example, $184 = 3 \cdot 61 + 1$, so use (i); $196 = 4 \cdot 7 \cdot 7$, so use (ii); $304 = 60 \cdot 5 + 0 \cdot 12 + 4$ so use (vii). The three exceptions require special constructions which we omit.

(ix) Suppose that the existence of a resolvable $(12u + 4, 4, 1)$BIBD has been established for all $u < 30k$, $k \geqslant 3$. Then establish existence for $30k \leqslant u < 30(k + 1)$ as follows.

For $30k \leqslant u \leqslant 30k + 4$ write $12u + 4 = 12(30k + h) + 4 = 60(6k - 1) + 12(5 + h) + 4$.

For $30k + 5 \leqslant u \leqslant 30k + 24$ write $12u + 4 = 60(6k + 1) + 12h + 4$.

For $30k + 25 \leqslant u \leqslant 30k + 29$ write $12u + 4 = 60(6k + 5) + 12h + 4$.

Apply (vii) in each case. (Hanani *et al.*, 1972)

7. *Existence of* PBD$(v, \{4, 5\}, 1)$ *whenever* $v \equiv 0$ *or* $1$ (mod $4$)$, v \geqslant 13$. Make use of the existence of the resolvable designs of the previous problem as follows.

 (i) $v = 13, \ldots, 25$ can be dealt with by adapting finite planes.

 (ii) Use a resolvable $(28, 4, 1)$ design to deal with $28 \leqslant v \leqslant 33$.

 (iii) For $v = 36$ take a $(41, 5, 1)$ design and delete all elements of one block.

 (iv) For $v = 48$ see Exercises 2, number 21.

 (v) The case $v = 60$. The sets $\{0, 1, 3, 21, 55\}$, $\{0, 12, 36, 8, 50\}$, $\{0, 5, 22, 35, 51\}$ yield a cyclic $(61, 5, 1)$ design. Delete one element.

 (vi) The case $v = 72$. Remove 13 elements from one group of a TD$(5, 17)$ to get a PBD$(72, \{4, 5, 17\}, 1)$; then replace blocks of size 17 by using a PBD$(17, \{4, 5\}, 1)$.

 (vii) The case $v \equiv 1$ (mod $12$) is dealt with by Exercises 6, number 13.

 (viii) The cases $4, 5, 8, 9$ (mod $12$) are dealt with by starting with a resolvable $(12u + 4, 4, 1)$ design and adding new elements to the blocks of $0, 1, 4, 5$ classes respectively.

 (ix) To deal with $v \equiv 0$ (mod $12$)$, v \geqslant 84$, take a resolvable $(12t + 4, 4, 1)$ design with $t \geqslant 5$, and, for each $i \leqslant 20$, add $\infty_i$ to each block of the $i$th class. This gives a PBD$(12t + 24, \{4, 5, 20\}, 1)$; replace each block of size 20 by using a PBD$(20, \{4, 5\}, 1)$.

8. Use question 7 to prove the existence of doubles schedules (as described at the end of section 1.4) for all $v \neq 0$ or $1$ (mod $4$)$, v \geqslant 13$.

# 8

# League schedules

## 8.1 Round robin tournaments

In a round robin tournament, every team plays every other team once. If there are $2n$ teams, and the $n(2n-1)$ games are to be arranged into $2n-1$ rounds, with each team playing in one game in each round, we seek a resolvable $(2n, 2, 1)$ design. As we have already seen in sections 1.2 and 2.3, such league schedules can be obtained elegantly by the 'circle method' and, more generally, by the use of a difference system. For example, the Scottish Rugby Union used the design of Example 1.2.3 for its 1995 premier league, altering the home and away venues slightly (we shall discuss this further in section 8.2).

The circle method gives a cyclic schedule for teams $1, 2, \ldots, 2n-1, \infty$, with round $i$ consisting of games $i$ v $\infty$ and $a$ v $b$ where $a + b \equiv 2i \pmod{2n-1}$. Who first discovered this construction is unknown, but, remarkably, in 1847 Kirkman gave a construction which appears to be completely different but which leads to the same structure with rounds in a different order (Anderson 1991).

Kirkman's idea was to take teams $1, 2, \ldots, 2n$, to consider the pairs $(i, j)$, $1 \leqslant i < j \leqslant 2n$, in their lexicographic order (i.e. $(i, j)$ before $(i, k)$ if $j < k$, and $(i, j)$ before $(k, l)$ if $i < k$), and to consider the rounds in cyclic order. Start by assigning $(1, 2)$ to round 1. In general, after assigning a pair to a round, assign the next pair to the first subsequent round in the cyclic order which can accept it without causing any repeated appearance of a team in that round.

**Example 8.1.1** $2n = 8$ teams, labelled $1, \ldots, 8$.
The order in which the pairs are assigned is obtained by going down the first column, then the second, and so on.

| Round 1: | 1,2 | | | 3,7 | 4,6 | | 5,8 | |
|---|---|---|---|---|---|---|---|---|
| Round 2: | 1,3 | | 2,8 | | 4,7 | 5,6 | | |
| Round 3: | 1,4 | 2,3 | | | | 5,7 | | 6,8 |
| Round 4: | 1,5 | 2,4 | | 3,8 | | | 6,7 | |
| Round 5: | 1,6 | 2,5 | 3,4 | | | | | 7,8 |
| Round 6: | 1,7 | 2,6 | 3,5 | | 4,8 | | | |
| Round 7: | 1,8 | 2,7 | 3,6 | 4,5 | | | | |

It is not immediately obvious that this method will always work. But remarkably it does, giving round $i$ consisting of pairs $(a, b)$ with $a + b \equiv i + 2 \pmod{2n - 1}$, $1 \leqslant a < b \leqslant 2n - 1$, and with $2n$ taking the place of $\infty$.

Let us call 'stage $k$' the part of the construction dealing with pairs $(k, j)$, $k < j \leqslant 2n$. To show that our claim is valid, we prove $P(k)$ by induction, where $P(k)$ is the statement:

At the end of stage $k$, each pair $(i, 2n)$, $1 \leqslant i \leqslant k$, is in round $2i - 2 \pmod{2n - 1}$, and, for $i < j \leqslant 2n$, each pair $(i, j)$ with $i \leqslant k$ is in round $i + j - 2 \pmod{2n - 1}$.

$P(1)$ is trivially true, so now consider the induction step from $k$ to $k + 1$. In completing step $k$ we have just assigned $(k, 2n)$ to round $2k - 2$. Next consider $(k + 1, k + 2)$. Round $2k - 1$ contains $(k, k + 1)$ and round $2k$ contains $(k, k + 2)$, so put $(k + 1, k + 2)$ in round $2k + 1$. This is permitted since a pair $(i, k + 1)$ could already occur in that round only if $i + k - 1 \equiv 2k + 1 \pmod{2n - 1}$, i.e. $i \equiv k + 2$, which is impossible; and similarly no $(i, k + 2)$ could have already appeared there.

So we put $(k + 1, k + 2)$ in round $2k + 1$. We then place $(k + 1, j)$ in round $k + j - 1$ for each $j = k + 3, \ldots, 2n - 1$. This gives no repeat of $k + 1$ in any round since $(i, k + 1)$ could be in round $k + j - 1$ only if $k + i - 1 \equiv k + j - 1$, i.e. $i \equiv j$. Similarly there are no repetitions of any $j$.

Finally, we have to place $(k + 1, 2n)$. The first round we come to is round $k + 2n - 1$, i.e. round $k$; but this already contains $(1, k + 1)$. The next rounds $k + 1, \ldots, 2k - 1$ already contain $(2, k + 1), \ldots, (k, k + 1)$ so we put $(k + 1, 2n)$ into round $2k$, noting that $2n$ has not yet appeared in that round. So $P(k + 1)$ is true, and the result is proved.

The arrangement of the games into rounds as given by the circle method (or by Kirkman's version) is known to graph theorists as the one-factorization $GK_{2n}$. In terms of graphs, we have the complete graph $K_{2n}$, with $2n$ vertices, such that every pair of vertices is joined by an edge, and where a round, or a one-factor, is a pairing of the vertices into $n$ pairs, the $n$ edges between the pairs corresponding to the games in a round. Other well-known one-factorizations are described in the exercises (question 7).

## 8.2 Venues

We now investigate the possibility of arranging the games of a league schedule so that each team plays alternatively home and away. We shall use the convention that in the game *a* v *b*, *a* is the home team. If we consider the schedule of Example 1.2.3 and interpret each arrow in Figure 1.3 as pointing from the home team to the away team, we obtain the following fixture list.

**Example 8.2.1**

| | | | | |
|---|---|---|---|---|
| Day 1: | ∞ v 1 | 7 v 2 | 3 v 6 | 5 v 4 |
| Day 2: | 2 v ∞ | 1 v 3 | 4 v 7 | 6 v 5 |
| Day 3: | ∞ v 3 | 2 v 4 | 5 v 1 | 7 v 6 |
| Day 4: | 4 v ∞ | 3 v 5 | 6 v 2 | 1 v 7 |
| Day 5: | ∞ v 5 | 4 v 6 | 7 v 3 | 2 v 1 |
| Day 6: | 6 v ∞ | 5 v 7 | 1 v 4 | 3 v 2 |
| Day 7: | ∞ v 7 | 6 v 1 | 2 v 5 | 4 v 3. |

From this we obtain a *venue sequence* for each team, indicating which games are at home (H) and which are away (A):

| | | | | | | | |
|---|---|---|---|---|---|---|---|
| ∞: | H | A | H | A | H | A | H |
| 1: | A | H | A | H | A | H | A |
| 2: | A | H | H | A | H | A | H |
| 3: | H | A | A | H | A | H | A |
| 4: | A | H | A | H | H | A | H |
| 5: | H | A | H | A | A | H | A |
| 6: | A | H | A | H | A | H | H |
| 7: | H | A | H | A | H | A | A. |

If we call a repetition of H or of A in consecutive positions in a venue sequence a *break*, then we note that teams ∞ and 1 have no venue breaks, but that each other team has one. This is in fact the best that we can do, for there are only two venue sequences with no break, and no two teams can have the same venue sequence (otherwise they would not meet). Notice also that teams ∞ and 1 have *complementary* venue sequences in the sense that one is at home precisely when the other is away, as do teams 2 and 3, 4 and 5, 6 and 7. This property is useful if we have two neighboring teams and we wish to avoid having them both playing at home on the same day.

If a fixture list is required for an *odd* number of teams, we can simply remove all the games involving ∞. For example, for 5 teams we can use the circle method to obtain the following schedule.

**Example 8.2.2** Schedule for 5 teams.

| Day 1: | 1 v 4 | 3 v 2 |
|--------|-------|-------|
| Day 2: | 2 v 5 | 4 v 3 |
| Day 3: | 3 v 1 | 5 v 4 |
| Day 4: | 4 v 2 | 1 v 5 |
| Day 5: | 5 v 3 | 2 v 1. |

This schedule has been used in the Five Nations Rugby Championship involving England, France, Ireland, Scotland, and Wales. The assignment of numbers to countries varies from year to year. Note that each team has an alternating venue sequence.

Return now to the case of $2n$ teams, and suppose that each pair of teams is to meet twice in a season. The simplest way to schedule this is to make the second half of the fixture list a repetition of the first half, but with home and away venues interchanged. How many breaks do we get? We claim that there have to be at least $6n - 6$ breaks altogether, no matter which schedule we use. For if team $i$ has $x_i$ breaks in the first half, it will have a break between halves if $x_i$ is odd; so if there are $s$ values of $i$ for which $x_i$ is odd, and $2n - s$ values of $i$ for which $x_i$ is even, the total number of breaks will be $2 \sum x_i + s$. But at most two of the $x_i$ can be zero, so

$$\sum x_i \geqslant \begin{cases} s + 2(2n - s - 2) & \text{if } s \leqslant 2n - 2 \\ s & \text{if } s \geqslant 2n - 2. \end{cases}$$

Thus the total number of breaks is

$$\geqslant \begin{cases} 8n - 8 - s & \geqslant 6n - 6 \quad \text{if } s \leqslant 2n - 2 \\ 3s & \geqslant 6n - 6 \quad \text{if } s \geqslant 2n - 2. \end{cases}$$

The circle construction gives precisely $6n - 6$ breaks, the best possible. However, in Example 8.2.1 we find that team 7 is away on three consecutive days (days 6, 7, 8) and so has two consecutive venue breaks; similarly for team 6. We can avoid consecutive breaks by reversing the venues in the games $6 \text{ v } \infty$ and $\infty \text{ v } 7$. This is what the Scottish Rugby Union did. A similar switch works in the general case of $2n$ teams, and we have the following result.

**Theorem 8.2.1** (de Werra, 1981) *A league schedule can be constructed for $2n$ teams, with $4n - 2$ rounds, each pair of teams meeting twice, such that*:

(a) *the second half of the schedule is the same as the first half with venues interchanged*;

(b) *there are $6n - 6$ venue breaks*;

(c) *no team has consecutive venue breaks*.

## 8.3 Carry-over effects

Example 8.2.1 has one disadvantage which we now discuss. There are 5 teams which play team 1 immediately after playing team 6 (namely teams 2, 3, 4, 5, 7). If team 6 were a weak team, team 1 could justifiably complain that on five occasions its opponents were 'on a high', with confidence boosted by a big victory. On the other hand, if team 6 were a strong physical team, then it would be to team 1's advantage to have its opponents suffering from a bad experience so often. Accordingly, it is advisable to balance out the 'carry-over' effects'; indeed, it would be best, if possible, if the following could be achieved:

> There are no choices of $x$, $y$, $z$, $w$ such that teams $x$ and $y$
> both play team $z$ immediately after playing team $w$.               (8.1)

It so happens that it is possible to achieve this in a league of 8 teams.

### Example 8.3.1

| | | | |
|---|---|---|---|
| Day 1: | $0 \, v \, \infty$ | $1 \, v \, 5$ | $3 \, v \, 2$ | $6 \, v \, 4$ |
| Day 2: | $\infty \, v \, 1$ | $2 \, v \, 6$ | $4 \, v \, 3$ | $5 \, v \, 0$ |
| Day 3: | $2 \, v \, \infty$ | $3 \, v \, 0$ | $5 \, v \, 4$ | $1 \, v \, 6$ |
| Day 4: | $\infty \, v \, 3$ | $4 \, v \, 1$ | $6 \, v \, 5$ | $0 \, v \, 2$ |
| Day 5: | $\infty \, v \, 4$ | $5 \, v \, 2$ | $0 \, v \, 6$ | $3 \, v \, 1$ |
| Day 6: | $5 \, v \, \infty$ | $6 \, v \, 3$ | $1 \, v \, 0$ | $4 \, v \, 2$ |
| Day 7: | $\infty \, v \, 6$ | $0 \, v \, 4$ | $2 \, v \, 1$ | $3 \, v \, 5.$ |

If we list each team's opponents in order, we get:

| | | | | | | | |
|---|---|---|---|---|---|---|---|
| $\infty$: | 0 | 1 | 2 | 3 | 4 | 5 | 6 |
| 0: | $\infty$ | 5 | 3 | 2 | 6 | 1 | 4 |
| 1: | 5 | $\infty$ | 6 | 4 | 3 | 0 | 2 |
| 2: | 3 | 6 | $\infty$ | 0 | 5 | 4 | 1 |
| 3: | 2 | 4 | 0 | $\infty$ | 1 | 6 | 5 |
| 4: | 6 | 3 | 5 | 1 | $\infty$ | 2 | 0 |
| 5: | 1 | 0 | 4 | 6 | 2 | $\infty$ | 3 |
| 6: | 4 | 2 | 1 | 5 | 0 | 3 | $\infty$. |

Observe that no number follows any other number twice. The venues have been arranged so that if the league is followed by a second half with venues reversed, a total of 20 breaks will occur. This is not quite the minimum of 18, but there is the benefit of no carry-over effects (apart from those resulting in the repetition of games in the second half).

There are few cases where schedules with a balance of carry-over effects have been constructed other than the case of $2n$ teams where $n$ is a power of 2. We present the construction.

**Theorem 8.3.1** (Russell 1980; Hwang 1989)  *If $n$ is a power of* 2, *a cyclic league schedule for* $2n$ *teams exists satisfying* (8.1).

**Proof**  Let $2n = 2^\alpha = q$, and let $\theta$ be a primitive element of $\mathrm{GF}(q)$. For $1 \leqslant j \leqslant q - 1$, define round $j$ to consist of the games
$$\theta^i \, \mathrm{v} \, (\theta^i + \theta^j).$$
(Note that, since we are in $\mathrm{GF}(2^\alpha)$, if $\theta^i + \theta^j = \theta^k$ then $\theta^k + \theta^j = \theta^i$.) Suppose that (8.1) does not hold. Then, for some $j, J$,
$$x = w + \theta^{j-1}, \ x = z + \theta^j, \ y = w + \theta^{J-1}, \ y = z + \theta^J.$$
But then
$$y - x = \theta^{J-1} - \theta^{j-1} = \theta^J - \theta^j,$$
so that
$$\theta^{j-1}(\theta - 1) = \theta^{J-1}(\theta - 1),$$
whence $j = J$; thus $x = y$. So (8.1) is satisfied.

Note that if $\theta^k$ plays $\theta^l$ in round $j$, then $\theta^k = \theta^l + \theta^j$ and so $\theta^{k+1} = \theta^{l+1} + \theta^{j+1}$, so that $\theta^{k+1}$ plays $\theta^{l+1}$ in round $j + 1$. Thus, if we replace $\theta^i$ by $i$ and $0$ by $\infty$, we obtain a cyclic schedule.

**Example 8.3.2**  Balanced schedule for 16 teams. Take the primitive element $\theta$ of $\mathrm{GF}(16)$ satisfying $\theta^4 = \theta + 1$. Then $\theta^5 = \theta^2 + \theta$, $\theta^6 = \theta^3 + \theta^2$, $\theta^7 = \theta^3 + \theta + 1$, $\theta^8 = \theta^2 + 1$, $\theta^9 = \theta^3 + \theta$, $\theta^{10} = \theta^2 + \theta + 1$, $\theta^{11} = \theta^3 + \theta^2 + \theta$, $\theta^{12} = \theta^3 + \theta^2 + \theta + 1$, $\theta^{13} = \theta^3 + \theta^2 + 1$, $\theta^{14} = \theta^3 + 1$, $\theta^{15} = 1$. The first round games are of the form $\theta^i \, \mathrm{v} \, (\theta^i + \theta)$, i.e. $0 \, \mathrm{v} \, \theta$, $\theta^2 \, \mathrm{v} \, \theta^5$, $\theta^3 \, \mathrm{v} \, \theta^9$, $\theta^4 \, \mathrm{v} \, \theta^{15}$, $\theta^6 \, \mathrm{v} \, \theta^{11}$, $\theta^7 \, \mathrm{v} \, \theta^{14}$, $\theta^8 \, \mathrm{v} \, \theta^{10}$, $\theta^{12} \, \mathrm{v} \, \theta^{13}$. So we obtain the schedule

| | | | | | | | |
|---|---|---|---|---|---|---|---|
| Day 1 : | $\infty$ v 1 | 2 v 5 | 3 v 9 | 4 v 15 | 6 v 11 | 7 v 14 | 8 v 10 | 12 v 13 |
| Day 2 : | $\infty$ v 2 | 3 v 6 | 4 v 10 | 5 v 1 | 7 v 12 | 8 v 15 | 9 v 11 | 13 v 14 |

$$\vdots$$

| | | | | | | | |
|---|---|---|---|---|---|---|---|
| Day 15: | $\infty$ v 15 | 1 v 4 | 2 v 8 | 3 v 14 | 5 v 10 | 6 v 13 | 7 v 9 | 11 v 12. |

## 8.4  Bipartite tournaments

Suppose now that two teams each consist of $n$ players, and that each player has to compete once against each of the $n$ players in the opposing team (e.g. at chess or tennis). The games have to be arranged into $n$ rounds, each of $n$ games, with each player playing in each round.

**Example 8.4.1**  Teams $T_1 = \{x_1, x_2, x_3, x_4\}$, $T_2 = \{y_1, y_2, y_3, y_4\}$.

| | | | | |
|---|---|---|---|---|
| Round 1: | $x_1 \, \mathrm{v} \, y_1$ | $x_2 \, \mathrm{v} \, y_2$ | $x_3 \, \mathrm{v} \, y_3$ | $x_4 \, \mathrm{v} \, y_4$ |
| Round 2: | $x_1 \, \mathrm{v} \, y_3$ | $x_2 \, \mathrm{v} \, y_1$ | $x_3 \, \mathrm{v} \, y_4$ | $x_4 \, \mathrm{v} \, y_2$ |
| Round 3: | $x_1 \, \mathrm{v} \, y_2$ | $x_2 \, \mathrm{v} \, y_4$ | $x_3 \, \mathrm{v} \, y_1$ | $x_4 \, \mathrm{v} \, y_3$ |
| Round 4: | $x_1 \, \mathrm{v} \, y_4$ | $x_2 \, \mathrm{v} \, y_3$ | $x_3 \, \mathrm{v} \, y_2$ | $x_4 \, \mathrm{v} \, y_1$. |

It is clear that if we write $x_i$'s opponents as the $i$th column of a $4 \times 4$ matrix, we obtain a Latin square:

$$\begin{matrix} 1 & 2 & 3 & 4 \\ 3 & 1 & 4 & 2 \\ 2 & 4 & 1 & 3 \\ 4 & 3 & 2 & 1 \end{matrix} \qquad (8.2)$$

Conversely, any $4 \times 4$ Latin square on $\{1,2,3,4\}$ will uniquely determine a fixture list. So bipartite tournaments are equivalent to Latin squares.

Another way to represent the above tournament by a Latin square is to define

$$l_{ij} = \text{round in which } x_i \text{ plays } y_j.$$

In the above example we obtain the Latin square

$$L = \begin{matrix} 1 & 3 & 2 & 4 \\ 2 & 1 & 4 & 3 \\ 3 & 4 & 1 & 2 \\ 4 & 2 & 3 & 1. \end{matrix}$$

## Court balance

Suppose now that $T_1$, $T_2$ are tennis teams, each of $n$ players, that there are $n$ courts available, and that it is desired to have each player play once on each court. The way to solve this problem is to take two MOLS $A, B$ on $\{1, \ldots, n\}$ and consider their join; take rows as corresponding to rounds and columns to courts, and interpret each ordered pair $(u, v)$ in the join as giving the game $x_u$ v $y_v$.

**Example 8.4.2** Take two MOLS of order 3 as in Example 1.3.6 to obtain the following schedule.

|  | Court 1 | Court 2 | Court 3 |
|---|---|---|---|
| Round 1: | $x_2$ v $y_3$ | $x_1$ v $y_2$ | $x_3$ v $y_1$ |
| Round 2: | $x_3$ v $y_2$ | $x_2$ v $y_1$ | $x_1$ v $y_3$ |
| Round 3: | $x_1$ v $y_1$ | $x_3$ v $y_3$ | $x_2$ v $y_2$ |

Clearly, if such a schedule exists, two MOLS can be read off, so we can assert that such a court-balanced schedule exists if and only if $n \neq 2$ or 6.

## Colour balance

If $T_1$ and $T_2$ are two chess teams, there is a new balance problem to be considered. It is advantageous to play white rather than black, so it is desirable to use a bipartite tournament in which each player plays an equal number of games with black and with white, and, in each round, each team plays an equal number of games with black and white. For this to be possible, $n$ must be even, so we now assume that it is.

One approach is again to consider the join of two MOLS $A, B$ on $\{1,\ldots,n\}$. This time let rows correspond to members of team $T_1$ and columns to members of $T_2$. We shall interpret $a_{ij} = k$ to mean that $x_i$ plays $y_j$ in round $k$, and $b_{ij} = h$ to tell us that $x_i$ plays white against $y_j$ if $h$ is even, and plays black if $h$ is odd.

**Example 8.4.3** Take the second of the MOLS of Example 1.3.4 to be $A$, and the third to be $B$. We obtain the following schedule in which games are underlined when team $T_1$ plays white.

Round 1:    $x_1 \vee y_1$    $\underline{x_2 \vee y_4}$    $\underline{x_3 \vee y_2}$    $\underline{x_4 \vee y_3}$

Round 2:    $x_1 \vee y_2$    $x_2 \vee y_3$    $x_3 \vee y_1$    $x_4 \vee y_4$

Round 3:    $x_1 \vee y_3$    $\underline{x_2 \vee y_2}$    $x_3 \vee y_4$    $\underline{x_4 \vee y_1}$

Round 4:    $\underline{x_1 \vee y_4}$    $x_2 \vee y_1$    $x_3 \vee y_3$    $x_4 \vee y_2$.

Since there do not exist two MOLS of order 6, this approach cannot be used when $n = 6$. However, a schedule can be constructed, with the extra property that no player plays the same color more than twice in succession.

**Example 8.4.4** Schedule for $n = 6$.

Round 1: $\underline{x_1 \vee y_1}$   $x_2 \vee y_5$   $x_3 \vee y_4$   $x_4 \vee y_2$   $x_5 \vee y_3$   $\underline{x_6 \vee y_6}$

Round 2: $\underline{x_1 \vee y_2}$   $x_2 \vee y_1$   $x_3 \vee y_3$   $\underline{x_4 \vee y_4}$   $x_5 \vee y_6$   $x_6 \vee y_5$

Round 3: $\underline{x_1 \vee y_3}$   $x_2 \vee y_2$   $x_3 \vee y_1$   $\underline{x_4 \vee y_6}$   $x_5 \vee y_5$   $x_6 \vee y_4$

Round 4: $x_1 \vee y_4$   $x_2 \vee y_6$   $x_3 \vee y_5$   $x_4 \vee y_1$   $x_5 \vee y_2$   $x_6 \vee y_3$

Round 5: $x_1 \vee y_5$   $x_2 \vee y_4$   $x_3 \vee y_6$   $x_4 \vee y_3$   $x_5 \vee y_1$   $x_6 \vee y_2$

Round 6: $x_1 \vee y_6$   $x_2 \vee y_3$   $\underline{x_3 \vee y_2}$   $x_4 \vee y_5$   $\underline{x_5 \vee y_4}$   $\underline{x_6 \vee y_1}$.

## Carry-over effects

Finally, corresponding to (8.1), we consider the following requirement for bipartite tournaments:

> There is no choice of $i$, $j$, $k$, $l$ such that $x_i$ and $x_j$ both
>
> play $y_k$ immediately after $y_l$.         (8.3)

If we use matrix formulation as in (8.2), the requirement (8.3) is that no number follows any other number in more than one column.

**Definition 8.4.1**  A Latin square $A = (a_{ij})$ on $\{1, \ldots, n\}$ is *row complete* if the pairs $(a_{ij}, a_{i,j+1})$ are all distinct, and is *column complete* if the pairs $(a_{ij}, a_{i+1,j})$ are all distinct. $A$ is *complete* if it is both row and column complete.

**Example 8.4.5**  A symmetric complete Latin square of order 6.

$$
\begin{array}{cccccc}
1 & 2 & 6 & 3 & 5 & 4 \\
2 & 3 & 1 & 4 & 6 & 5 \\
6 & 1 & 5 & 2 & 4 & 3 \\
3 & 4 & 2 & 5 & 1 & 6 \\
5 & 6 & 4 & 1 & 3 & 2 \\
4 & 5 & 3 & 6 & 2 & 1.
\end{array}
$$

This square is constructed as in the following theorem.

**Theorem 8.4.1** (Williams 1949)  *Let $n = 2m$. Take the first row and column of an $n \times n$ array $L$ to be*

$$1 \quad 2 \quad 2m \quad 3 \quad 2m-1 \quad 4 \quad \cdots \quad m \quad m+2 \quad m+1$$

*and, for $i > 1$, $j > 1$, define $l_{ij} = l_{i1} + l_{1j} - 1 \pmod{2m}$. Then $L$ is a symmetric complete Latin square.*

**Proof**  It is straightforward to check that $L$ is Latin. The $j$th column is obtained from the first by adding $l_{1j} - 1$ to each entry, and so, as the $l_{1j}$ are all distinct, we obtain all possible shifts of the first column. So we need only check that the differences $l_{i+1,1} - l_{i,1}$ give every non-zero element of $Z_{2m}$ exactly once. But the differences are

$$1, 2m - 2, 3, 2m - 4, 5, \ldots, 2m - 3, 2, 2m - 1.$$

It is harder to construct complete Latin squares of odd orders; indeed, none of orders 3, 5, or 7 exists. But see Exercises 8, number 4 for *quasicomplete* Latin squares.

It is tempting to think that if a column complete Latin square representing the opponents of each $x_i$ is also row complete, then not only are the opponents of the $x_i$ balanced, but so also are the opponents of the $y_i$. But of course the rows do not represent the opponents of the $y_i$ in order. If, however, the complete Latin square is as given in the above theorem, the corresponding bipartite tournament *is* balanced for both teams.

**Theorem 8.4.2**  *Let $L = (l_{i,j})$ be the $2m \times 2m$ matrix defined in the previous theorem. Construct a bipartite tournament by:*

$$\text{if } l_{ij} = u, \text{ then } x_j \text{ plays } y_u \text{ in round } i.$$

*Then the tournament is balanced for carry-over effects for both teams.*

**Proof** We already know that there is carry-over balance for the $x_i$. Suppose there exist $i, j, k, h$ such that $y_i$ and $y_j$ both play $x_k$ immediately after $x_h$. Then $x_h$ plays $y_i$ in round $I$, $x_k$ plays $y_i$ in round $I + 1$; $x_h$ plays $y_j$ in round $J$, $x_k$ plays $y_j$ in round $J + 1$ for some $I, J$. Thus $l_{Ih} = l_{I+1,k}$ and $l_{Jh} = l_{J+1,k}$. So

$$l_{1I} + l_{1h} = l_{1,I+1} + l_{1k} \quad \text{and} \quad l_{1J} + l_{h1} = l_{1,J+1} + l_{1k}.$$

Thus

$$l_{1,I+1} - l_{1I} = l_{1,J+1} - l_{1J}.$$

But all the differences $l_{1,i+1} - l_{1i}$ are distinct, so $I = J$; so $y_i = y_j$, whence $i = j$.

**Example 8.4.6** Let $L$ be the matrix given in Example 8.4.5. Then column $j$ gives the opponents of $x_j$ in order. The opponents of the $y_i$ are, in order,:

$$
\begin{aligned}
y_1&: \quad 1, 3, 2, 5, 4, 6 \\
y_2&: \quad 2, 1, 4, 3, 6, 5 \\
y_3&: \quad 4, 2, 6, 1, 5, 3 \\
y_4&: \quad 6, 4, 5, 3, 2, 1 \\
y_5&: \quad 5, 6, 3, 4, 1, 2 \\
y_6&: \quad 3, 5, 1, 6, 2, 4.
\end{aligned}
$$

This completes our discussion of court and carry-over balance (but see exercise 9 below). Further topics on balance are discussed in the following two chapters.

## 8.5 Exercises 8

1. Use Kirkman's method to construct a schedule for 10 teams.
2. Show that the construction of Theorem 8.3.1 in the case $q = 8$ leads to the schedule of Example 8.3.1. (Use $\theta$ satisfying $\theta^3 = \theta^2 + 1$.)
3. Construct a court balanced bipartite tournament for two teams of five players.
4. A Latin square in which each *unordered* pair of elements appears twice as successive elements in rows, and twice in columns, is said to be *quasicomplete*. Show that quasicomplete Latin squares of all odd orders $2m + 1$ exist by imitating the proof of Theorem 8.4.1, starting with first row and column given by

$$1 \ 2 \ 2m + 1 \ 3 \ 2m \ 4 \ 2m - 1 \ \cdots \ m \ m + 3 \ m + 1 \ m + 2.$$

Construct such squares in the cases $m = 2, 3, 4$.

5. Suppose that the first $r$ rounds of a bipartite tournament have been constructed by an *ad hoc* method. Use Theorem 1.5.3 to show that these $r$ rounds can always be extended to a full bipartite tournament.

6. In an attempt to construct a league schedule for $2n$ teams, the teams are labelled $x_1, \ldots, x_n, y_1, \ldots, y_n$, and the first $n$ rounds take the form of a bipartite tournament between teams $\{x_1, \ldots, x_n\}$ and $(y_1, \ldots, y_n)$. Show that these rounds can be extended to a league schedule if and only if $n$ is even.

7. A construction of league schedules for $2n$ teams (known to graph theorists as $GA_{2n}$).
   (i) *$n$ odd.* Label the teams $x_1, \ldots, x_n, y_1, \ldots, y_n$ and introduce two new teams $\infty_x, \infty_y$. Use the circle construction to obtain schedules $S_1, S_2$ on $\{x_1, \ldots, x_n, \infty_x\}$ and $\{y_1, \ldots, y_n, \infty_y\}$ in which $x_i$ plays $\infty_x$ and $y_i$ plays $\infty_y$ in round $i$. For the $i$th round of the required schedule, $i \leqslant n$, take the $i$th round games of $S_1$ and $S_2$, replacing the two games $x_i \vee \infty_x$ and $y_i \vee \infty_y$ by $x_i \vee y_i$. Then construct a bipartite tournament for the remaining rounds.
   (ii) *$n$ even.* As in question 6 above, using the circle construction on $\{x_1, \ldots, x_n\}$ and $\{y_1, \ldots, y_n\}$ for the last $n-1$ rounds.

8. *Completely balanced round robin tournaments.* A league schedule is balanced for carry-over effects at level $k$ if the condition (8.1) holds with the phrase 'immediately after' being replaced by 'exactly $k$ rounds after', and is completely balanced if it is balanced at level $k$ for all $k$. (The rounds are considered in cyclic order.) Show that the schedules of Theorem 8.3.1 are completely balanced. (Beintema *et al.*)

9. (i) An ordering $\alpha: g_1, \ldots, g_n$ of the elements of a (multiplicative) group $G$ of order $n$ is called a *directed terrace* if $g_1^{-1}g_2, \ldots, g_{n-1}^{-1}g_n$ are all distinct. Define the ordering $\bar{\alpha}$ to be $g_1^{-1}, \ldots, g_n^{-1}$. Show that if $G$ is abelian then $\bar{\alpha}$ is a directed terrace if and only if $\alpha$ is.
   (ii) Verify that if $g_1, \ldots, g_n$ is a directed terrace in $G$ and $L = (l_{ij})$ is the Latin square defined by $l_{ij} = g_i g_j$ then $L$ is complete.
   (iii) If $L$ is a Latin square, the conjugate Latin square $M = L^{(23)}$ is defined by $m_{ij} = k \Leftrightarrow l_{ik} = j$. Verify that a bipartite tournament represented by $L$ as in (8.2) is balanced for carry-over effects for both teams if and only if $L$ and $M$ are both column complete.
   (iv) Show that if $L$ is obtained from an abelian group $G$ as in (ii) then $L$ and $L^{(23)}$ are both column complete.
   (v) Use the directed terrace 1 2 4 7 3 8 6 5 in the additive group $Z_8$ to construct a bipartite tournament balanced with respect to carry-over effects for both teams.
   (This example is largely due to Rosemary Bailey.)

10. *Mixed doubles tournaments.* A tournament of mixed doubles tennis matches is to be arranged for $n$ men and $n$ women. No two players

are to play in the same game more than once. Show that if $f(n)$ denotes the maximum possible number of matches in such a tournament, then $f(n) \leqslant \left[ \dfrac{n}{2} \left[ \dfrac{n}{2} \right] \right]$.

(a) Suppose $n = 12k \pm 2$. Construct a tournament with $\left[ \dfrac{n}{2} \left[ \dfrac{n}{2} \right] \right]$ games in $N = \frac{1}{2}n$ rounds as follows. Label the men $M_i$ and the women $W_i$, $1 \leqslant i \leqslant n$, and for each $i, j \in I_N$ arrange a match $M_i$, $W_{i+j}$ v $M_{i+2j}$, $W_{i+3j}$. These games can be arranged into rounds by taking the $j$th round to consist of all those matches given by that particular $j$.

(b) Represent the match involving $M_i$, $M_j$, $W_k$, $W_l$ by $(i, j, k, l)$. Show that $(0, 1, 0, 2)$ and $(0, 2, 5, 6)$ mod 8 generate 16 matches for eight men and eight women. What is needed here is that all $i - j$ are different, all $k - l$ are different, and all $i - k$, $i - l$, $j - k$, $j - l$ are different. Use the same two starting matches mod 9, 10, 11.

(c) Show that the difference method in (b) can achieve the maximum number of games only if $n \equiv 0$ or $1 \pmod 4$. As examples, consider $(0, 1, 0, 2)$, $(0, 2, 7, 8)$, $(0, 5, 3, 9)$ mod 13, and $(0, 1, 0, 6)$, $(0, 4, 12, 13)$, $(0, 6, 1, 10)$, $(0, 5, 3, 7)$ mod 16. (Gilbert, 1961)

# 9

# Room squares and bridge tournaments

## 9.1 Room squares

In the first chapter it was shown how to construct a league schedule for $2n$ teams in $2n - 1$ rounds, each round consisting of $n$ games, with each pair of teams meeting exactly once. Here, for example, is a possible league schedule for eight teams $\infty, 0, 1, \ldots, 6$.

| | | | |
|---|---|---|---|
| Day 1: | $\infty$ v 0 | 1 v 6 | 2 v 5 | 3 v 4 |
| Day 2: | $\infty$ v 1 | 2 v 0 | 3 v 6 | 4 v 5 |
| Day 3: | $\infty$ v 2 | 3 v 1 | 4 v 0 | 5 v 6 |
| Day 4: | $\infty$ v 3 | 4 v 2 | 5 v 1 | 6 v 0 |
| Day 5: | $\infty$ v 4 | 5 v 3 | 6 v 2 | 0 v 1 |
| Day 6: | $\infty$ v 5 | 6 v 4 | 0 v 3 | 1 v 2 |
| Day 7: | $\infty$ v 6 | 0 v 5 | 1 v 4 | 2 v 3. |

Suppose now that seven pitches are available, of different quality. Can the games be allocated to the pitches in such a way that each team will play exactly once on each pitch? If football is replaced by bridge and pitches by tables, the problem is to arrange for a bridge tournament in which each of $2n$ partnerships play in each round, and overall each partnership plays each other once, and each plays each table once. Here is a possible solution for $n = 4$ using the above league schedule.

**Example 9.1.1**  A Room square of order 8. Rows correspond to rounds and columns to pitches or tables.

| | 1 | 2 | 3 | 4 | 5 | 6 | 7 |
|---|---|---|---|---|---|---|---|
| 1 | $\infty$ 0 | — | — | 2 5 | — | 1 6 | 3 4 |
| 2 | 4 5 | $\infty$ 1 | — | — | 3 6 | — | 2 0 |
| 3 | 3 1 | 5 6 | $\infty$ 2 | — | — | 4 0 | — |
| 4 | — | 4 2 | 6 0 | $\infty$ 3 | — | — | 5 1 |
| 5 | 6 2 | — | 5 3 | 0 1 | $\infty$ 4 | — | — |
| 6 | — | 0 3 | — | 6 4 | 1 2 | $\infty$ 5 | — |
| 7 | — | — | 1 4 | — | 0 5 | 2 3 | $\infty$ 6 |

Such an array is known as a *Room square*, after T.G. Room who, in 1955, pointed out the existence of the following array which corresponds to a different (non-cyclic) league schedule for eight teams $1, \ldots, 8$.

**Example 9.1.2**

| | | | | | | |
|---|---|---|---|---|---|---|
| 12 | — | 34 | — | 56 | — | 78 |
| — | 37 | 25 | — | — | 48 | 16 |
| 47 | 15 | — | — | 38 | 26 | — |
| — | — | — | 68 | 14 | 57 | 23 |
| 58 | — | 67 | 24 | — | 13 | — |
| — | 46 | 18 | 35 | 27 | — | — |
| 36 | 28 | — | 17 | — | — | 45 |

(a) Room's square of order 8.

| | | | | | | |
|---|---|---|---|---|---|---|
| — | — | — | hi | kl | mn | op |
| — | il | mo | — | np | hk | — |
| — | no | hl | mp | — | — | ik |
| lp | — | in | ko | hm | — | — |
| im | — | kp | — | — | lo | hn |
| ho | km | — | ln | — | ip | — |
| kn | hp | — | — | io | — | lm |

(b) Kirkman's Room square of order 8.

**Definition 9.1.1** A *Room square of order* $2n$ (or of side $2n - 1$) is a $(2n - 1) \times (2n - 1)$ array based on $2n$ symbols $x_1, \ldots, x_{2n}$ such that

(i) each cell is empty or contains an *unordered* pair of distinct symbols;
(ii) each row and each column contains each $x_i$ exactly once;
(iii) each of the $n(2n - 1)$ unordered pairs of distinct symbols occurs in exactly one cell of the array.

It is natural to ask if Room squares of all orders $2n$ can be constructed. We shall see that indeed they can, except for two exceptions, $n = 2$ and $n = 3$.

Room squares had in fact been discussed (although not under that name!) long before 1955. Away back in 1850, Kirkman, while discussing the schoolgirls problem, had exhibited the above 'curious arrangement' which is just another Room square of order 8.

Kirkman observed that if he adjoined 1 to each pair in the first column, 2 to each pair in the second column, and so on, then the triples in each row, along with the triple of missing numbers, form one resolution class of a KTS(15). In this way he obtained a solution to the 15 schoolgirls problem:

| | | | | | |
|---|---|---|---|---|---|
| Day 1: | 123 | hi4 | kl5 | mn6 | op7 |
| Day 2: | 147 | il2 | mo3 | np5 | hk6 |
| Day 3: | 156 | no2 | hl3 | mp4 | ik7 |
| Day 4: | 267 | lp1 | in3 | ko4 | hm5 |
| Day 5: | 245 | im1 | kp3 | lo6 | hn7 |
| Day 6: | 357 | ho1 | km2 | ln4 | ip6 |
| Day 7: | 346 | kn1 | hp2 | io5 | lm7. |

Anstice (1853) extended this idea to construct $KTS(2p+1)$ for all primes $p \equiv 1 \pmod 6$.

Later on, Room squares were studied in the context of bridge tournaments; solutions were known to E.C. Howell by 1897 for all $2n$ from 8 to 30, and were known as *Howell mastersheets*. More recently, in the 1970s, connections have been discovered with other combinatorial structures such as Latin squares, and it is now known that Room squares exist for all even orders except 4 and 6.

We now exhibit Room squares of orders 10 and 12; in the second of these, $X$ stands for 10. Each is based on the numbers $\infty; 0, \dots, 2n-2$ ($n = 5, 6$) and has the property shared by that of Example 9.1.1 that the pair $\{\infty, i\}$ occurs in the $(i+1, i+1)$ position in the main diagonal.

**Example 9.1.3**   Room square of order 10.

| | | | | | | | | |
|---|---|---|---|---|---|---|---|---|
| ∞0 | 58 | 37 | — | — | — | — | 46 | 12 |
| 23 | ∞1 | 60 | 48 | — | — | — | — | 57 |
| 68 | 34 | ∞2 | 71 | 50 | — | — | — | — |
| — | 70 | 45 | ∞3 | 82 | 61 | — | — | — |
| — | — | 81 | 56 | ∞4 | 03 | 72 | — | — |
| — | — | — | 02 | 67 | ∞5 | 14 | 83 | — |
| — | — | — | — | 13 | 78 | ∞6 | 25 | 04 |
| 15 | — | — | — | — | 24 | 80 | ∞7 | 36 |
| 47 | 26 | — | — | — | — | 35 | 01 | ∞8 |

**Example 9.1.4**   Room square of order 12.

| | | | | | | | | | | |
|---|---|---|---|---|---|---|---|---|---|---|
| ∞0 | 48 | — | 12 | 5X | 97 | — | — | — | 36 | — |
| — | ∞1 | 59 | — | 23 | 60 | X8 | — | — | — | 47 |
| 58 | — | ∞2 | 6X | — | 34 | 71 | 09 | — | — | — |
| — | 69 | — | ∞3 | 70 | — | 45 | 82 | 1X | — | — |
| — | — | 7X | — | ∞4 | 81 | — | 56 | 93 | 20 | — |
| — | — | — | 80 | — | ∞5 | 92 | — | 67 | X4 | 31 |
| 42 | — | — | — | 91 | — | ∞6 | X3 | — | 78 | 05 |
| 16 | 53 | — | — | — | X2 | — | ∞7 | 04 | — | 89 |
| 9X | 27 | 64 | — | — | — | 03 | — | ∞8 | 15 | — |
| — | X0 | 38 | 75 | — | — | — | 14 | — | ∞9 | 26 |
| 37 | — | 01 | 49 | 86 | — | — | — | 25 | — | ∞X |

By interchanging rows and columns, any Room square on the $2n$ symbols $\infty, 0, \dots, 2n-2$ can be transformed into one with the pair $\{i, \infty\}$ in position $(i+1, i+1)$; a Room square with this property is said to be *standardized*. Each of the examples 9.1.1, 9.1.3, and 9.1.4 has a further property: any row can be obtained from the previous row by adding 1 mod $2n-1$ to each entry other than $\infty$ and moving one step diagonally towards the bottom right. This 'cyclic' property is possessed by many of the squares to be constructed, and clearly depends on a clever choice of the first row; this will be explained in the next section.

Having obtained Room squares for small orders, it is possible to combine them to form Room squares of larger orders. This depends on the following composition or multiplication theorem, the first satisfactory proof of which was given by Stanton and Horton (1972). Their proof uses the existence of two MOLS of order $n$ whenever $n$ is odd.

**Theorem 9.1.1** (Composition theorem for Room squares)  *If Room squares of sides $m, n$ exist, then a Room square of side $mn$ also exists.*

**Proof** Let $M$ and $N$ be Room squares on $\{0, \ldots, m\}$ and $\{0, \ldots, n\}$ respectively, and let $L_1$ and $L_2$ be two MOLS of order $n$ on $\{1, \ldots, n\}$. Alter $M$ by replacing each of its entries by an $n \times n$ array as follows.

  (i) If cell $(i, j)$ of $M$ is empty, place an empty $n \times n$ array in it.
 (ii) If cell $(i, j)$ of $M$ has $\{0, k\}$ in it, place in that cell $N + kn$, which is obtained from $N$ by adding to $kn$ to each *non-zero* entry. (Note that $N + kn$ is a Room square of side $n$ on the symbols $0$; $kn + 1$, $\ldots, kn + n$.)
(iii) If cell $(i, j)$ of $M$ has $\{u, v\}$ with $0 < u < v$, add $un$ to each entry of $L_1$ and $vn$ to each entry of $L_2$ and place the join of the resulting MOLS in that cell, with ordered pairs replaced by unordered ones.

When this is done, clearly an $mn \times mn$ square array $R$ is obtained on the symbols $0$; $n + 1, n + 2, \ldots, n + mn$; further, each cell of $R$ is either empty or contains a pair of symbols. It is also clear that each row and column of $R$ contains each symbol exactly once; so it remains to prove that each pair of symbols occurs exactly once in $R$. Certainly there are $m \cdot \frac{1}{2}n(n + 1)$ pairs arising from (ii) and $\frac{1}{2}m(m - 1)n^2$ pairs arising from (iii), i.e. $\frac{1}{2}mn(mn + 1)$ pairs altogether; so we need only show that all pairs are distinct. Let $P(i, j)$ denote the collection of all pairs of $R$ arising from the $(i, j)$ cell of $M$. Then certainly all pairs in $P(i, j)$ are distinct; either they are all pairs in a Room square or they are all pairs in the join of two MOLS. Further, if $(i, j) \neq (h, k)$ then $P(i, j)$ and $P(h, k)$ have no pairs in common; for if, say, $P(i, j)$ contained the pair $\{un + l_1, vn + l_2\}$ which is also in $P(h, k)$, the the join of $L_1$ and $L_2$ would have $(l_1, l_2)$ occurring in two different positions, contradicting orthogonality.

**Example 9.1.5**  Since Room squares of orders 8 and 12, i.e. of sides 7, 11, exist, a Room square of side 77, i.e. order 78, exists.

## 9.2 The use of starters

We have already seen examples of standardized Room squares of orders 8, 10, and 12 with cyclic structure, each row being determined by the first row. What special properties does the first row have to possess to enable it to generate a cyclic Room square in this way? Since every pair of distinct symbols is to appear in the square just once, the pairs in the first row must form a difference system with $\lambda = 1$. For example, in the square of order 10, the differences arising from the pairs in the first row are $\pm 3$, $\pm 4$, $\pm 2$, $\pm 1$, so that each non-zero member of $Z_9$ crops up exactly once as a difference. Further, no two pairs in the difference system can have an element in common (this of course is to enable the rows of the square to form resolution classes of a resolvable $(2n, 2, 1)$ design). In other words, for the first row pairs we need a starter, as defined in Section 2.3.

Having obtained a starter, we are faced with a further problem: in which places in the first row should the pairs of the starter be placed so that, if the row is developed cyclically, it will generate a Room square? Label the rows and columns by the elements of the group $G = \{g_0 = 0, g_1, \ldots, g_{2n-2}\}$ (so that the 'first' row is now row 0), and place the pair $\{0, \infty\}$ in position $(g_0, g_0)$. Suppose we decide to place the pair $\{x_i, y_i\}$ in position $(g_0, -a_i)$ for some $a_i \in G$, $1 \leqslant i \leqslant n - 1$, where $a_i \neq a_j$ unless $i = j$, and where $a_i \neq g_0 = 0$. If we then develop this first row cyclically by placing $\{g_i, \infty\}$ in position $(g_i, g_i)$ and placing $\{x_i + g, y_i + g\}$ in position $(g, -a_i + g)$, then the following pairs will occur in column $g_0$: $\{0, \infty\}$, $\{x_i + a_i, y_i + a\}$, $1 \leqslant i \leqslant n - 1$. Since these pairs have to contain in their union each element of $G$ exactly once, we make the following definition.

**Definition 9.2.1** An *adder* for a starter $\{x_1, y_1\}, \ldots, \{x_{n-1}, y_{n-1}\}$ in $G$ is a set of $n - 1$ distinct non-zero elements $a_1, \ldots, a_{n-1}$ of $G$ such that $x_1 + a_1, y_1 + a_1, \ldots, x_{n-1} + a_{n-1}, y_{n-1} + a_{n-1}$ are precisely all the non-zero elements of $G$.

**Example 9.2.1** The starter $\{5, 8\}$, $\{3, 7\}$, $\{4, 6\}$, $\{1, 2\}$ in $Z_9$ has an adder $8, 7, 2, 1$ since $5 + 8 = 4$, $8 + 8 = 7$, $3 + 7 = 1$, $7 + 7 = 5$, $4 + 2 = 6$, $6 + 2 = 8$, $1 + 1 = 2$, $2 + 1 = 3$ are precisely the non-zero elements of $Z_9$. So, for example, the pair $\{5, 8\}$ is placed in position $(0, -8)$, i.e. the position $(0, 1)$ in row 0. The square of Example 9.1.3 is obtained in this way.

Note that it is not always possible to find an adder for a given starter. For example, in $Z_5$ the starter $\{1, 4\}$, $\{2, 3\}$ has no adder, as can readily be verified. However, if an adder *can* be found, then a Room square will result.

**Theorem 9.2.1** *If there exists a starter with adder in an abelian group $G$ of order $2n - 1$ then a Room square of order $2n$ exists.*

**Proof** Let the starter be $\{x_1, y_1\}, \ldots, \{x_{n-1}, y_{n-1}\}$, with adder $a_1, \ldots, a_{n-1}$. Take a square array of order $2n - 1$, with rows and columns labelled by the elements $g_0 = 0, g_1, \ldots, g_{2n-2}$ of $G$; place $\{\infty, g_i\}$ in position $(g_i, g_i)$ and place $\{x_i + g, y_i + g\}$ in position $(g, g - a_i)$. Then each row contains $\infty$ and each element of $G$ once, and the entries in column 0 are $\infty, 0, x_i + a_i, y_i + a_i$ ($1 \leq i \leq n - 1$), so that column 0, and hence each other column, contains each of the $2n$ symbols once. Finally, every pair occurs in the square because the pairs $\{x_i, y_i\}$ form a difference system.

The starter–adder method was essentially used by Anstice in 1853, but his work remained unrecognised. The method was re-introduced by Stanton and Mullin in 1968; they gave starters and adders in $Z_k$ for all $k$ from 7 to 47. Mullin and Nemeth gave infinite families of starters and adders in the following year (see Exercises 9, number 4 as well as the following theorems). Their adders were of a particularly simple form.

**Definition 9.2.2** A starter $\{x_1, y_1\}, \ldots, \{x_{n-1}, y_{n-1}\}$ is a *strong* starter if the sums $x_i + y_i$ are all distinct and non-zero.

**Lemma 9.2.2** *If $\{x_1, y_1\}, \ldots, \{x_{n-1}, y_{n-1}\}$ is a strong starter in $G$, then the elements $a_i = -(x_i + y_i)$ form an adder.*

**Proof** Certainly the $a_i$ are all distinct and non-zero. Also, $x_i + a_i = -y_i$ and $y_i + a_i = -x_i$, so condition (i) for a starter yields the required fact that these sums are all the non-zero elements of $G$.

**Corollary 9.2.3** *If a strong starter exists in an abelian group $G$ of order $2n - 1$, then a Room square of order $2n$ exists.*

**Example 9.2.2** The pairs $\{1, 2\}$, $\{4, 8\}$, $\{5, 10\}$, $(9, 7)$, $\{3, 6\}$ form a strong starter in $Z_{11}$. They form a starter because the *differences* are $\pm 1$, $\pm 4$, $\pm 5$, $\pm 2$, $\pm 3$; and the starter is strong because the *sums* $1 + 2 = 3$, $4 + 8 = 1$, $5 + 10 = 4$, $9 + 7 = 5$, $3 + 6 = 9$ are all distinct and non-zero. So a corresponding adder is $-3, -1, -4, -5, -9$, i.e. $8, 10, 7, 6, 2$ and a cyclic Room square of order 12 can be constructed; the reader should check that the square is precisely the one exhibited in Example 9.1.4.

Before proving the more general Theorem 9.2.5, we first present the simplest case, where $2n - 1$ is a prime $p \equiv 3 \pmod 4$.

**Theorem 9.2.4**  *Let $p \equiv 3 \pmod 4$ be a prime, $p > 3$. Then there exists a strong starter in $Z_p$, and hence there exists a Room square of order $p + 1$.*

**Proof**  Let $\theta$ be a primitive root mod $p$, where $p = 4m + 3$, $m \geqslant 1$. Then $\theta^{4m+2} = 1$, $\theta^{2m+1} = -1$. We claim that the pairs

$$\{\theta^0, \theta^1\}, \{\theta^2, \theta^3\}, \ldots, \{\theta^{4m}, \theta^{4m+1}\}$$

form a strong starter. Their differences are

$$\pm(\theta - 1), \ \pm\theta^2(\theta - 1), \ldots, \ \pm\theta^{4m}(\theta - 1),$$

i.e. $\theta - 1$ times the elements $\theta^{2i}$ and $\theta^{2i+2m+1}$, $0 \leqslant i \leqslant 2m$; i.e. the differences are all the non-zero elements of $Z_p$ exactly once. Similarly, their sums are the elements $\theta^{2i}(\theta + 1)$, and since $\theta + 1 \neq 0$ (because $m \geqslant 1$), these are again all distinct and non-zero.

**Example 9.2.3**  Since 2 is a primitive root of 11, we see now where the pairs of the strong starter exhibited in Example 9.2.2 came from!

**Example 9.2.4**  Since 2 is a primitive root of 19, we can take the pairs $\{1, 2\}$, $\{4, 8\}$, $\{16, 13\}$, $\{7, 14\}$, $\{9, 18\}$, $\{17, 15\}$, $\{11, 3\}$, $\{6, 12\}$, $\{5, 10\}$ as a starter, with adder 16, 7, 9, 17, 11, 6, 5, 1, 4. So as the first row of a cyclic Room square of order 20 we can take

$$\{\infty, 0\}{-}\{7, 14\}\{1, 2\}{-}{-}{-}{-}\{9, 18\}{-}\{16, 13\}$$

$$-\{4, 8\}\{17, 15\}\{3, 11\}\{5, 10\}{-}{-}\{6, 12\}.$$

Now we present the more general result.

**Theorem 9.2.5**  (Anstice 1853; Mullin and Nemeth 1969)  *If $p$ is prime, and $p^n = 2^k t + 1$ where $t > 1$ is odd and $k > 0$, then there exists a strong starter in $\mathrm{GF}(p^n)$ and hence a Room square of order $p^n + 1$.*

**Proof**  Let $\theta$ be a primitive element of $\mathrm{GF}(q)$ where $q = p^n$, and let $d = 2^{k-1}$, so that $\theta^{2d t} = 1$ and $\theta^{d t} = -1$. Write down the elements of $G^* = \mathrm{GF}(q) - \{0\}$ in pairs as follows.

$$\{\theta^0, \theta^d\} \qquad \{\theta^{2d}, \theta^{3d}\} \qquad \cdots \qquad \{\theta^{(2t-2)d}, \theta^{(2t-1)d}\}$$

$$\{\theta^1, \theta^{d+1}\} \qquad \{\theta^{2d+1}, \theta^{3d+1}\} \qquad \cdots \qquad \{\theta^{(2t-2)d+1}, \theta^{(2t-1)d+1}\}$$

$$\vdots$$

$$\{\theta^{d-1}, \theta^{2d-1}\} \quad \{\theta^{3d-1}, \theta^{4d-1}\} \quad \cdots \quad \{\theta^{(2t-1)d-1}, \theta^{2td-1}\}$$

Note that, taking a column at a time, we obtain all the elements of $G^*$ in their 'natural' order as powers of $\theta$.

We now assert that these pairs form a strong starter! First consider differences: they are $\pm\theta^{2id+j}(1 - \theta^d)$, $0 \leqslant i \leqslant t - 1$, $0 \leqslant j \leqslant d - 1$, where $1 - \theta^d \neq 0$ since $t \geqslant 2$. Now we cannot have $\theta^{2id+j}(1 - \theta^d) = \theta^{2Id+J}(1 - \theta^d)$ unless $i = I$ and $j = J$; for cancelling by $(1 - \theta^d)$ gives $\theta^{2id+j} = \theta^{2Id+J}$, i.e. $\theta^{2d(i-I)} = \theta^{J-j}$, which, due to the ranges of possible values of $i, j, I, J$, can only occur if $i = I$ and $j = J$. Suppose next that $\theta^{2id+j}(1 - \theta^d) = -\theta^{2Id+J}(1 - \theta^d)$; then we would have $\theta^{2id+j} + \theta^{2Id+J} = 0$. If $2id + j < 2Id + J$ then we would have $\theta^{2id+j}(1 + \theta^{(2I-2i)d+(J-j)}) = 0$ so that $(2I - 2i)d + J - j = dt$. But $J - j$ lies in the interval $[-d + 1, d - 1]$ and so, being a multiple of $d$, must be 0. So $j = J$; so $2I - 2i = t$, contradicting the oddness of $t$.

Thus the pairs do indeed form a starter. To show that the starter is strong, we have to show that the *sums* of pairs are all distinct. But the above argument applies with $1 + \theta^d$ in place of $1 - \theta^d$, where we note that $1 + \theta^d$ is non-zero. This completes the proof of the theorem.

Note that the condition $t > 1$ is used to ensure that the starter is strong. If $t = 1$ then $p = 2^k + 1$, and the sums $\theta^i + \theta^{d+i}$ are in fact all zero since $\theta^d = \theta^{dt} = -1$.

**Example 9.2.5** A Room square of order 14. Consider $p = 13 = 2^2 3 + 1$. Here we take $t = 3$, $k = 2$, $d = 2$, $\theta = 2$, and we obtain the pairs

$$\{1, 4\} \qquad \{3, 12\} \qquad \{9, 10\}$$
$$\{2, 8\} \qquad \{6, 11\} \qquad \{5, 7\}.$$

These yield the following Room square.

| | | | | | | | | | | | | |
|---|---|---|---|---|---|---|---|---|---|---|---|---|
| 13,0 | — | 3,12 | — | 6,11 | 1,4 | 9,10 | — | — | — | 2,8 | — | 5,7 |
| 6,8 | 13,1 | — | 4,0 | — | 7,12 | 2,5 | 10,11 | — | — | — | 3,9 | — |
| — | 7,9 | 13,2 | — | 5,1 | — | 8,0 | 3,6 | 11,12 | — | — | — | 4,10 |
| 5,11 | — | 8,10 | 13,3 | — | 6,2 | — | 9,1 | 4,7 | 12,0 | — | — | — |
| — | 6,12 | — | 9,11 | 13,4 | — | 7,3 | — | 10,2 | 5,8 | 0,1 | — | — |
| — | — | 7,0 | — | 10,12 | 13,5 | — | 8,4 | — | 11,3 | 6,9 | 1,2 | — |
| — | — | — | 8,1 | — | 11,0 | 13,6 | — | 9,5 | — | 12,4 | 7,10 | 2,3 |
| 3,4 | — | — | — | 9,2 | — | 12,1 | 13,7 | — | 10,6 | — | 0,5 | 8,11 |
| 9,12 | 4,5 | — | — | — | 10,3 | — | 0,2 | 13,8 | — | 11,7 | — | 1,6 |
| 2,7 | 10,0 | 5,6 | — | — | — | 11,4 | — | 1,3 | 13,9 | — | 12,8 | — |
| — | 3,8 | 11,1 | 6,7 | — | — | — | 12,5 | — | 2,4 | 13,10 | — | 0,9 |
| 1,10 | — | 4,9 | 12,2 | 7,8 | — | — | — | 0,6 | — | 3,5 | 13,11 | — |
| — | 2,11 | — | 5,10 | 0,3 | 8,9 | — | — | — | 1,7 | — | 4,6 | 13,12 |

Note that the starter in 9.2.5 is *skew* in the following sense.

**Definition 9.2.3** A strong starter is *skew* if the sums $x_i + y_i$ and their negatives are all distinct.

**Definition 9.2.4**  A Room square is *skew* if, whenever $i \neq j$, precisely one of the $(i, j)$th and $(j, i)$th cells is occupied.

It can be shown (Exercises 9, number 11) that a Room square derived from a skew strong starter is always skew. Further, the Mullin–Nemeth starters are always skew. The following definition will be useful.

**Definition 9.2.5**  If $R$ is a skew Room square on $\{\infty, 1, \ldots, r\}$, with pairs involving $\infty$ on the main diagonal, the *sum* of $R$ and $R'$ is the array obtained by deleting all diagonal entries, adding $r$ to each entry of $R'$, and superimposing the resulting arrays.

It follows from Theorem 9.2.5 that starters and adders exist in $Z_p$ for all odd primes $p$ not of the form $2^k + 1$. Primes of the form $2^k + 1$ are known as *Fermat primes*. At present only five such primes are known: 3, 15, 17, 257, 65537. Of these, 17 has been dealt with already by Mullin and Stanton (see Exercises 9, number 1), and it is easy to check that no starters and adders can exist for 3 or 5. The remaining Fermat primes (and there may be infinitely many of them!) have to be dealt with separately. Observe that such primes $2^k + 1$ must in fact have $k$ a power of 2; for if $k = 2^a m$ where $m \geq 3$ is odd, then $2^k + 1 = b^m + 1$ (where $b = 2^{2^a}$) is divisible by $b + 1$ and so cannot be prime.

**Theorem 9.2.6** (Chong and Chan, 1974)   *If $p \geq 17$ is a Fermat prime, then there is a strong starter in $Z_p$ and hence there exists a Room square of side $p$.*

**Proof**   Let $p = 2^{2d} + 1$ where $d = 2^{m-1}$, and consider the following pairs of elements of $Z_p$:

(i) $\{i + (r-1)2^d, i2^d - (r-1)\}$;
(ii) $\{(2^d - i)2^d + r, (2^{d-1} - r)2^d + 2^{d-1} - i + 1\}$;
(iii) $\{(2^{d-1} + r - 1)2^d + 2^{d-1} + i, (2^{d-1} + i - 1)2^d + 2^{d-1} - (r-1)\}$;
(iv) $\{(2^{d-1} - i)2^d + 2^{d-1} + r, (2^d - r + 1)2^d - i + 1\}$

where, in each case, $1 \leq r \leq 2^{d-2}$, $1 \leq i \leq 2^{d-1}$. It is left to the reader to check that these pairs form a strong starter.

**Example 9.2.6**   Take $p = 17 = 2^4 + 1$, so that $m = 2$ and $d = 2$. The starter pairs are

(i) $\{1, 4\}, \{2, 8\}$;          (iii) $\{11, 10\}, \{12, 14\}$;
(ii) $\{13, 6\}, \{9, 5\}$;        (iv) $\{7, 16\}, \{3, 15\}$.

The adders are (i) 12, 7; (ii) 15, 3; (iii) 13, 8; (iv) 11, 16. All of this gives a cyclic Room square of order 18 whose first row is
$\{\infty, 0\}\{3, 15\}\{13, 6\} — \{11, 10\}\{1, 4\}\{7, 16\} — — \{12, 14\}\{2, 8\}$

$— — — \{9, 5\} — —$.

From the results of this section and the composition theorem we now have the following.

**Theorem 9.2.7** *There exists a Room square of side $\prod_i p_i^{\alpha_i}$ whenever all the primes $p_i$ are $\geqslant 7$.*

## 9.3 Triplication and quintuplication

Our aim is now to complete the proof of the existence of a Room square of side $2n - 1$ for all $n \geqslant 4$. Since Room squares of side 9, 15, and 25 all exist (see Example 9.1.3, Exercises 9, number 1, and Theorem 9.2.5) and since Theorem 9.2.7 has been established, we have in fact dealt with every $2n - 1$ of the form $3^\alpha 5^\beta p_1^{\alpha_1} \cdots p_r^{\alpha_r}$ where each $p_i$ is $\geqslant 7$ and where $\alpha + \beta = 0$ or 2. The general existence result will therefore follow once the following results of Horton and Wallis have been established.

**Theorem 9.3.1** (Wallis, 1973) *If a Room square of side $r$ exists, then a Room square of side $3r$ also exists.*

**Theorem 9.3.2** (Horton, 1971) *If a Room square of side $r$ exists, then a Room square of side $5r$ also exists.*

**Proof of Theorem 9.3.1** (The triplication theorem)  Suppose we have a Room square $R$ of side $r = 2s + 1$ with entries $\infty, 0, 1, \ldots, r - 1$. By permuting rows and/or columns we can assume that the square is standardized, with the pairs $\{\infty, i\}$ occurring in order on the main diagonal. For each $i = 1, 2, 3$ let $x_i = x + r(i - 1)$, and $y_i = y + r(i - 1)$, and define $R_{ij}$ to be the array obtained from $R$ by deleting all main diagonal entries and replacing each remaining entry $\{x, y\}$, $x < y$, by $\{x_i, y_j\}$. For example, if $R$ is the Room square of Example 9.1.1, then $R_{12}$ has first row

$$————\{2, 12\}—\{1, 13\}\{3, 11\}.$$

Note that the arrays $R_{ij}$, $1 \leqslant i \leqslant 3$, $1 \leqslant j \leqslant 3$, will contain among them all unordered pairs of numbers from 1 to $3r$ apart from the pairs $\{x_i, x_j\}$. Putting them together is therefore a hopeful method of attempting to form a Room square of side $3r$. Suppose then that we consider the following $3r \times 3r$ array:

$$
\begin{array}{ccc}
R_{11} & R_{22} & R_{33} \\
R_{32} & R_{13} & R_{21} \\
R_{23} & R_{31} & R_{12}
\end{array}
$$

Each of the first $r$ rows contain each of $0, \ldots, 3r - 1$ exactly once, except that, in the $(i + 1)$th row, the numbers $i, i + r, i + 2r$ are missing. Similarly, these are missing from rows $i + r + 1$ and $i + 2r + 1$. A similar observation can be made about the columns. Note also that every unordered pair of numbers from 0 to $3r - 1$ occurs exactly once, except for the pairs $x, y$ where $x$ and $y$ differ by a multiple of $r$; these pairs do not occur at all, like the pairs involving $\infty$. Now if we try to fit these missing pairs into the array, we find that we have problems. There are, for example, the nine top left corner positions of the $R_{ij}$ in which to place the pairs $\{\infty, 0\}$, $\{\infty, r\}$, $\{\infty, 2r\}$, $\{0, r\}$, $\{r, 2r\}$, $\{0, 2r\}$; but this fitting in just cannot be done. Clearly the above approach needs to be modified.

So let us now consider the empty cells of $R$. Define the $r \times r$ $(0, 1)$ matrix $M = (m_{ij})$ by

$$m_{ij} = \begin{cases} 1 & \text{if cell } (i, j) \text{ of } R \text{ is empty,} \\ 0 & \text{otherwise.} \end{cases}$$

Then, since $r = 2s + 1$, $M$ has $s$ 1s in each row and column, and so, by Theorem 1.5.4,

$$M = P_1 + \cdots + P_s,$$

where each $P_i$ is a permutation matrix. Let $\phi$ be the permutation corresponding to $P_1$; i.e. $\phi(k) = l$ if and only if $P_1$ has entry 1 in the $(k, l)$ position: and note that each cell $(k, \phi(k))$ in $R$ is empty.

**Example 9.3.1** The matrix $M$ corresponding to the Room square of Example 9.1.1 is

$$\begin{bmatrix} 0 & 1 & 1 & 0 & 1 & 0 & 0 \\ 0 & 0 & 1 & 1 & 0 & 1 & 0 \\ 0 & 0 & 0 & 1 & 1 & 0 & 1 \\ 1 & 0 & 0 & 0 & 1 & 1 & 0 \\ 0 & 1 & 0 & 0 & 0 & 1 & 1 \\ 1 & 0 & 1 & 0 & 0 & 0 & 1 \\ 1 & 1 & 0 & 1 & 0 & 0 & 0 \end{bmatrix} = \begin{bmatrix} 0 & 1 & 0 & 0 & 0 & 0 & 0 \\ 0 & 0 & 1 & 0 & 0 & 0 & 0 \\ 0 & 0 & 0 & 1 & 0 & 0 & 0 \\ 0 & 0 & 0 & 0 & 1 & 0 & 0 \\ 0 & 0 & 0 & 0 & 0 & 1 & 0 \\ 0 & 0 & 0 & 0 & 0 & 0 & 1 \\ 1 & 0 & 0 & 0 & 0 & 0 & 0 \end{bmatrix}$$

$$+ \begin{bmatrix} 0 & 0 & 1 & 0 & 0 & 0 & 0 \\ 0 & 0 & 0 & 1 & 0 & 0 & 0 \\ 0 & 0 & 0 & 0 & 1 & 0 & 0 \\ 0 & 0 & 0 & 0 & 0 & 1 & 0 \\ 0 & 0 & 0 & 0 & 0 & 0 & 1 \\ 1 & 0 & 0 & 0 & 0 & 0 & 0 \\ 0 & 1 & 0 & 0 & 0 & 0 & 0 \end{bmatrix} + \begin{bmatrix} 0 & 0 & 0 & 0 & 1 & 0 & 0 \\ 0 & 0 & 0 & 0 & 0 & 1 & 0 \\ 0 & 0 & 0 & 0 & 0 & 0 & 1 \\ 1 & 0 & 0 & 0 & 0 & 0 & 0 \\ 0 & 1 & 0 & 0 & 0 & 0 & 0 \\ 0 & 0 & 1 & 0 & 0 & 0 & 0 \\ 0 & 0 & 0 & 1 & 0 & 0 & 0 \end{bmatrix}$$

so that $\phi$ is given by $\phi(1) = 2$, $\phi(2) = 3, \ldots, \phi(6) = 7$, $\phi(7) = 1$.

Consider now the following $3r \times 3r$ array:

$$
\begin{array}{ccc}
R_{11} & \phi R_{22} & \phi R_{33} \\
\phi R_{32} & R_{13} & R_{21} \\
\phi R_{23} & R_{31} & R_{12}
\end{array}
$$

where $\phi R_{ij}$ denotes the array obtained from $R_{ij}$ by permuting the *columns* by $\phi$. Because $R_{ij}$ is always in the same column of the array as $R_{ij}$, and either both or neither have their columns permuted by $\phi$, this array still has every row and column containing every one of $0, \ldots, 3r - 1$, except now that, for each $i < r$, $i, i + r, i + 2r$ are missing from rows $i + 1$, $i + r + 1$, $i + 2r + 1$; $i$ is still missing from columns $i + 1$, $i + r + 1$, $i + 2r + 1$; while $i + r$ is missing from columns $\phi(i + 1)$, $r + \phi(i + 1)$, $i + 2r + 1$, and $i + 2r$ is missing from columns $\phi(i + 1)$, $i + r + 1$, $\phi(i + 1) + 2r$. So we can now place

$\{\infty, i\}$ in cell $(i + 1, i + 1)$;
$\{\infty, i + r\}$ in cell $(i + r + 1, \phi(i + 1) + r)$;
$\{\infty, i + 2r\}$ in cell $(i + 2r + 1, \phi(i + 1) + 2r)$;
$\{i, i + r\}$ in cell $(i + 2r + 1, i + 2r + 1)$;
$\{i, i + 2r\}$ in cell $(i + r + 1, i + r + 1)$;
$\{i + r, i + 2r\}$ in cell $(i + 1, \phi(i + 1))$.

The resulting array is then a Room square of side $3r$.

**Proof of Theorem 9.3.2** (The quintuplication theorem)  The proof follows the lines of the proof of the triplication theorem, except that this time we use the following $5r \times 5r$ array

$$
\begin{array}{ccccc}
R_{11} & \phi R_{22} & \phi R_{33} & \phi R_{44} & \phi R_{55} \\
\phi R_{52} & R_{13} & \theta R_{24} & \theta R_{35} & R_{41} \\
\theta R_{43} & \theta R_{54} & R_{15} & R_{21} & \theta R_{32} \\
\theta R_{34} & \theta R_{45} & R_{51} & R_{12} & \theta R_{23} \\
\phi R_{25} & R_{31} & \theta R_{42} & \theta R_{53} & R_{14}
\end{array}
$$

where $\phi, \theta$ are the permutations associated with the permutation matrices $P_1, P_2$ which arise in the decomposition $M = P_1 + \cdots + P_s$ given by Theorem 1.5.4.

It is worth remarking that this method can be generalized to yield the following result (Wallis; see Wallis *et al.* (1972)).

**Theorem 9.3.3** *If there exists a Room square of side $r$, and if $n < r$ is odd, then a Room square of side $n$ exists.*

However, Theorems 9.3.1 and 9.3.2 are sufficient to enable us to complete the existence proof for Room squares.

**Theorem 9.3.4** *If $n$ is any odd number $\geqslant 7$, there exists a Room square of side $n$.*

**Proof** Let $n = 3^{2a+b}5^{2c+d}p_1^{\alpha_1} \cdots p_r^{\alpha_r}$ where $b = 0$ or $1, d = 0$ or $1$, and each $p_i \geqslant 7$. If $b = d = 0$ then $n$ is a product of prime powers each at least 7, so a Room square of side $n$ exists. If $b = d = 1$, then $n = 15m$ where a Room square of side 15 exists and, if $m > 1$, a Room square of side $m$ exists. Finally, the cases $b = 1$, $d = 0$ and $b = 0$, $d = 1$, yield to the triplication and quintuplication theorems.

It should, perhaps, be pointed out that, historically, this was not the way in which the existence of Room squares was first established. The case of Fermat primes was dealt with differently, and at one stage it was known that Room squares existed for all odd $n \geqslant 7$ except possibly 257. See Wallis *et al.* (1972) for further details.

## 9.4 Balanced Room squares for bridge tournaments

Room squares of order $2n$ provide a solution to the problem of constructing a round robin tournament involving $2n$ teams on $2n - 1$ days, each team playing once on each of $2n - 1$ pitches. The pairs of teams in the cells of the Room square are *unordered*, since it is irrelevant which team in any given game is named first. In bridge tournaments, however, with boards in place of games, *order does matter*. When two teams (each with two players) play each other, one team plays North–South (NS) and the other team plays East–West (EW), and we shall follow the convention that the first named team in any pair is the one playing NS. Now, in playing a board NS, a team is not just competing against its (EW) opponents for that particular game, but it is also competing against all the other teams which will play that board NS in a different round, since their relative performances on that board will be taken into account. We are therefore led to the following definitions.

### Definition 9.4.1

(a) If teams $i, j$ oppose each other on a board, one playing NS and the other playing EW, we say that $i$ *plays with* $j$ on that board.

(b) If $i$ and $j$ both play a board in the same direction (i.e. both play NS or both play EW, necessarily in different rounds), we say that $i$ *plays against* $j$ on that board.

For fairness, we want a bridge tournament to have the property that, for some integers $\mu$, $\lambda$, every pair of teams play with each other $\mu$ times and play against each other $\lambda$ times.

**Example 9.4.1** Consider the following Room square.

$$
\begin{bmatrix}
81 & - & - & 36 & - & 27 & 45 \\
56 & 82 & - & - & 47 & - & 31 \\
42 & 67 & 83 & - & - & 51 & - \\
- & 53 & 71 & 84 & - & - & 62 \\
73 & - & 64 & 12 & 85 & - & - \\
- & 14 & - & 75 & 23 & 86 & - \\
- & - & 25 & - & 16 & 34 & 87
\end{bmatrix}
$$

(Observe that this is really just the square of Example 9.1.1 in disguise.) Take the rounds to be given by the rows, and the boards given by the columns. Note that team 1 plays against team 3 thrice: both play NS on board 4 and both play EW on boards 1 and 3. It is easy to check that each pair play against each other thrice; and of course because the array is a Room square, each pair plays with each other once. So the required balance conditions are satisfied.

Note also that, in terms of block designs, we can take the $7 \times 2 = 14$ columns as blocks of size 4 to obtain a design with $\lambda = 3$.

We extract from this example the following ideas.

**Definition 9.4.2** An *ordered Room square* is obtained from a Room square by replacing each unordered pair $\{x, y\}$ by the ordered pair $(x, y)$ or $(y, x)$.

**Definition 9.4.3** The *block design of an ordered Room square* of order $2n$ is the design whose $4n - 2$ blocks are given by the first (or second) entries of the pairs in the $2n - 1$ columns.

**Example 9.4.2** The ordered Room square of Example 9.4.1 yields a design whose blocks are $\{8, 5, 4, 7\}$, $\{1, 6, 2, 3\}$, $\{8, 6, 5, 1\}$, $\{2, 7, 3, 4\}$, $\{8, 7, 6, 2\}$, $\{3, 1, 4, 5\}$, $\{7, 8, 3, 1\}$, $\{4, 5, 2, 6\}$, $\{4, 8, 2, 1\}$, $\{7, 5, 3, 6\}$, $\{2, 5, 8, 3\}$, $\{7, 1, 6, 4\}$, $\{4, 3, 6, 8\}$, $\{5, 1, 2, 7\}$.

Note that each column of the Room square gives two complementary blocks.

**Definition 9.4.4** A *balanced Room square* (BRS) of order $2n$ is an ordered Room square whose block design is balanced.

For example, we have already observed that the ordered Room square of Example 9.4.1 is balanced.

**Definition 9.4.5** The bridge tournament determined by a balanced Room square is called a *complete balanced Howell rotation*.

Complete balanced Howell rotations can be presented in a different way. Consider again Example 9.4.1. Since four games are played in each round, we can use four tables, and move the boards around from table to table. The schedule of Example 9.4.1 can then be presented as follows.

| Round | Table 1 Teams | Board | Table 2 Teams | Board | Table 3 Teams | Board | Table 4 Teams | Board |
|---|---|---|---|---|---|---|---|---|
| 1 | 8,1 | 1 | 3,6 | 4 | 2,7 | 6 | 4,5 | 7 |
| 2 | 8,2 | 2 | 4,7 | 5 | 3,1 | 7 | 5,6 | 1 |
| 3 | 8,3 | 3 | 5,1 | 6 | 4,2 | 1 | 6,7 | 2 |
| 4 | 8,4 | 4 | 6,2 | 7 | 5,3 | 2 | 7,1 | 3 |
| 5 | 8,5 | 5 | 7,3 | 1 | 6,4 | 3 | 1,2 | 4 |
| 6 | 8,6 | 6 | 1,4 | 2 | 7,5 | 4 | 2,3 | 5 |
| 7 | 8,7 | 7 | 2,5 | 3 | 1,6 | 5 | 3,4 | 6 |

Parker and Mood (1955) credit a General Gruenther with this rotation.

Before we proceed with the actual construction of particular balanced Room squares, we investigate their block design properties.

**Lemma 9.4.1** *The parameters of the block design $D$ of a balanced Room square of order $2n$ are $v = 2n$, $b = 2(2n - 1)$, $r = 2n - 1$, $k = n$, $\lambda = n - 1$. Further, $D$ is self-complementary.*

**Proof** Certainly $v = 2n$, and $b = 2(2n - 1)$ since each of the $2n - 1$ columns of the BRS yields two blocks. Also $k = n$ is obvious. Then $bk = vr$ gives $r = 2n - 1$ and $\lambda(v - 1) = r(k - 1)$ gives $\lambda = n - 1$. Finally $D$ is self-complementary since if a block consists of, say, the teams playing a given board NS, then the complement of the block is the set of teams playing that board EW, and hence is also a block.

**Lemma 9.4.2** (Schellenberg, 1973) *The block design D of a BRS of order 2n is a 3-design; every triple of elements is contained in exactly $\frac{1}{2}(n-2)$ blocks.*

**Proof** Let $S_{i,...,j}$ denote the set of blocks containing $i,...,j$. Then $|S_u - (S_v \cup S_w)|$—the number of blocks containing $u$ but neither of $v,w$— is also the number of blocks containing $v,w$ but not $u$ (since $D$ is self-complementary), $|S_{vw} - S_u|$. Thus

$$|S_u| - |S_{uv}| - |S_{uw}| + |S_{uvw}| = |S_{vw}| - |S_{uvw}|,$$

i.e.

$$2|S_{uvw}| = |S_{vw}| + |S_{uv}| + |S_{uw}| - |S_u|$$
$$= 3\lambda - r = 3(n-1) - (2n-1) = n-2.$$

**Example 9.4.3** Example 9.4.2 is just Example 6.3.1!

Note that this result shows that $n$ must be even.

**Corollary 9.4.3** *A balanced Room square of order 2n can exist only if n is even.*

But a stronger result is possible. Since $n$ is even, put $n = 2m$, so that the order of the square is $2n = 4m$.

**Theorem 9.4.4** *A balanced Room square of order 4m can exist only if a Hadamard matrix of order 4m exists.*

**Proof** Suppose that a BRS of order $4m$ exists. Then its block design $D$ is a $(4m, 8m - 2, 4m - 1, 2m, 2m - 1)$ design. For any element $u$ let $S_u$ denote the set of all blocks containing $u$. Remove $u$ from each block in $S_u$ and throw away all other blocks of $D$: this gives a design $E$ with $v = 4m - 1$, $b = 4m - 1$, $r = 2m - 1$, $k = 2m - 1$. If $x, y$ are any two elements, the number of blocks of $E$ containing both $x$ and $y$ is just $|S_{uxy}| = \frac{1}{2}(2m - 2) = m - 1$ by Lemma 9.4.2. So $E$ is balanced, with $\lambda = m - 1$, and so is a Hadamard design of order $4m - 1$. The result now follows by Theorem 3.2.4.

In the light of this result it comes as no surprise to learn that the problem of determining those values of $m$ for which a BRS of order $4m$ exists is far from being solved. However, some infinite families of balanced Room squares are known. For example, Hwang (1970) showed that a BRS of order $4m$ exists whenever $4m - 1 > 3$ is a prime power $q$; Schellenberg (1973) gave a starter–adder proof of this result which we now describe. For this, we need a balanced version of a starter.

**Definition 9.4.6** A *balanced starter* $X$ in an abelian group $G$ of order $4m - 1$ is a set $X = \{(x_1, y_1), \ldots, (x_{2m-1}, y_{2m-1})\}$ of ordered pairs of elements of $G$ such that;

(i) the unordered pairs $\{x_i, y_i\}$ form a starter, and
(ii) the blocks $\{0, x_1, \ldots, x_{2m-1}\}$ and $\{y_1, \ldots, y_{2m-1}\}$ form a difference system in $G$.

Note that the concurrence number $\lambda$ associated with the difference system must satisfy $2m(2m - 1) + (2m - 1)(2m - 2) = \lambda(4m - 2)$ (compare Lemma 2.2.1), i.e. $\lambda = 2m - 1$.

**Definition 9.4.7** A balanced starter $X$ is *strong* if the element $a_i = -x_i - y_i$ form an adder for $X$.

**Theorem 9.4.5** *If a strong balanced starter in an abelian group $G$ of order $4m - 1$ exists then a balanced Room square of order $4m$ exists.*

**Proof** Let $\{(x_1, y_1), \ldots, (x_{2m-1}, y_{2m-1})\}$ be a balanced starter with adder $a_i = -x_i - y_i$. Form a square array of side $4m - 1$, with rows and columns labelled by the elements of $G = \{0 = g_0, g_1, \ldots, g_{4m-2}\}$. Put $(\infty, g_i)$ in position $(g_i, g_i)$ and $(x_i + g, y_i + g)$ in position $(g, -a_i + g)$. Certainly, as in Theorem 9.2.1, we obtain an unordered Room square. But now consider the blocks given by the first and second entries of the columns. The first entries in the pairs in column $g_0$ form a block $\{\infty, x_1 + a_1, \ldots, x_{2m-1} + a_{2m-1}\} = \{\infty, -y_1, \ldots, -y_{2m-1}\}$, and the second entries form the block $\{0, y_1 + a_1, \ldots, y_{2m-1} + a_{2m-1}\} = \{0, -x_1, \ldots, -x_{2m-1}\}$. Thus the $8m - 2$ blocks obtained from the pairs in a column are

$$C_i = \{\infty, g_i - y_1, \ldots, g_i - y_{2m-1}\} \quad (0 \leqslant i \leqslant 4m - 2)$$

and

$$D_i = \{g_i - x_0, g_i - x_1, \ldots, g_i - x_{2m-1}\} \quad (0 \leqslant i \leqslant 4m - 2)$$

where we have written $x_0$ for 0. So altogether we obtain these $2(4m - 1)$ blocks of size $2m$, and we have to verify balance, i.e., show that each pair of elements occurs in precisely $2m - 1$ of these blocks. Clearly $\infty$ occurs with each element of $G$ in precisely $2m - 1$ of the blocks $C_i$. Next consider the pair $g_l, g_k$, and let $g_l - g_k = g$. Then $g_l, g_k$ occur together in a block whenever there exist $i, j, h$ such that $g_i - y_j = g_l, g_i - y_h = g_k$ or $g_i - x_j = g_l$,

$g_i - x_h = g_k$. But by property (ii) of a balanced starter, there are $2m - 1$ ways of representing $g$ as $g = z_h - z_j$ ($z = x$ or $y$), i.e. $g_l - g_k = z_h - z_j$, i.e. $g_l + z_j = g_k + z_k$; so there are indeed $2m - 1$ choices of $i, j, h$ such that $g_l + z_j = g_i = g_k + z_h$, $z = x$ or $y$.

This result yields an easy proof of Hwang's theorem.

**Theorem 9.4.6** (Hwang) *If* $q = p^\alpha \equiv 3$ (mod 4), $q > 3$, *then a balanced Room square of order* $q + 1$ *exists.*

**Proof** (Schellenberg) Let $\theta$ be a primitive element of $GF(q)$, and let $X = \{(\theta^{2i+1}, \theta^{2i}): 0 \leqslant i \leqslant \frac{1}{2}(q - 3)\}$. Then, as in Theorem 9.2.4, $X$ is a strong starter. To show that it is balanced we have to verify that $\{0, \theta, \theta^3, \ldots, \theta^{q-2}\}$ and $\{1, \theta^2, \theta^4, \ldots, \theta^{q-3}\}$ form a difference system. But this was essentially shown in the proof of Theorem 2.2.7. So $X$ is a strong balanced starter, and so, by the previous theorem, a BRS of order $q + 1$ exists.

**Example 9.4.4** Take $q = 7$, $\theta = 3$; then $X = \{(3, 1), (6, 2), (5, 4)\}$ is a strong balanced starter, with adder $3, 6, 5$, leading to the following BRS of order 8.

$$
\begin{bmatrix}
\infty 0 & 62 & 54 & - & 31 & - & - \\
- & \infty 1 & 03 & 65 & - & 42 & - \\
- & - & \infty 2 & 14 & 06 & - & 53 \\
64 & - & - & \infty 3 & 25 & 10 & - \\
- & 05 & - & - & \infty 4 & 36 & 21 \\
32 & - & 16 & - & - & \infty 5 & 40 \\
51 & 43 & - & 20 & - & - & \infty 6
\end{bmatrix}
$$

**Example 9.4.5** Take $q = 11$, $\theta = 2$: then $X = \{(2, 1), (8, 4), (10, 5), (7, 9), (6, 3)\}$ is a strong balanced starter, yielding a BRS of order 12 which is essentially the one exhibited in Example 9.1.4.

Note that Theorem 9.4.6 gives balances Room squares of orders 8, 12, 20, 24, 28, 32 but not 16. We can deal with 16 by using a 'doubling' construction due to Schellenberg.

**Theorem 9.4.7** *If* $q \equiv 3$ (mod 4), $q > 3$, *is a prime power, then there exists a BRS of order* $2(q + 1)$.

**Proof** Let $R$ be the skew BRS of order $q + 1$ construction as in Theorem 9.4.6, and let $R^*$ denote the array obtained from it by removing the main diagonal entries and replacing each remaining pair $(a, b)$ by $(a_1, b_1)$. Let

$S^*$ be the array obtained from $R^*$ by placing $(b_2, a_2)$ in position $(h, g)$ whenever $R^*$ has $(a_1, b_1)$ in position $(g, h)$. Since $R$ is skew, the resulting array $A$ has an ordered pair in each non-diagonal cell. For example, with $R$ as in Example 9.4.4, we obtain

$$
A = \begin{bmatrix}
- & 6_1,2_1 & 5_1,4_1 & 4_2,6_2 & 3_1,1_1 & 2_2,3_2 & 1_2,5_2 \\
2_2,6_2 & - & 0_1,3_1 & 6_1,5_1 & 5_2,0_2 & 4_1,2_1 & 3_2,4_2 \\
4_2,5_2 & 3_2,0_2 & - & 1_1,4_1 & 0_1,6_1 & 6_2,1_2 & 5_1,3_1 \\
6_1,4_1 & 5_2,6_2 & 4_2,1_2 & - & 2_1,5_1 & 1_1,0_1 & 0_2,2_2 \\
1_2,3_2 & 0_1,5_1 & 6_2,0_2 & 5_2,2_2 & - & 3_1,6_1 & 2_1,1_1 \\
3_1,2_1 & 2_2,4_2 & 1_1,6_1 & 0_2,1_2 & 6_2,3_2 & - & 4_1,0_1 \\
5_1,1_1 & 4_1,3_1 & 3_2,5_2 & 2_1,0_1 & 1_2,2_2 & 0_2,4_2 & -
\end{bmatrix}.
$$

Next obtain $B$ from $A$ by putting $(g_1, g_2)$ in position $(g, g)$ for each $g \in \mathrm{GF}(q)$, and by replacing each $(a_1, b_1)$ by $(b_2, a_1)$ and each $(b_2, a_2)$ by $(b_1, a_2)$. Since the two half-column blocks obtained from the $i$th column of $R^*$ are the same as those of $S^*$, the balance property of $R^*$ goes over to give balance between elements of differing suffices. To fit in pairs involving two further elements $\infty_1$ and $\infty_2$, we proceed as follows.

Consider the array

Take a pair $(x_1, y_2)$ in row $0_2$ of $B$. If it is in cell $(0_2, h_2)$, consider all the cells in the transversal $T_1 = \{(0 + g)_2, (h + g)_2 : g \in \mathrm{GF}(q)\}$ of $B$. If the cell $(i_2, j_2)$ in $T_1$ contains $(n_1, k_2)$, $n \neq k$, move $(n_1, k_2)$ down to cell $(\infty, j_2)$, and place $(\infty_1, n_1)$ and $(\infty_2, k_2)$ in cells $(i_2, n_1)$ and $(i_2, k_1)$ respectively.

Similarly, consider all cells of a transversal $T_2$ of $B$ consisting of cells $((0 + g)_2, (l + g)_2)$ where cell $(0_2, l_2)$ contains a pair of the form $(x_2, y_1)$. If cell $(i_2, j_2)$ in $T_2$ contains $(m_2, h_1)$, $m \neq h$, move it along to cell $(i_2, \infty)$, and place $(m_2, \infty_1)$ and $(\infty_2, h_1)$ in cells $(m_1, j_2)$ and $(h_1, j_2)$ respectively.

Finally, place $(\infty_2, \infty_1)$ in cell $(\infty, \infty)$. A balanced design results.

**Example 9.4.6** A BRS of order 16.

```
 —    6₁2₁  5₁4₁  4₂6₂  3₁1₁  2₂3₂  1₂5₂ │          ∞₂0₁          0₂∞₁
2₂6₂   —    0₁3₁  6₁5₁  5₂0₂  4₁2₁  3₂4₂ │               ∞₂1₁          1₂∞₁
4₂5₂  3₂0₂   —    1₁4₁  0₁6₁  6₂1₂  5₁3₁ │                    ∞₂2₁          2₂∞₁
6₁4₁  5₂6₂  4₂1₂   —    2₁5₁  1₁0₁  0₂2₂ │                         ∞₂3₁          3₂∞₁
1₂3₂  0₁5₁  6₂0₂  5₂2₂   —    3₁6₁  2₁1₁ │ 4₂∞₁                         ∞₂4₁
3₁2₁  2₂4₂  1₁6₁  0₂1₂  6₂3₂   —    4₁0₁ │      5₂∞₁                         ∞₂5₁
5₁1₁  4₁3₁  3₂5₂  2₁0₁  1₂2₂  0₂4₂   —   │ ∞₂6₁      6₂∞₁
────────────────────────────────────────┼──────────────────────────────────────────────
      ∞₁1₁                    ∞₁5₂       │ 0₁0₂  2₂6₁  4₂5₁  4₁6₂   —    2₁3₂   —   │ 1₂3₁
           ∞₁2₁                    ∞₁6₂  │  —    1₁1₂  3₂0₁  5₂6₁  5₁0₂   —    3₁4₂ │ 2₂4₁
∞₂0₂            ∞₁3₁                     │ 4₁5₂   —    2₁2₂  4₂1₁  6₂0₁  6₁1₂   —   │ 3₂5₁
      ∞₁2₂                 ∞₁4₁          │  —    5₁6₂   —    3₁3₂  5₂2₁  0₂1₁  0₁2₂ │ 4₂6₁
           ∞₂2₂                 ∞₁5₁     │ 1₁3₂   —    6₁0₂   —    4₁4₂  6₂3₁  1₂2₁ │ 5₂0₁
                ∞₂3₂                 ∞₁6₁│ 2₂3₁  2₁4₂   —    0₁1₂   —    5₁5₂  0₂4₁ │ 6₂1₁
∞₁0₁                      ∞₂4₂           │ 1₂5₁  3₂4₁  3₁5₂   —    1₁2₂   —    6₁6₂ │ 0₂2₁
─────────────────────────────────────────┼──────────────────────────────────────────────
                                          │ 2₁6₂  3₁0₂  4₁1₂  5₁2₂  6₁3₂  0₁4₂  1₁5₂ │ ∞₂∞₁
```

More recent progress in the construction of BRS has been based on the use of a special class of starters called *symmetric skew balanced starters*.

**Definition 9.4.8** Let $q = 4m + 1$ be a prime power. Then an ordered starter $(x_1, y_1), \ldots, (x_{2m}, y_{2m})$ in $GF(q)$ is called a *symmetric skew balanced starter* (SSBS) if

(i) $\{\pm(x_i + y_i)\} = GF(q) - \{0\}$;
(ii) $\{x_1, \ldots, x_{2m}\} = \{-x_1, \ldots, -x_{2m}\}$;
(iii) $\{x_1, \ldots, x_{2m}\}, \{y_1, \ldots, y_{2m}\}$ form a difference system.

Note that (i) is the skew property, (ii) is the symmetric property, and (iii) is a balance property slightly different from that in definition 9.4.6.

**Example 9.4.7** $(1, 11), (3, 7), (4, 2), (9, 8), (10, 5), (12, 6)$ is a SSBS in $Z_{13}$.

It is now known (see Du and Hwang (1988)) that an SSBS exists for all prime powers $q \equiv 1 \pmod 4$ which are not Fermat primes. We do not give an account of this work, which depends on the theory of character sums, but limit ourselves to constructing SSBS for primes $p \equiv 5 \pmod 8$.

**Theorem 9.4.8** Let $p = 8k + 5$ be a prime, $p \neq 5$. Then there exists a primitive root $\theta$ of $p$ such that the pairs $(\theta^{4i}, -\theta^{4i+1})$, $(\theta^{4i+2}, \theta^{4i+1})$, $0 \leqslant i \leqslant 2k$, form an SSBS.

**Proof** It can be shown that for all such $p$ there exists a primitive root $\theta$ such that $\theta^2 - 1$ is a square (mod $p$). Now the differences arising from the given pairs are $\pm\theta^{4i}(\theta + 1)$ and $\pm\theta^{4i+1}(\theta - 1)$, where $\theta^{4k+2} = -1$; so we have a starter provided $\theta^2 - 1$ is a square. A similar argument with the sums confirms condition (i). Condition (ii) is satisfied since $\theta^{4k+2} = -1$, and condition (iii) is satisfied as in Theorem 2.2.5.

**Example 9.4.8**  Take $p = 13$, $\theta = 2$ to obtain the SSBS of Example 9.4.7.

A BRS of order 28 can be obtained from the SSBS of Example 9.4.7 in a way similar to Example 9.4.6, taking $A$ to have first row

$$-2_2 12_2\ 10_1 5_1\ 6_2 10_2\ 9_1 8_1\ 12_1 6_1\ 4_1 2_1\ 11_2 9_2\ 7_2 1_2\ 5_2 4_2\ 3_1 7_1\ 8_2 3_2\ 1_1 11_1$$

and $B$ to have first row

$$0_2 0_1\ 12_2 2_1\ 10_1 5_2\ 10_2 6_1\ 9_2 8_1\ 12_1 6_2\ 4_1 2_2\ 9_1 11_2\ 1_1 7_2\ 4_2 5_1\ 3_2 7_1\ 3_1 8_2\ 1_2 11_1,$$

and then rearranging two transversals of $B$ as before. The resulting array takes advantage of the balance properties of the squares and the non-squares as in Theorem 2.2.5. Note that $A$ uses the skew Room square obtained from a skew starter, and its transpose.

## 9.5 Exercises 9

1. Find an adder for each of the following starters, and hence construct Room squares of orders 16 and 18:
   (a) $\{2,7\}$, $\{10,13\}$, $\{6,12\}$, $\{4,5\}$, $\{14,3\}$, $\{9,11\}$, $\{1,8\}$ in $Z_{15}$;
   (b) $\{1,16\}$, $\{2,15\}$, $\{3,14\}$, $\{4,13\}$, $\{5,12\}$, $\{6,11\}$, $\{7,10\}$, $\{8,9\}$ in $Z_{17}$.
2. Use Theorem 9.2.4 to find strong starters in $Z_{23}$ and $Z_{31}$.
3. Suppose that a Room square of side $2n - 1$ is obtained from a starter in $Z_{2n-1}$ with adder $a_1, \ldots, a_{n-1}$. Show that $\sum_{i=1}^{n-1} a_i \equiv 0 \pmod{2n-1}$. Show also that $\sum_{i=1}^{n-1} a_i^2 \equiv -\sum_{i=1}^{n-1} a_i(x_i + y_i) \pmod{2n-1}$. Does this help in positioning $\{1,6\}$, $\{2,5\}$, and $\{3,4\}$ in the first row of a Room square of order 8?
4. Let $p^n = 6t + 1$ and let $\theta$ be a primitive element of $GF(q^n)$. Show that the pairs $\{\theta^{2t+i} - \theta^i, \theta^{4t+i} - \theta^i\}$, $\{\theta^i - \theta^{2t+i}, \theta^{4t+i} - \theta^{2t+i}\}$, $\{\theta^i - \theta^{4t+i}, \theta^{2t+i} - \theta^{4t+i}\}$, $0 \le i \le t-1$, form a strong starter in $GF(p^n)$. Hence find Room squares of orders 8 and 14. (Mullin and Nemeth, 1969)
5. Suppose that $\{\{x_1, y_1\}, \ldots, \{x_s, y_s\}\}$ is a strong starter in $GF(2s + 1)$. Show that the elements $a_i = -\frac{1}{2}(x_i + y_i)$ form an adder.
6. The starter in $Z_{17}$ exhibited in exercise 1(b) above is an example of a *patterned* starter, i.e. one whose pairs are all of the form $\{x, -x\}$. (Compare Exercises 2, number 11.) Given a strong starter in $GF(2s + 1)$, construct a Room square of order $2s + 2$ by using the adder of the previous question, and show that the transpose of this Room square has a patterned starter. Deduce that there exists a patterned starter

with adder in GF($q$) for all prime powers $q$ covered by Theorem 9.2.5 and 9.2.6. (Byleen, 1970)

7. *Orthogonal starters.* Note that if $X = \{\{x_1, y_1\}, \ldots, \{x_s, y_s\}\}$ and $Y = \{\{u_1, v_1\}, \ldots, \{u_s, v_s\}\}$ are two starters in $Z_{2s+1}$, then, for each $i$, there is a unique $j$ such that $x_i - y_i = \pm(u_j - v_j)$. Without loss of generality, we can suppose that $x_i - y_i = u_j - v_j$. Let $d_i = u_j - x_i = v_j - y_i$. Then if the $d_i$ are all distinct and non-zero we say that $X$ and $Y$ are *orthogonal* starters.

   (i) Show that $X, Y$ are orthogonal then $A = \{d_1, \ldots, d_s\}$ is an adder for $X$. Illustrate with $X = \{\{1, 3\}, \{2, 6\}, \{4, 5\}\}$.

   (ii) Show that if $Z_{2s+1}$ has a strong starter then there are three pairwise orthogonal starters. (If $X = \{\{x_i, y_i\}\}$ is a strong starter consider $Y = \{\{-x_1, -y_1\}\}$ and $H = \{\{i, -i\}\}$.)

   (iii) See number 9 for application. (Horton, 1981)

8. *Orthogonal league schedules.* Let $R$ be a Room square of side $r$ based on $\{0, 1, \ldots, r\}$. Then the pairs in the $r$ rows form a league schedule for $r + 1$ teams, called the *row schedule* of $R$. Similarly the columns give a league schedule called the *column schedule* of $R$. These two schedules are orthogonal, i.e., any resolution class of one schedule has at most one pair in common with any class of the other schedule. Conversely, any two orthogonal schedules for $r + 1$ teams yield a Room square of side $r$.

9. *Orthogonal symmetric Latin squares.*

   (i) Let $\mathscr{F} = \{F_1, \ldots, F_r\}$ be the set of resolution classes of a resolvable $(r + 1, 2, 1)$ design $\{0, 1, \ldots, r\}$, where $\{0, i\} \in F_i$. Define an $r \times r$ array $F = (f_{ij})$ by

$$f_{ii} = i;$$
$$f_{ij} = k \text{ where } \{i, j\} \in F_k.$$

   Then $F$ is a symmetric Latin square called the *Latin square of $\mathscr{F}$*.

   (ii) Let $F, G$ be the Latin squares of $\mathscr{F}, \mathscr{G}$, where $\mathscr{F}, \mathscr{G}$ are the row and column schedules of a Room square $R$ of side $r$ with $\{0, i\}$ in the $i$th diagonal place. Show that $F, G$ have the following properties:

      (a) for each $i$, there is just one position in which $F$ and $G$ both have $i$, and this is on the main diagonal;

      (b) if $i \neq j$, there is at most one position above the main diagonal in which $F$ has $i$ and $G$ has $j$.

   (iii) Any two symmetric Latin squares with properties (a) and (b) are called *orthogonal symmetric Latin squares* (OSLS). It follows that there exist OSLS of all odd orders $r \geq 7$. Show that there can exist at most $r - 2$ mutually OSLS of order $r$.

   (iv) Prove that orthogonal starters give rise to orthogonal schedules and hence to OSLS. Deduce from question 7 that there exist at least three MOSLS of all odd prime orders $q \geq 7$.

10. Show that if $q = p^n \equiv 3 \pmod 4$ then there exist at least $\frac{1}{2}(q-1)$ MOSLS of order $q$. Hint: let $Q$ and $N$ denote the sets of non-zero squares and non-squares of GF($q$) and let $X_t = \{\{s, st\}: s \in Q\}$ for each $t \in N$; verify that each $X_t$ is a starter and that they are orthogonal. (Horton, 1981)

11. Prove that a Room square arising from a skew starter is skew, by observing that $(x_i, y_i)$ in cell $(0, x_i + y_i)$ leads to $(-y_i, -x_i)$ in cell $(-x_i - y_i, 0)$.

12. Skew Room squares proved useful in the search for constructions of Room squares of all odd sides of at least 7. As an example, we show now that if there exists a skew Room square $R$ of side $r$, then there exists a Room square of side $2r - 1$. Let $R$ be on $\{\infty, 1, 2, \ldots, r\}$.

    (i) Take two MOLS $L, M$ of size $r - 1$ with diagonal $1, 2, \ldots, r - 1$ (e.g. a SOLS $L, M = L'$). Obtain $N$ from $M$ by adding $r$ to each entry, and let $B$ denote the join of $L$ and $N$ with all diagonal entries deleted.

    (ii) Take the sum of $R$ and its transpose, delete every entry involving $r$ or $2r$, and introduce $\{i, r + i\}$ in the $i$th diagonal cell, $1 \leq i \leq r$. Call the resulting array $A$.

    (iii) Let $C = \begin{bmatrix} A & \overline{\phantom{x}} \\ \overline{\phantom{x}} & B \end{bmatrix}$. Then $C$ contains every unordered pair of elements of $S = \{1, \ldots, 2r\}$ except those pairs involving one of $r, 2r$. Finally, if $\{i, r\}$ is in cell $(x_i, y_i)$ of $R$, now insert $\{i, r\}$ in cell $\{x_i, i + r\}$, $\{i, 2r\}$ in cell $(i + r, y_i)$, $\{i + r, r\}$ in cell $(i + r, x_i)$ and $\{i + r, 2r\}$ in cell $(y_i, i + r)$. The resulting array is a Room square of order $2r$. (W. D. Wallis, 1972)

13. If $R$ is a Room square of side $r$ containing an $s \times s$ subarray $S$ which is itself a Room square, we call $S$ a *subsquare* of $R$. Show that if a subsquare exists with $r > s$, then in fact $r \geq 3s + 2$, as follows. Suppose that $R = \begin{bmatrix} S & A \\ B & C \end{bmatrix}$ where $R, S$ are based on sets $X, Y$ with $Y \subset X$. Let $x \in X - Y$. If pair $\{x, x_i\}$ occurs in row $i$ of $R$ and pair $\{x, y_i\}$ occurs in column $i$, then none of the elements $x_i, y_i, i \leq s$, can be in $S$. So $r \geq 3s + 1$; but $r, s$ are odd, so $r \geq 3s + 2$. (Note: this bound $3s + 2$ is the best possible for infinitely many values of $s$.)

14. Show that no skew Room square of side $3k$ can be obtained from a patterned starter. (Note that by exercise 3 the adder would satisfy $\Sigma_i a_i^2 \equiv 0 \pmod{3k}$.) (Constable, 1974).

15. Use Theorem 9.4.6 to construct BRS of orders 20 and 24.

16. Use Theorem 9.4.7 to construct a BRS of order 24.

17. Construct an SSBS in $Z_{29}$. (Take $\theta = 8$ in Theorem 9.4.8.)

# 10

# Balanced tournament designs

## 10.1 Factored balanced tournament designs

The existence of Room squares of order $2n$ for all $n \geqslant 4$ permits the construction of a league schedule for $2n \geqslant 8$ teams on $2n - 1$ days with each team playing exactly once on each of $2n - 1$ grounds. On any given day only $n$ of the $2n - 1$ grounds are used. Suppose now that just $n$ grounds are available. Then, since $2n - 1$ is not a multiple of $n$, it is certainly not possible to assign games to grounds so that each team plays the same number of games on each ground. But is it possible to arrange for each team to play *at most twice* on each ground? If so, then each team will play once on one ground and twice on all other grounds.

**Definition 10.1.1** A *balanced tournament design* for $2n$ teams (BTD($2n$)) is a league schedule in which the same $n$ grounds are used in each round such that each team plays at most twice on each ground.

**Example 10.1.1** A BTD(8) due to E. Nemeth.

|  |  | Grounds | | | |
|---|---|---|---|---|---|
|  |  | 1 | 2 | 3 | 4 |
|  | 1 | 1,2 | 3,5 | 4,7 | 6,0 |
|  | 2 | 6,7 | 1,3 | 4,0 | 2,5 |
|  | 3 | 0,3 | 5,7 | 2,6 | 1,4 |
| Rounds | 4 | 7,0 | 4,6 | 1,5 | 2,3 |
|  | 5 | 3,4 | 1,6 | 2,7 | 5,0 |
|  | 6 | 4,5 | 2,0 | 3,6 | 1,7 |
|  | 7 | 5,6 | 2,4 | 1,0 | 3,7 |

Clearly, a BTD($2n$) can be considered as a $(2n - 1) \times n$ array in which each position contains an unordered pair of elements from $\{0, 1, \ldots, 2n - 1\}$ such that

(i)   each element occurs once in each row;
(ii)  each element occurs at most twice in each column;
(iii) each unordered pair occurs in exactly one position.

It is the aim of this section to prove that a BTD($2n$) exists for all $n \neq 2$. The first BTDs to be constructed were for $n \equiv 0$ or $1 \pmod 3$; these were obtained by Haselgrove and Leech in 1977. The existence for all $n \neq 2$ was shortly afterwards established by Schellenberg *et al.* (1977). In this section we present a simplification of their proof, due to Lamken and Vanstone (1985); this proof involves the idea of a factored BTD.

**Definition 10.1.2** A balanced tournament design is *factored* if, for each of the $n$ grounds, there are $n$ games played on that ground which together involve each of the $2n$ teams. The $n$ pairs of teams are said to form a *factor*.

**Example 10.1.2** The BTD(8) of Example 10.1.1 is a factored balanced tournament design (FBTD): the pairs in the factors are underlined below.

| | | | |
|---|---|---|---|
| 1,2 | 3,5 | 4,7 | 6,0 |
| 6,7 | 1,3 | 4,0 | 2,5 |
| 0,3 | 5,7 | 2,6 | 1,4 |
| 7,0 | 4,6 | 1,5 | 2,3 |
| 3,4 | 1,6 | 2,7 | 5,0 |
| 4,5 | 2,0 | 3,6 | 1,7 |
| 5,6 | 2,4 | 1,0 | 3,7 |

**Example 10.1.3** (Lamken and Vanstone)   An FBTD(12). In this example, X and E stand for 10 and 11; factors are again indicated.

| | | | | | |
|---|---|---|---|---|---|
| E,8 | 6,9 | X,7 | 4,0 | 5,3 | 1,2 |
| 4,6 | E,9 | 7,0 | X,8 | 5,1 | 2,3 |
| 6,2 | 5,7 | E,0 | 8,1 | X,9 | 3,4 |
| X,0 | 7,3 | 6,8 | E,1 | 9,2 | 4,5 |
| 0,3 | X,1 | 8,4 | 7,9 | E,2 | 5,6 |
| E,3 | 1,4 | X,2 | 9,5 | 8,0 | 6,7 |
| 9,1 | E,4 | 2,5 | X,3 | 0,6 | 7,8 |
| 1,7 | 0,2 | E,5 | 3,6 | X,4 | 8,9 |
| X,5 | 2,8 | 1,3 | E,6 | 4,7 | 9,0 |
| 5,8 | X,6 | 3,9 | 4,2 | E,7 | 0,1 |
| 2,7 | 3,8 | 4,9 | 5,0 | 6,1 | X,E |

Instead of just proving that a BTD($2n$) exists for all $n \neq 2$, we shall in fact prove that an FBTD($2n$) exists for all $n \neq 2$. The first step is to deal with the odd values of $n$.

**Theorem 10.1.1** *If $n$ is odd, an FBTD($2n$) exists.*

**Proof** Let $n = 2k + 1$ so that there are $4k + 2$ teams, $4k + 1$ rounds, and $2k + 1$ grounds. We shall construct a $(4k + 1) \times (2k + 1)$ array whose rows are labelled by $0, 1, \ldots, 4k$ (considered as elements of $Z_{4k+1}$) and whose columns are labelled by $0, 1, \ldots, 2k$ (considered as elements of $Z_{2k+1}$). The $4k + 2$ teams will be labelled by $0_1, 0_2, 1_1, 1_2, \ldots, (2k)_1, (2k)_2$, where $0, \ldots, 2k$ are again considered as elements of $Z_{2k+1}$. We start by assigning pairs to column 0 as follows, and then obtain column $i$ by replacing each pair $x_u, y_v$ by $(x + i)_u, (y + i)_v$.

|        | 0 | 1 | $\cdots$ | $2k$ |
|--------|---|---|----------|------|
| 0      | $0_1, 0_2$ | $1_1, 1_2$ | $\cdots$ | $(2k)_1, (2k)_2$ |
| 1      | $1_1, (2k)_1$ | $2_1, 0_1$ | | $0_1, (2k-1)_1$ |
| 2      | $2_1, (2k-1)_1$ | $3_1, (2k)_1$ | | $1_1, (2k-2)_1$ |
| $\vdots$ | | | | |
| $k$    | $k_1, (k+1)_1$ | $(k+1)_1, (k+2)_1$ | | $(k-1)_1, k_1$ |
| $k+1$  | $1_2, (2k)_2$ | $2_2, 0_2$ | | $0_2, (2k-1)_2$ |
| $\vdots$ | | | | |
| $2k$   | $k_2, (k+1)_2$ | $(k+1)_2, (k+2)_2$ | | $(k-1)_2, k_2$ |
| $2k+1$ | $1_1, (2k)_2$ | $2_1, 0_2$ | | $0_1, (2k-1)_2$ |
| $\vdots$ | | | | |
| $3k$   | $k_1, (k+1)_2$ | $(k+1)_1, (k+2)_2$ | | $(k-1)_1, k_2$ |
| $3k+1$ | $1_2, (2k)_1$ | $2_2, 0_1$ | | $0_2, (2k-1)_1$ |
| $\vdots$ | | | | |
| $4k$   | $k_2, (k+1)_1$ | $(k+1)_2, (k+2)_1$ | $\cdots$ | $(k-1)_2, k_1$ |

Certainly no column contains any team more than twice. As far as the rows are concerned, row 0 and rows $2k + 1$ to $4k$ each contain each team once, but rows 1 to $2k$ are unsatisfactory and require alteration. We show how to move entries around so as to obtain the required properties.

Consider the pairs in row $i$, $1 \leqslant i \leqslant k$; they are of the form $\{x_1, y_1\}$ where $y - x = 2k + 1 - 2i \equiv -2i \pmod{2k + 1}$. So row $i$ contains the pairs $\{x_1, (x - 2i)_1\}, \{(x - 2i)_1, (x - 4i)_1\}, \ldots, \{(x - (t - 1)2i)_1, x_1\}$ where $t$ is the period of $-2i \pmod{2k + 1}$. Now the period $t$ must divide $2k + 1$ and hence must be odd; and the pairs in row $i$ will form $(2k + 1)/t$ cosets each of $t$ pairs. The same happens in rows $k + i$ and $2k + i$.

We break off the proof of the theorem to illustrate the above by an example.

**Example 10.1.4** Take $n = 2k + 1 = 9$. Row 3 in the array is

$$\{3_1, 6_1\} \{4_1, 7_1\} \{5_1, 8_1\} \{6_1, 0_1\} \{7_1, 1_1\} \{8_1, 2_1\} \{0_1, 3_1\} \{1_1, 4_1\} \{2_1, 5_1\}.$$

Here $-2i = -6 = 3$ which has period 3; so the pairs fall into three cosets:

$$\{\{3_1, 6_1\}, \{6_1, 0_1\}, \{0_1, 3_1\}\}, \{\{4_1, 7_1\}, \{7_1, 1_1\}, \{1_1, 4_1\}\},$$
$$\{\{5_1, 8_1\}, \{8_1, 2_1\}, \{2_1, 5_1\}\}.$$

A similar thing happens in row $3 + 4 = 7$ and $3 + 8 = 11$. We thus obtain sets of cosets such as:

$$
\begin{array}{ccc}
\{3_1, 6_2\} & \{6_1, 0_2\} & \{0_1, 3_2\} \\
\updownarrow & & \\
\{3_2, 6_2\} & \{6_2, 0_2\} & \{0_2, 3_2\} \\
\updownarrow & & \\
\{3_1, 6_1\} & \{6_1, 0_1\} & \{0_1, 3_1\}
\end{array}
$$

Suppose we interchange $\{6_1, 0_1\}$ and $\{6_2, 0_2\}$, interchange $\{0_1, 3_1\}$ and $\{0_1, 3_2\}$, and finally interchange $\{3_2, 6_2\}$ and $\{3_1, 6_2\}$. This is shown by the arrows, and leads to

$$
\begin{array}{ccc}
\{3_1, 6_1\} & \{6_2, 0_2\} & \{0_1, 3_2\} \\
\{3_1, 6_2\} & \{6_1, 0_1\} & \{0_2, 3_2\} \\
\{3_2, 6_2\} & \{6_1, 0_2\} & \{0_1, 3_1\}
\end{array}
$$

which has each row containing each element once. Since interchanges are all within columns, the column properties of the array remain unaltered.

We can now complete the proof of the theorem by imitating this process. If there is a coset.

$$\{x, x - 2i\} \{x - 2i, x - 4i\} \{x - 4i, x - 6i\} \cdots \{x - (t - 1)2i, x\}$$

where $t$ is necessarily odd, then, for each odd $r$, interchange

$$\{(x - 2ri)_1, (x - 2ri - 2i)_1\} \quad \text{and} \quad \{(x - 2ri)_2, (x - 2ri - 2i)_2\}.$$

Also interchange

$$\{x_2, (x - 2i)_2\} \quad \text{and} \quad \{x_1, (x - 2i)_2\}$$

and finally interchange

$$\{(x - (t - 1)2i)_1, x_1\} \quad \text{and} \quad \{(x - (t - 1)2i)_1, x_2\}.$$

If we do this for each coset, the columns continue to contain each element at most twice, and the new rows contain each exactly once. So a BTD$(4k + 2)$ is obtained. Further, it is factored, the factors being given by the pairs in rows 0 to $2k$ in the array initially written down.

**Example 10.1.5** An FBTD(6). Here $n = 2k + 1 = 3, k = 1$. The period of $-2 \equiv 1 \pmod 3$ is 3, so we obtain

$$
\begin{array}{ccc}
0_1,0_2 & 1_1,1_2 & 2_1,2_2 \\
1_1,2_1 & 2_2,0_2 & 0_1,1_2 \\
1_1,2_2 & 2_1,0_1 & 0_2,1_2 \\
1_2,2_2 & 2_1,0_2 & 0_1,1_1 \\
1_2,2_1 & 2_2,0_1 & 0_2,1_1
\end{array}
$$

which, on replacing $x_i$ by $x + (i - 1)3$, can be represented as

$$
\begin{array}{ccc}
\underline{0,3} & \underline{1,4} & 2,5 \\
\underline{1,2} & \underline{5,3} & 0,4 \\
1,5 & 2,0 & \underline{3,4} \\
\underline{4,5} & 2,3 & \underline{0,1} \\
4,2 & 5,0 & 3,1
\end{array}
$$

**Example 10.1.6** An FBTD(10). Here $n = 2k + 1 = 5, k = 2$. This time two values of $i$ are relevant: (a)$i = 1$; $-2i = -2 = 3$ has period 5, (b)$i = 2$; $-2i = -4 = 1$ has period 5. The two cycles of period 5 are 1,4 4,2 2,0 0,3 3,1 and 2,3 3,4 4,0 0,1 1,2. The construction in the proof of the theorem therefore gives

$$
\begin{array}{ccccc}
0_1,0_2 & 1_1,1_2 & 2_1,2_2 & 3_1,3_2 & 4_1,4_2 \\
1_1,4_1 & 2_1,0_1 & 3_1,1_2 & 4_2,2_2 & 0_2,3_2 \\
2_1,3_1 & 3_2,4_2 & 4_1,0_1 & 0_2,1_2 & 1_1,2_2 \\
1_1,4_2 & 2_2,0_2 & 3_2,1_2 & 4_1,2_1 & 0_1,3_1 \\
2_1,3_2 & 3_1,4_1 & 4_2,0_2 & 0_1,1_1 & 1_2,2_2 \\
1_2,4_2 & 2_1,0_2 & 3_1,1_1 & 4_1,2_2 & 0_1,3_2 \\
2_2,3_2 & 3_1,4_2 & 4_1,0_2 & 0_1,1_2 & 1_1,2_1 \\
1_2,4_1 & 2_2,0_1 & 3_2,1_1 & 4_2,2_1 & 0_2,3_1 \\
2_2,3_1 & 3_2,4_1 & 4_2,0_1 & 0_2,1_1 & 1_2,2_1
\end{array}
$$

which can be re-expressed as

| | | | | |
|---|---|---|---|---|
| <u>0,5</u> | <u>1,6</u> | <u>2,7</u> | <u>3,8</u> | <u>4,9</u> |
| <u>1,4</u> | <u>2,0</u> | <u>3,6</u> | <u>9,7</u> | <u>5,8</u> |
| <u>2,3</u> | <u>8,9</u> | <u>4,0</u> | <u>5,6</u> | <u>1,7</u> |
| <u>1,9</u> | <u>7,5</u> | <u>8,6</u> | <u>4,2</u> | <u>0,3</u> |
| <u>2,8</u> | <u>3,4</u> | <u>9,5</u> | <u>0,1</u> | <u>6,7</u> |
| <u>6,9</u> | <u>2,5</u> | <u>3,1</u> | <u>4,7</u> | <u>0,8</u> |
| <u>7,8</u> | <u>3,9</u> | <u>4,5</u> | <u>0,6</u> | <u>1,2</u> |
| 6,4 | 7,0 | 8,1 | 9,2 | 5,3 |
| 7,3 | 8,4 | 9,0 | 5,1 | 6,2. |

To deal with $n$ even, we provide a method of doubling.

**Theorem 10.1.2** *If an* FBTD($2n$) *exists,* $n > 3$, *then an* FBTD($4n$) *also exists.*

**Proof** Let $A$ be an FBTD($2n$) on $X = \{1, \ldots, 2n\}$. Obtain $A_1$ from $A$ as follows: in each of the columns of $A$, replace each $m$ in the factor of that column by $m_1$, and replace each other element $m$ by $m_2$. Similarly, obtain $A_2$ from $A$ by, in each column, replacing each $m$ in the factor of that column by $m_2$, and each other $m$ by $m_1$. Then let $B = [A_1, A_2]$. Thus $B$ is a $(2n - 1) \times 2n$ array on $X_1 \cup X_2$, where $X_i = \{1_i, \ldots, 2n_i\}$, and all pairs of the form $\{x_1, y_1\}$ or $\{x_2, y_2\}$ $(x \neq y)$ occur precisely once in $B$. Further, each of the rows of $B$ contains each of the $4n$ elements of $X_1 \cup X_2$ exactly once, and the columns of $B$ contain each element at most once.

Now pairs of the form $\{x_1, y_2\}$ are so far missing. Take two MOLS of order $2n$, one on $X_1$ and the other on $X_2$, and let $C$ denote their join. Then $C$ is a $2n \times 2n$ array of pairs $\{x_1, y_2\}$, containing each such pair exactly once. So consider the array

$$D = \begin{bmatrix} B \\ C \end{bmatrix}.$$

$D$ is a $(4n - 1) \times 2n$ array on $X_1 \cup X_2$; it contains all of the $\binom{4n}{2} = 2n(4n - 1)$ pairs of elements of $X_1 \cup X_2$ once each; its rows contain each element once, and its columns contain each element at most twice. Thus $D$ is a BTD($4n$). But, further, it is factored, since the pairs in $C$ give a factor in each column.

**Example 10.1.7** An FBTD(16) obtained from an FBTD(8). Here we take

$n = 4$. We shall use the FBTD(8) of Example 10.1.2, and the following Graeco-Latin square of order 8 which can be obtained by the method presented in the proof of Theorem 4.1.1.

$$
\begin{array}{cccccccc}
83 & 38 & 64 & 17 & 52 & 46 & 25 & 71 \\
36 & 84 & 48 & 75 & 21 & 63 & 57 & 12 \\
61 & 47 & 85 & 58 & 16 & 32 & 74 & 23 \\
15 & 72 & 51 & 86 & 68 & 27 & 43 & 34 \\
54 & 26 & 13 & 62 & 87 & 78 & 31 & 45 \\
42 & 65 & 37 & 24 & 73 & 81 & 18 & 56 \\
28 & 53 & 76 & 41 & 35 & 14 & 82 & 67 \\
77 & 11 & 22 & 33 & 44 & 55 & 66 & 88
\end{array}
$$

We obtain the following FBTD(16):

$$
\begin{array}{cccccccc}
1_12_1 & 3_25_2 & 4_27_2 & 6_18_1 & 1_22_2 & 3_15_1 & 4_17_1 & 6_28_2 \\
6_27_2 & 1_13_1 & 4_18_1 & 2_15_1 & 6_17_1 & 1_23_2 & 4_28_2 & 2_25_2 \\
8_23_2 & 5_17_1 & 2_26_2 & 1_14_1 & 8_13_1 & 5_27_2 & 2_16_1 & 1_24_2 \\
7_18_1 & 4_16_1 & 1_15_1 & 2_23_2 & 7_28_2 & 4_26_2 & 1_25_2 & 2_13_1 \\
3_14_1 & 1_26_2 & 2_17_1 & 5_28_2 & 3_24_2 & 1_16_1 & 2_27_2 & 5_18_1 \\
4_25_2 & 2_18_1 & 3_16_1 & 1_27_2 & 4_15_1 & 2_28_2 & 3_26_2 & 1_17_1 \\
5_16_1 & 2_24_2 & 1_28_2 & 3_17_1 & 5_26_2 & 2_14_1 & 1_18_1 & 3_27_2 \\
8_13_2 & 3_18_2 & 6_14_2 & 1_17_2 & 5_12_2 & 4_16_2 & 2_15_2 & 7_11_2 \\
3_16_2 & 8_14_2 & 4_18_2 & 7_15_2 & 2_11_2 & 6_13_2 & 5_17_2 & 1_12_2 \\
6_11_2 & 4_17_2 & 8_15_2 & 5_18_2 & 1_16_2 & 3_12_2 & 7_14_2 & 2_13_2 \\
1_15_2 & 7_12_2 & 5_11_2 & 8_16_2 & 6_18_2 & 2_17_2 & 4_13_2 & 3_14_2 \\
5_14_2 & 2_16_2 & 1_13_2 & 6_12_2 & 8_17_2 & 7_18_2 & 3_11_2 & 4_15_2 \\
4_12_2 & 6_15_2 & 3_17_2 & 2_14_2 & 7_13_2 & 8_11_2 & 1_18_2 & 5_16_2 \\
2_18_2 & 5_13_2 & 7_16_2 & 4_11_2 & 3_15_2 & 1_14_2 & 8_12_2 & 6_17_2 \\
7_17_2 & 1_11_2 & 2_12_2 & 3_13_2 & 4_14_2 & 5_15_2 & 6_16_2 & 8_18_2
\end{array}
$$

We can now complete the proof of the existence of an FBTD($2n$) for each $n \neq 2$.

**Theorem 10.1.3** *An FBTD($2n$) exists for all $n \neq 2$.*

**Proof** For odd $n$, existence is given by Theorem 10.1.1. Then Theorem 10.1.2 gives existence for all $n = 2^r m$, $m$ odd, $m > 3$, and, in view of Example 10.1.1, for all $n = 2^r, r \geqslant 2$. Finally consider $n = 3 \cdot 2^r$; for $r = 1$ use Example 10.1.3, and then all $r > 1$ yield to Theorem 10.1.2.

This completes the proof that a BTD($2n$) exists for all $n \neq 2$. The corresponding problem for an odd number of teams is much more easily dealt with: see Exercises 10, number 3.

## 10.2 Partitioned balanced tournament designs

It has been shown that, provided $n \neq 2$, it is possible to construct a balanced tournament design for $2n$ teams. Each team plays in each of the $2n - 1$ rounds, and plays on all but one of the $n$ grounds twice. Clearly, for each ground, there are two teams which only play on that ground once. In other words, in each column $C$ of a BTD($2n$) there are two elements which occur only once in $C$. We call these the *deficient* elements of $C$. If, for each column $C$, the two deficient elements actually occur as a pair in $C$, we say that the design is *linked*; then in the corresponding tournament, the two teams which play only once on a particular ground actually play against each other on that ground.

**Definition 10.2.1** A BTD is *linked* if each column contains its two deficient elements as a pair.

**Example 10.2.1** (a) The FBTD(6) of Example 10.1.5 is linked: the deficient pairs in the three columns are $\{0, 3\}$, $\{1, 4\}$, $\{2, 5\}$.
(b) On the other hand, the BTD(8) of Example 10.1.1 is not linked; for example, the deficient elements of the second column are 0 and 7, and the pair $\{0, 7\}$ does *not* occur in that column.

Here is another linked design, a BTD(10), in which the deficient pairs occur in the middle row.

**Example 10.2.2** (Stinson, 1985)

| | | | | |
|---|---|---|---|---|
| 0,7 | 1,2 | 5,6 | 8,3 | 9,4 |
| 1,6 | 4,7 | 0,3 | 9,5 | 8,2 |
| 2,5 | 8,0 | 9,7 | 4,6 | 1,3 |
| 4,3 | 9,6 | 8,1 | 0,2 | 5,7 |
| 8,9 | 5,3 | 4,2 | 1,7 | 0,6 |
| 1,5 | 9,0 | 6,7 | 8,4 | 2,3 |
| 3,7 | 8,6 | 0,1 | 9,2 | 4,5 |
| 2,6 | 1,4 | 9,3 | 0,5 | 8,7 |
| 0,4 | 7,2 | 8,5 | 6,3 | 9,1 |

This design has a further remarkable property: the middle row and the rows above it form factors in each column, and (hence!) also the middle row and the rows below it form factors in each column. The design is therefore said to be *partitioned*.

**Definition 10.2.2** A BTD($2n$) is said to be a *partitioned balanced tournament design* (PBTD($2n$)) if the rows can be permuted so that in each column the first $n$ pairs form a factor, as do the last $n$ pairs.

Clearly, any PBTD is automatically an FBTD; but the converse does not hold, since an FBTD need not even be linked (for example, the FBTD(8) of Example 10.1.2). No PBTD(8) in fact exists. It is now known (see Lamken, 1996) that a PBTD($2n$) exists for all $n \geqslant 5$ except possibly for $n = 9, 11, 15$. We do not prove this result here, but simply indicate an interesting connection between PBTDs and Room squares.

In any Room square of order $2n$ there are $2n - 1$ rows and columns, and in each row or column precisely $n - 1$ of the cells are empty. The biggest square array of empty cells in any Room square of order $2n$ is therefore of size $m$ where $m \leqslant n - 1$. Can a Room square of order $2n$ with as large a 'hole' as a square of size $n - 1$ be constructed? It was Stinson who, in 1985, first showed the connection between this problem and that of constructing PBTDs: a PBTD($2n$) exists if and only if a Room square of order $2n$ with an empty $(n - 1) \times (n - 1)$ array of cells exists. We show the connection clearly by means of an example.

Consider for example the PBTD(10) of Example 10.2.2. Just as the middle row is used twice there in providing two factors given by the top 'half' and the bottom 'half', so is it used twice in obtaining the resolution classes in the following Room square design.

**Example 10.2.3** A Room square of order 10 with a $4 \times 4$ hole.

$$
\begin{bmatrix}
8,9 & & & & & 1,5 & 3,7 & 2,6 & 0,4 \\
& 5,3 & & & & 9,0 & 8,6 & 1,4 & 7,2 \\
& & 4,2 & & & 6,7 & 0,1 & 9,3 & 8,5 \\
& & & 1,7 & & 8,4 & 9,2 & 0,5 & 6,3 \\
& & & & 0,6 & 2,3 & 4,5 & 8,7 & 9,1 \\
4,3 & 9,6 & 8,1 & 0,2 & 5,7 & & & & \\
2,5 & 8,0 & 9,7 & 4,6 & 1,3 & & & & \\
1,6 & 4,7 & 0,3 & 9,5 & 8,2 & & & & \\
0,7 & 1,2 & 5,6 & 8,3 & 9,4 & & & &
\end{bmatrix}
$$

**Definition 10.2.3** A Room square of size $2n - 1$ which contains an empty $(n - 1) \times (n - 1)$ subarray of cells is called a *maximum empty subarray Room square* (MESRS($2n - 1$)).

As suggested by the above example, we have the following theorem.

**Theorem 10.2.1** (Stinson, 1985)  *A PBTD($2n$) exists if and only if a MESRS($2n - 1$) exists.*

**Proof**  Given a PBTD($2n$) with middle row $\{a_1, b_1\}, \ldots, \{a_n, b_n\}$, let $A$ be the array obtained by writing the last $n - 1$ rows as columns, and let $B$ be the array obtained by writing the first $n - 1$ rows in reverse order. Then

$$D = \begin{bmatrix} C & A \\ B & — \end{bmatrix}$$

is a MESRS($2n - 1$), where $C$ is the $n \times n$ array with $\{a_i, b_i\}$ in the $i$th diagonal place and with every non-diagonal cell empty.

Conversely, given a MESRS($2n - 1$) we can permute the rows and columns so that the empty $(n - 1) \times (n - 1)$ array of cells is at the bottom right. The form of the MESRS is then like $D$ above, where $A$ and $B$ have no empty cells and where in $C$ there is precisely one non-empty cell in each row and column. By permuting rows and columns we can bring these cells on to the diagonal, and then read off a PBTD($2nm$) by reversing the process described in the first part of the proof.

Note that the theorem establishes the following fact about designing league schedules. Suppose a league schedule for $2n$ teams in $2n - 1$ rounds is to be formed. Then we can arrange such a schedule using $n$ grounds with each team appearing once on each ground in each 'half' of the tournament if and only if we can arrange a schedule using $2n - 1$ grounds in which $n - 1$ of the grounds are used only on the first $n$ days.

### 10.3  Howell designs

The subarray of the PBTD(10) of Example 10.2.2, obtained by taking just the first five rows, is an example of a Howell design H(5, 10).

**Definition 10.3.1**  A *Howell design* H($s, 2n$) is an $s \times s$ array in which each cell either is empty or contains an unordered pair of elements of a $2n$-set $X$, such that:

 (i) each row and each column contains each element of $X$ exactly once;
(ii) each unordered pair of elements of $X$ occurs in at most one cell.

**Lemma 10.3.1**  *If an* H($s, 2n$) *exists,* $n \leqslant s \leqslant 2n - 1$.

**Proof**  Since $n$ pairs must appear in row, we must have $s \geqslant n$. Also, the $s$ rows contain between them $sn$ pairs, so we must have

$$sn \leqslant \binom{2n}{2} = n(2n - 1); \qquad \text{so } s \leqslant 2n - 1.$$

At one extreme, where $s = n$, there are no empty cells; at the other extreme, where $s = 2n - 1$, the Howell design $H(2n - 1, 2n)$ is just a Room square of order $2n$.

Here are some examples

**Example 10.3.1**   An $H(6, 12)$, due to Hung and Mendelsohn (1974).

$$\begin{bmatrix} 1,7 & 2,8 & 3,9 & 4,X & 5,E & 6,0 \\ 2,3 & 0,7 & 4,5 & 8,9 & 6,1 & X,E \\ E,0 & 3,4 & 7,8 & 5,6 & 9,X & 1,2 \\ 5,X & 6,E & 1,0 & 2,7 & 3,8 & 4,9 \\ 6,9 & 1,X & 2,E & 3,0 & 4,7 & 5,8 \\ 4,8 & 5,9 & 6,X & 1,E & 2,0 & 3,7 \end{bmatrix}$$

**Example 10.3.2**   An $H(4, 6)$.

$$\begin{bmatrix} - & 4,5 & 2,3 & 1,6 \\ 2,6 & - & 1,5 & 3,4 \\ 4,1 & 3,6 & - & 2,5 \\ 3,5 & 1,2 & 4,6 & - \end{bmatrix}$$

(Note how 5, 6 behave like $\infty$, while $1, \ldots, 4$ change cyclically mod 4.)

**Example 10.3.3**   An $H(7, 10)$.

$$\begin{bmatrix} 0,7 & 4,6 & 3,8 & 2,5 & 1,9 & - & - \\ - & 1,7 & 5,0 & 4,8 & 3,6 & 2,9 & - \\ - & - & 2,7 & 6,1 & 5,8 & 4,0 & 3,9 \\ 4,9 & - & - & 3,7 & 0,2 & 6,8 & 5,1 \\ 6,2 & 5,9 & - & - & 4,7 & 1,3 & 0,8 \\ 1,8 & 0,3 & 6,9 & - & - & 5,7 & 2,4 \\ 3,5 & 2,8 & 1,4 & 0,9 & - & - & 6,7 \end{bmatrix}$$

(Note how 7, 8, 9 behave like $\infty$, while $0, \ldots, 6$ move cyclically mod 7.)

It is now known that an $H(s, 2n)$ exists for all $n \leqslant s \leqslant 2n - 1$ except for the cases $(s, n) = (3, 2), (2, 2), (5, 3), (5, 4)$. Of these, $(3, 2)$ and $(5, 3)$ correspond to the non-existing Room squares of orders 4, 6.

The case $s = n$ is easy.

**Theorem 10.3.2**   *An* $H(n, 2n)$*exists for all* $n \neq 2$.

**Proof**   The case $n = 6$ is dealt with by Example 10.3.1. For $n \neq 2, 6$, simply take the join of two MOLS of order $n$, one on $\{1, \ldots, n\}$ and the other on $\{n + 1, \ldots, 2n\}$.

We do not intend to give a proof of the existence of all the possible $H(s, 2n)$; details can be found for odd $s$ in Stinson (1982) and for even $s$ in Anderson *et al.* (1984). The methods used are basically variations on the methods already used in other constructions—for example, using starters and adders, pairwise balanced designs, and MOLS. Some starter methods are given in the exercises, as are some examples of the use of MOLS. For some of these, the following definition is useful.

**Definition 10.3.2** An $H^*(s, 2n)$ is an $H(s, 2n)$ such that among its $2n$ elements are $2n - s$ elements $\infty_i$ no two of which occur as a pair. If its elements are $x_1, \ldots, x_s, \infty_1, \ldots, \infty_{2n-s}$, it is *standardized* w.r.t. $\infty_1$ if the pair $\{x_j, \infty_1\}$ occurs in the $(j, j)$ position for each $j \leqslant s$.

These ideas generalize the situation which occurs when a Room square is obtained by the starter–adder method: there $2n - s = 1$.

**Example 10.3.4**

(a) The $H(4, 6)$ of Example 10.3.2 is an $H^*(4, 6)$; 5 and 6 act as $\infty_1$ and $\infty_2$. We can standardize it w.r.t. 5 by permuting rows and columns to obtain

$$\begin{bmatrix} 1,5 & 3,4 & 2,6 & - \\ - & 2,5 & 4,1 & 3,6 \\ 4,6 & - & 3,5 & 1,2 \\ 2,3 & 1,6 & - & 4,5 \end{bmatrix}$$

(b) The $H(7, 10)$ of Example 10.3.3 is an $H^*(7, 10)$ which is standardized w.r.t. 7.

(c) All designs $H(n, 2n)$ obtained for MOLS as in Theorem 10.3.2 are $H^*(n, 2n)$ designs.

## 10.4 Exercises 10

1. Use Theorem 10.1.1 to construct an FBTD(14).
2. Use the FBTD(10) of Example 10.1.6 and the Graeco-Latin square of order 10 exhibited in Example 4.2.2 to construct an FBTD(20).
3. An *odd BTD* for $2n + 1$ teams (an OBTD$(2n + 1)$) is a league schedule for $2n + 1$ teams in $2n + 1$ rounds, with one team sitting out in each round, every pair of teams meeting exactly once, with games assigned to $n$ grounds, each ground being used in each round, such that each team plays exactly twice on each ground. An OBTD(7) can be obtained by removing games involving $\infty$ from the schedule of Example 1.2.3. Show that the method used there gives an OBTD$(2n + 1)$ for all $n$.

4. *Using skew Room squares* to construct BTDs. Let $R$ be a skew Room square of size $n$, with $\{i,\infty\}$ in position $(i,i)$. Form the sum of $R$ and $R'$, as described in Definition 9.2.5; then put $\{i, n+i\}$ in position $\{i,i\}$ to obtain the array $B$. Also let $C$ be the join of two MOLS of order $n$, one on $\{1,\ldots,n\}$ and the other on $\{n+1,\ldots,2n\}$, with $\{i, n+i\}$ in the $(1,i)$ position. Remove the first row of $C$ to obtain the array $D$. Then the array $\begin{bmatrix} B \\ D \end{bmatrix}$ is a BTD($2n$). (Schellenberg *et al.*, 1977)

5. *The starter–adder method.* Here we construct an H($n+1, 2n$) for each odd integer $n \geqslant 3$. The $2n$ entries will be $1,\ldots,n+1$ (considered mod $n+1$) and $\infty_1,\ldots,\infty_{n-1}$. For the first row take $\{n+1,\infty_1\}$ $\cdots$ $\{m+3,\infty_{m-2}\} - \{m+2,\infty_{m-1}\}$ $\cdots$ $\{4,\infty_{n-2}\}$ $\{2,3\}$ $\{1,\infty_{n-1}\}$, where $n+1 = 2m$, and obtain the other rows by developing cyclically mod $n+1$. The case $n=3$ gives the H($4,6$) of Example 10.3.2 with $5,6$ in place of $\infty_1,\infty_2$. To show that this works, verify (i) that all elements occur in the first row once; (ii) differences are all different; (iii) each pair is positioned so as to give no repetitions in any column. (Note: no such method works when $n$ is even.) (Hung and Mendelsohn, 1974)

6. Starter–adder construction of an H($n+2, 2n$) for all $n \geqslant 5$.
   (a) If $n$ is odd, take elements $0,1,\ldots,n+1$ (considered mod $n+2$) and $\infty_1,\ldots,\infty_{n-2}$ and first row

   $$\{0,\infty_1\} \; \{n-1, n+1\} \; \{n-2,\infty_2\} \cdots \{3,\infty_{n-3}\} \; \{2,n\} \; \{1,\infty_{n-2}\}\text{——}.$$

   (The case $n=5$ gives the H($7,10$) of Example 10.3.3.)
   (b) If $n$ is even, $n = 2m \geqslant 8$, take elements $0,\ldots,n+1$ (considered mod $n+2$) and $\infty_1,\ldots,\infty_{n-2}$, and first row.

   $$\{0,\infty_1\} \; \{n-1,n+1\} \; \{n-2,\infty_2\} \cdots$$
   $$\{m+1,\infty_{m-1}\}-\{m,\infty_m\} \cdots \{3,\infty_{n-3}\} \; \{2,n\} \; \{1,\infty_{n-2}\}\text{—}.$$

   (Hung and Mendelsohn, 1974)

7. Show that if $s = 2n-1$ or $2n-2$ then every H($s,2n$) is an H*($s,2n$).

8. A *multiplication theorem* for Howell designs. Prove that if an H*($s,2m$) and an H($t,2n$) both exist, and two MOLS of order $s$ exist, then an H($st, (2n-1)s + (2m-s)$) exists. Take an H*($s,2m$) $H_1$ on $\{1,\ldots,2m\}$, and an H($t,2n$) $H_2$ on $\{1,\ldots,2n-1,\infty\}$, with $\{i,\infty\}$ in position $(i,i)$. Let $L, M$ be two MOLS on $\{1,\ldots,s\}$. Then replace each cell of $H_2$ by an $s \times s$ array as follows.
   (i) Replace each empty cell of $H_2$ by an $s \times s$ empty square.
   (ii) Replace the diagonal entry $\{i,\infty\}$ by $A_i$, where $A_i$ is obtained from $H_1$ by replacing each $x \leqslant s$ by $x_i$.
   (iii) If $j \neq k$, the cell of $H_2$ containing $\{j,k\}$ is replaced by $B_{jk}$ where $B_{jk}$ is the $s \times s$ array obtained by replacing each pair $(l,m)$ in the join of $L$ and $M$ by $\{l_j, m_k\}$.

The resulting array $H$ is the required design: it has $s(2n-1)$ elements with subscripts and $2m-s$ without subscripts. (Hung and Mendelsohn, 1974)

9. (a) Construct a PBTD(12) on $\{0,\ldots,9,\alpha,\beta\}$ as follows. Take as the first row $\{\alpha,4\}\,\{\beta,3\}\,\{1,2\}\,\{0,6\}\,\{7,9\}$ and develop it cyclically (mod 10) to obtain a $10\times 10$ array. Form an 11th column with $\{4+i,7+i\}$ in the $i$th row, and an 11th row $R_{11}$ with $\{i,5+i\}$ in the $i$th column $(i\leqslant 5)$. Finally place $\{\alpha,\beta\}$ in the new corner cell. Then alternate rows, together with $R_{11}$, form an H(6,12).

   (b) Use this PBTD(12) to construct a MESRS(11). (Lamken and Vanstone, 1985)

10. Construct a PBTD($2n$) for $n=8$, 10, 12 by proceeding as in the previous exercise, taking first rows as follows:

$n=8$:   $\alpha,11$   $1,4$   $5,7$   $\beta,9$   $8,13$   $0,10$   $6,12$

$n=10$:   $2,4$   $10,16$   $\beta,14$   $7,11$   $0,1$   $5,12$   $9,17$   $3,6$   $\alpha,15$

$n=12$:   $2,4$   $\alpha,17$   $\beta,15$   $10,20$   $7,16$   $8,13$   $14,21$   $12,18$

   $3,6$   $5,9$   $11,19$.

# 11

# Whist tournaments

## 11.1 The use of differences

The concept of a whist (or doubles tennis) tournament was introduced in Section 2.3. Recall that a Wh($4n$) consists of $4n - 1$ rounds of $n$ games; in each game two players play against two others, and altogether

(i) each player plays in one game per round;
(ii) each player partners every other player exactly once;
(iii) each player opposes every other player exactly twice.

Examples of Wh(12), Wh(16), and Wh(20) were given in Examples 2.3.2 and 2.3.3 and Exercises 2.17.

**Example 11.1.1** A cyclic Wh(8).

$$
\begin{array}{ll}
\infty,0 \text{ v } 4,5 & 1,3 \text{ v } 2,6 \\
\infty,1 \text{ v } 5,6 & 2,4 \text{ v } 3,0 \\
\infty,2 \text{ v } 6,0 & 3,5 \text{ v } 4,1 \\
\infty,3 \text{ v } 0,1 & 4,6 \text{ v } 5,2 \\
\infty,4 \text{ v } 1,2 & 5,0 \text{ v } 6,3 \\
\infty,5 \text{ v } 2,3 & 6,1 \text{ v } 0,4 \\
\infty,6 \text{ v } 3,4 & 0,2 \text{ v } 1,5
\end{array}
$$

Whist tournament designs Wh($4n$), $n = 1, \ldots, 10$, due to Howell, Mitchell Safford, and Whitfeld, were published in Volume 1 of *Whist*, 1891–2; most of the designs were cyclic like the one above.

From now on we shall use a different notation: thinking of the players seated round a table, with partners facing each other, we shall represent a game by

$$(a, b, c, d),$$

where $\{a, c\}$ and $\{b, d\}$ are partner pairs, and $\{a, b\}, \{a, d\}, \{b, c\}$ and $\{c, d\}$

are opponent pairs. Thus the first round of the Wh(8) above will be
denoted by

$$(\infty, 4, 0, 5) \qquad (1, 2, 3, 6). \tag{11.1}$$

**Example 11.1.2**  A cyclic Wh(24) can be obtained from the following first
round:

$$(\infty, 7, 0, 10), (1, 12, 8, 22), (5, 2, 6, 11), (3, 14, 9, 18),$$
$$(17, 4, 19, 16), (15, 13, 20, 21).$$

That this is a suitable first round can be verified by checking the conditions of Lemma 2.3.1 with $G = Z_{23}$.

Among the earliest mathematicians to work on whist tournaments was
E. H. Moore who, in 1896, showed that designs Wh($4n$) exist for infinitely
many values of $n$. However it was not until the 1970s that R. M. Wilson
and R. D. Baker established the existence of a Wh($4n$) for all $n \geqslant 1$. It is
the aim of this chapter to give a complete proof of this result, and to
consider also the problem with $4n + 1$ players. Our arguments will make
use of transversal designs, SOLS with symmetric mates, different systems,
and pairwise balanced designs, thereby showing the interconnections which
exist between many areas of design theory.

We saw in Chapter 2 that difference systems can be used to construct
related tournaments Wh($4n + 1$). These tournament designs were defined
in Definition 2.3.3: essentially there are $4n + 1$ rounds, with each player
sitting out in exactly one round. Differences were used in Theorem 2.3.3 to
construct a Wh($4n + 1$) whenever $4n + 1$ is a prime power. Take a primitive element $\theta$ of GF($4n + 1$); take the games

$$(\theta^i, \theta^{n+i}, \theta^{2n+i}, \theta^{3n+i}), \qquad 0 \leqslant i \leqslant n - 1, \tag{11.2}$$

for the first round. Then further rounds are obtained by adding a fixed
element of GF($4n + 1$) to each entry. See Example 2.3.5 for a Wh(13).

**Example 11.1.3**  Construction of a Wh(9). As a primitive element of GF(9)
take $\theta$ satisfying $\theta^2 = \theta + 1$ over GF(3), as in Example 1.6.3. Here $n = 2$,
$\theta^n = \theta + 1$, $\theta^{2n} = 2$, $\theta^{3n} = 2\theta + 2$. Using the multiplication table of
Example 1.6.3, we obtain the following Wh(9).

| Round 1: | $(1, \theta + 1, 2, 2\theta + 2)$, | $(\theta, 2\theta + 1, 2\theta, \theta + 2)$ |
|---|---|---|
| 2: | $(2, \theta + 2, 0, 2\theta)$, | $(\theta + 1, 2\theta + 2, 2\theta + 1, \theta)$ |
| 3: | $(0, \theta, 1, 2\theta + 1)$, | $(\theta + 2, 2\theta, 2\theta + 2, \theta + 1)$ |
| 4: | $(\theta + 1, 2\theta + 1, \theta + 2, 2)$, | $(2\theta, 1, 0, 2\theta + 2)$ |
| 5: | $(\theta + 2, 2\theta + 2, \theta, 0)$, | $(2\theta + 1, 2, 1, 2\theta)$ |
| 6: | $(\theta, 2\theta, \theta + 1, 1)$, | $(2\theta + 2, 0, 2, 2\theta + 1)$ |
| 7: | $(2\theta + 1, 1, 2\theta + 2, \theta + 2)$, | $(0, \theta + 1, \theta, 2)$ |
| 8: | $(2\theta + 2, 2, 2\theta, \theta)$, | $(1, \theta + 2, \theta + 1, 0)$ |
| 9: | $(2\theta, 0, 2\theta + 1, \theta + 1)$, | $(2, \theta, \theta + 2, 1)$ |

We can simplify this by first expressing each non-zero element as a power of $\theta$: $\theta^2 = \theta + 1, \theta^3 = 2\theta + 1, \theta^4 = 2, \theta^5 = 2\theta, \theta^6 = 2\theta + 2, \theta^7 = \theta + 2, \theta^8 = 1$. Making these substitutions and then replacing $\theta^i$ by $i$, we obtain

| Round 1: | $(8, 2, 4, 6)$, | $(1, 3, 5, 7)$ |
|---|---|---|
| 2: | $(4, 7, 0, 5)$, | $(2, 6, 3, 1)$ |
| 3: | $(0, 1, 8, 3)$, | $(7, 5, 6, 2)$ |
| 4: | $(2, 3, 7, 4)$, | $(5, 8, 0, 6)$ |
| 5: | $(7, 6, 1, 0)$, | $(3, 4, 8, 5)$ |
| 6: | $(1, 5, 2, 8)$, | $(6, 0, 4, 3)$ |
| 7: | $(3, 8, 6, 7)$, | $(0, 2, 1, 4)$ |
| 8: | $(6, 4, 5, 1)$, | $(8, 7, 2, 0)$ |
| 9: | $(5, 0, 3, 2)$, | $(4, 1, 7, 8)$. |

We shall see that the study of designs $\mathrm{Wh}(4n + 1)$ is necessary for the study of $\mathrm{Wh}(4n)$. One $\mathrm{Wh}(4n + 1)$ in particular which will be needed later on is a $\mathrm{Wh}(33)$. Since 33 is not a prime power, we need a separate construction.

**Example 11.1.4** A cyclic $\mathrm{Wh}(33)$ can be obtained by taking the following first round and developing mod 33:

$$(9, 28, 22, 30), (1, 19, 8, 18), (2, 3, 31, 24), (11, 20, 17, 5),$$
$$(13, 4, 21, 23), (12, 16, 15, 7), (27, 25, 32, 14), (6, 26, 29, 10).$$

Verification of this involves checking that the differences between partners give all the non-zero elements of $Z_{33}$ once, and the differences between opponents give all the non-zero elements of $Z_{33}$ twice. (Baker, 1975$b$)

The following recent example, due to Finizio, will also be useful.

**Example 11.1.5** A cyclic $\mathrm{Wh}(57)$ can be obtained by developing the following first round mod 57:

$$(25, 12, 26, 29), (41, 52, 2, 14), (7, 56, 28, 8), (44, 3, 49, 23),$$
$$(21, 40, 48, 46), (16, 15, 31, 22), (32, 30, 36, 54), (6, 37, 34, 53),$$
$$(4, 38, 18, 51), (50, 20, 5, 45), (24, 17, 35, 39), (10, 1, 13, 27),$$
$$(47, 19, 55, 42), (33, 9, 43, 11).$$

Tournaments Wh(4n + 1) will be studied further in a later section of this chapter, where it will be shown that a Wh(4n + 1) exists for all positive integers $n$.

## 11.2 Using pairwise balanced designs

As in the construction of so many other types of combinatorial structure, we can use a PBD(v, K, 1) to carry over existence from structures of size $k \in K$ to a structure of size $v$.

**Theorem 11.2.1** *Suppose that a* PBD(v, K, 1) *exists, where* $k \equiv 1$ (mod 4) *for all* $k \in K$, *and where a* Wh(k) *exists for all* $k \in K$. *Then a* Wh(v) *exists.*

**Proof** On each block $B$ of the PBD form a Wh(|B|). For any element $x$, take the blocks $B$ containing $x$, and take all the games of the round of the Wh(|B|) on $B$ in which $x$ sits out. These games will form a round of the required Wh(v). 

**Example 11.2.1** Since a (21, 5, 1) design exists and a Wh(5) exists, a Wh(21) also exists. Indeed we can take the cyclic (21, 5, 1) design derived from the difference set {1, 2, 5, 15, 17} in $Z_{21}$. The blocks containing 1 are {1, 2, 5, 15, 17}, {1, 6, 7, 10, 20}, {1, 4, 14, 16, 21}, {1, 11, 13, 18, 19}, {1, 3, 8, 9, 12}. Using the Wh(5) given in Section 1.4, we obtain a Wh(21) in which the games of the round in which 1 sits out are

$$(5, 2, 15, 17), (7, 6, 10, 20), (14, 4, 16, 0), (13, 11, 18, 19), (8, 3, 9, 12).$$

**Example 11.2.2** A PBD(69, {5, 17}, 1) can be constructed by starting with a resolvable (52, 4, 1) design (Exercises 2, number 24): adjoin $\infty_i$ to each block of the $i$th resolution class $(1 \leqslant i \leqslant 17)$ and form a new block of size 17. Since a Wh(5) and a Wh(17) both exist, it follows that a Wh(69) also exists.

We can also use resolvable (v, 4, 1) designs to construct some Wh(4n). Such designs can exist only if $v \equiv 4$ (mod 12), and it was shown in outline in Exercises 7, number 6 how to prove that they do indeed exist in all such cases. For our purposes here we only use their existence in the case $v = 3q + 1$ where $q \equiv 1$ (mod 4) is a prime power; a proof in this case is given in Exercise 2, number 23.

**Lemma 11.2.2** *If a resolvable* (12m + 4, 4, 1) *design exists, then a* Wh(12m + 4) *exists.*

**Proof** On each block of a given resolution class, construct a Wh(4). Taken together, these give three rounds of a Wh(12m + 4). Repeat for each class.

This construction goes back to Moore in 1896. It yields the following.

**Corollary 11.2.3** *A* Wh(4n) *exists in each of the following cases*: $4n = 28, 40, 52, 76, 88, 112, 124, 148, 160, 184, 220, 244, 268, 292, 304$.

**Proof** These values of $4n$ are all of the form $3q + 1$ where $q \equiv 1 \pmod 4$ is a prime power: $q = 9, 13, 17, 25, 29, 37, 41, 49, 53, 61, 73, 81, 89, 97, 101$.

Returning to Theorem 11.2.1 briefly, it is worth pointing out that use is made of the *absence* of one player from each round. Such an absence is also useful in the proof of the following important result.

**Theorem 11.2.4** *Let* $v \equiv 0 \pmod 4$, *and suppose that a* GDD $(v - 1, K, G)$ *exists where* $k \equiv 1 \pmod 4$ *and a* Wh(k) *exists for all* $k \in K$, *and where* $g \equiv 3 \pmod 4$ *and a* Wh(g + 1) *exists for all* $g \in G$. *Then a* Wh(v) *exists.*

**Proof** Construct a Wh(v) on the set $X \cup \{\infty\}$, where $X$ is the set of elements of the GDD, as follows. On each block of size $k$ form a Wh(k); do this for each $k \in K$. On each group of size $g$, with $\infty$ adjoined, form a Wh(g + 1) and label its rounds by the $g$ elements of the group; do this for each $g \in G$. Then, for any $x \in X$, take as the games of round $x$ of the required Wh(v) all games of round $x$ of the Wh(g + 1) constructed from the group containing $x$, and all the games in the rounds omitting $x$ in all the Wh(k) constructed from blocks containing $x$.

**Example 11.2.3**

(a) Since a TD(5, 11) exists on 55 elements, and a Wh(5) and a Wh(12) both exist, a Wh(56) exists.
(b) Since a TD(13, 19) exists on 247 elements, and a Wh(13) and a Wh(20) both exist, a Wh(248) also exists.

Theorem 11.2.4 is in fact the basis of a reduction of the problem of the existence of a Wh(4n) for all $n \geq 1$ to that of the existence of a Wh(4n) for all $n \leq 80$. Our task will thereafter be to construct a Wh(4n) in these first 80 cases.

**Theorem 11.2.5** *Suppose that a* Wh(4n) *exists for all* $n \leq 80$. *Then a* Wh(4n) *exists for all* $n \geq 1$.

**Proof** Let $n > 80$. We shall prove the existence of a GDD(4n − 1, {5, 17}, G) where $G = \{19, 23, 27, \ldots, 319\}$; it will then follow from Theorem 11.2.4 that a Wh(4n) exists.

Suppose that we have a TD$(17, g)$ on $17g$ elements: it will have 17 groups each of size $g$, and each block will have 17 elements, one from each group. Take $u$ elements of the first group and $v$ elements of the second group, $0 \leqslant u \leqslant g, 0 \leqslant v \leqslant g$, and replace each such element by five others, all allocated to the same group as the replaced element. This gives $17g + 4(u + v)$ elements, in 17 groups, 15 of which are of size $g$, one of size $g + 4u$, one of size $g + 4v$. If a block of the TD has no element replaced, leave it unaltered. If a block $B$ has one element replaced, consider the 16 unaltered elements and the five new ones—21 in all. On this set of size 21 construct a $(21, 5, 1)$ design, with the five new elements as one of the blocks, and replace $B$ by the other 20 blocks. If a block $B$ has two elements replaced, consider the 15 unaltered elements and the 10 new ones—25 in all. Form a $(25, 5, 1)$ design on them, in which the two sets of five new elements form disjoint blocks, and replace $B$ by the other 28 blocks. The total effect of all this is to construct a GDD$(17g + 4u + 4v, \{5, 17\}, H)$ where $H = \{g, g + 4u, g + 4v\}$.

First take $g = 19$. Then $17g = 323 = 4 \cdot 81 - 1$, and the choices $0, \ldots, 19$ for $u$ and $v$ give a GDD$(4n - 1, \{5, 17\}, G)$ for $n = 81, 82, \ldots, 119$. Next take $g = 27$. Then $17g = 459 = 4 \cdot 115 - 1$, and the choices $2, \ldots, 19$ for $u$ and $v$ give a GDD$(4n - 1, \{5, 17\}, G)$ for $n = 120, \ldots, 169$. After this, we take further choices of $g$ as described below: if the resulting elements of $H$ are not in $G$, replace each group of each such size $h$ by the groups and blocks of a GDD$(h, \{5, 17\}, G)$ which will have already been constructed. The following values of $g$ take us up to $n = 10494$: 31, 43, 47, 67, 83, 107, 139, 199, 243, 323, 443, 619, 863, 1207, 1679. (Note that for each such $g$, $g \equiv 3 \pmod 4$ and a TD$(17, g)$ exists.)

| For $n = 170$–194 | take $g =$ 31 | $n = 1245$–1519 | $g =$ 243 |
|---|---|---|---|
| 195–269 | 43 | 1520–2019 | 323 |
| 270–294 | 47 | 2020–2769 | 443 |
| 295–419 | 67 | 2770–3869 | 619 |
| 420–519 | 83 | 3870–5394 | 863 |
| 520–669 | 107 | 5395–7544 | 1207 |
| 670–869 | 139 | 7545–10 494 | 1679. |
| 870–1244 | 199 | | |

This deals with all $n$ up to $10\,494$, and in particular all $n$ with $4n - 1 < 17 \cdot 3^7$.

Suppose now that we have constructed a GDD$(4n - 1, \{5, 17\}, G)$ for all $n$ with $4n - 1 < 17 \cdot 3^s$ ($s \geqslant 7$). Then the successive choices

$$g = 3^s, 31 \cdot 3^{s-3}, 113 \cdot 3^{s-4}, 47 \cdot 3^{s-3}, 59 \cdot 3^{s-3}, 79 \cdot 3^{s-3}, 127 \cdot 3^{s-3},$$

$$19 \cdot 3^{s-1}, 23 \cdot 3^{s-1}, 3^{s+2}$$

enable us to deal with all $n$ for which $4n - 1 < 17 \cdot 3^{s+2}$; for, since $N(3^{s-4}) \geqslant N(27) > 15$, a TD$(17, g)$ exists in each case. This completes the proof of the theorem on using induction on $s$.

The existence problem for whist tournaments has therefore been reduced to that of constructing a Wh$(4n)$ for each $n \leqslant 80$. We close this section by listing those values of $n \leqslant 80$ for which a Wh$(4n)$ has already been constructed: they are

$$n = 1\text{–}7, 10, 13, 14, 19, 22, 28, 31, 37, 40, 46, 55, 61, 62, 67, 73, 76.$$

## 11.3 Product theorems

As for other combinatorial structures, it is possible to combine two whist tournament designs together to obtain a larger one.

**Theorem 11.3.1** (Baker) *Suppose that a* Wh$(v)$ *and a* Wh$(w)$ *exist, and that* $N(v) \geqslant 3$ *and* $N(w) \geqslant 3$. *Then a* Wh$(vw)$ *exists.*

**Note** In fact, $N(n) \geqslant 3$ for all $n \geqslant 4$, $n \neq 6$ or 10. We have shown that $N(4n) \geqslant 3$ for all $n$, and we will use Baker's theorem for odd $w$ only in cases where we know that $N(w) \geqslant 3$.

**Proof** There are really three theorems here, depending on whether $v$ and $w$ are both of the form $4n$, both of the form $4n + 1$, or one is of each form. Accordingly the proof deals with three separate cases.

**Case (i)** Suppose $v \equiv w \equiv 0 \pmod 4$. Given a Wh$(v)$ on $I_v = \{1, \ldots, v\}$ and a Wh$(w)$ on $I_w$, we show how to construct a Wh$(vw)$ on $S = I_v \times I_w$. This Wh$(vw)$ will contain $vw - 1$ rounds. Of these, $w - 1$ can be obtained very simply, involving games between elements of $S$ with the same first component. Suppose we take the given Wh$(w)$ and replace each $x \in I_w$ by $(j, x)$ for some fixed $j \in I_v$. If we do this for each $j$, then, for each $i \leqslant w - 1$, take all the $i$th rounds in the resulting $v$ tournaments, we obtain the $i$th round of the required Wh$(vw)$ as their union. In these $w - 1$ rounds, each pair of elements of $S$ with the same first component will partner each other once and oppose each other twice. It remains to construct rounds involving games in which the players have different first components. To do this, we make use of a resolvable TD$(4, w)$, which exists since $N(w) \geqslant 3$.

Take the games of any one round of the given Wh$(v)$. For each such game $(i, j, k, l)$ form a resolvable TD$(4, w)$ on $\{i, j, k, l\} \times I_w$, with the groups determined by the first components. A typical block in the TD will be $\{(i, p), (j, q), (k, r), (l, s)\}$; form from it the game $((i, p), (j, q), (k, r),$ $(l, s))$. If we take all such games formed from all the games of one round of the Wh($v$) and all the blocks of one resolution class of the TD, they will together form one round of the required Wh($vw$). Since the Wh($v$) has $v - 1$ rounds and the TD($4,w$) has $w$ resolution classes, we obtain $w(v - 1)$ rounds. Any two elements of $S$ with different first components, say $(i, m)$ and $(j, n)$, will play together once since there is just one block of the TD containing them both; and they will oppose each other in the games arising from the two games in Wh($v$) in which $i, j$ oppose each other.

Altogether we have $w(v - 1) + w - 1 = uw - 1$ rounds, and a Wh($vw$).

**Case (ii)** Suppose $v \equiv 0, w \equiv 1 \pmod 4$. Form $w(v - 1)$ rounds as in case (i) by making use of a resolvable TD($4,w$). These rounds contain all the required meetings of elements of $S$ with different first components. We now remove one of these rounds, and use its games, along with games involving elements with equal first components, to construct further rounds.

Let $((i, p), (j, q), (k, r), (l, s))$ be one of the games in the round removed. Along with it, take all the games of the round of a Wh($w$) on $\{i\} \times I_w$ which omits $(i, p)$, all the games of the round of the Wh($w$) on $\{j\} \times I_w$ which omits $(j, q)$, all the games of the round of the Wh($w$) on $\{k\} \times I_w$ which omits $(k, r)$, and all the games of the round of the Wh($w$) on $\{l\} \times I_w$ which omits $(l, s)$. If we do this for every game $(i, j, k, l)$ in one round of the Wh($v$) we obtain games which together form a round of the Wh($vw$). On repeating for each of the $w$ blocks of the chosen resolution class of the TD, we obtain $w$ rounds, giving $v(w - 1) - 1 + w = vw - 1$ altogether.

**Case (iii)** Suppose $v \equiv w \equiv 1 \pmod 4$. Form resolvable TD($4,w$) designs as before on $\{i, j, k, l\} \times I_w$, for each game $(i, j, k, l)$ of a Wh($v$) on $I_v$, and in each case label the parallel classes by the elements of $I_w$. Then construct a Wh($vw$) on $I_v \times I_w$ as follows: for the round omitting the player $(i, m)$, take all the games of the round omitting $(i, m)$ in a Wh($w$) on $\{i\} \times I_w$, and all the games obtained from the $m$th parallel classes of all the resolvable TD($4,w$) designs constructed from the games of the round of a Wh($v$) on $I_v$ which omits $i$.

Case (iii) is not in fact needed for the construction of tournaments Wh($4n$), but will be used in the construction of Wh($4n + 1$).

Using the product theorem, we can now deal with most values of $n \leqslant 80$.

**Corollary 11.3.2** *For all $n \leqslant 80$, excluding 11, 23, 38, 43, 44, 47, 59, 71, 77, 79, a Wh($4n$) exists.*

**Proof** Each such $n$ not listed at the end of section 11.2 can be dealt with using the product theorem. For example, $48 = 4 \times 12, 240 = 5 \times 48, 264 = 8 \times 33$.

It now remains only to deal with those values of $n$ excluded in the statement of the corollary. We show how to deal with these by means of resolvable SAMDRRs.

## 11.4 Using SAMDRR

We have already seen the connection between resolvable SAMDRR and SOLS with symmetric mates. Another unexpected connection is between resolvable SAMDRR and whist tournaments; this enables us to complete the existence proof for Wh$(4n)$.

**Theorem 11.4.1** *If a resolvable* SAMDRR$(n)$ *exists, then a* Wh$(4n)$ *exists.*

**Proof** Suppose first that $n$ is odd, and that, in a given resolvable SAMDRR$(n)$, round $t$ omits the pair $(H_t, W_t)$ and is made up of games $H_i W_l$ v $H_j W_k$. Then we obtain from round $t$ the following three rounds of the required Wh$(4n)$:

(i) all tables $(i_1, l_3, j_1, k_3)$, $(i_2, l_4, j_2, k_4)$, and $(t_1, t_2, t_3, t_4)$;
(ii) all tables $(i_4, j_4, l_1, k_1)$, $(i_3, j_3, l_2, k_2)$, and $(t_1, t_3, t_4, t_2)$;
(iii) all tables $(i_1, k_2, l_2, j_1)$, $(i_3, k_4, l_4, j_3)$, and $(t_1, t_4, t_2, t_3)$.

This gives $3n$ rounds. We obtain $n - 1$ further rounds by taking a resolvable TD$(4, n)$ on the set $\{m_i : 1 \leqslant m \leqslant n, 1 \leqslant i \leqslant 4\}$, with the $i$th group, $1 \leqslant i \leqslant 4$, consisting of all the elements with suffix $i$, and with one resolution class consisting of all the blocks $\{t_1, t_2, t_3, t_4\}$, $1 \leqslant t \leqslant n$. From each of the other $n - 1$ resolution classes obtain a round of the required Wh$(4n)$ by replacing each block $\{i_1, j_2, k_3, l_4\}$ by the game $(i_1, l_4, k_3, j_2)$.

We have to check that this construction works. For example, elements $a_1$ and $b_3$ appear as opponents in the tables arising from the game in which $H_a$ partners $W_b$ and the game in which $H_a$ opposes $W_b$; and as partners in the table arising from the block of the TD containing $a_1$ and $b_3$.

If $n$ is even, suppose round $t \leqslant n - 1$ of the SAMDRR$(n)$ is made up of games $H_i W_l$ v $H_j W_k$. Then we obtain four rounds as follows:

(i) all tables $(i_1, l_3, j_1, k_3)$, $(i_2, l_4, j_2, k_4)$;
(ii) all tables $(i_4, j_4, l_1, k_1)$, $(l_3, j_3, l_2, k_2)$;
(iii) all tables $(i_1, k_2, l_2, j_1)$, $(l_3, k_4, l_4, j_3)$;
(iv) all tables $(i_1, l_4, k_3, j_2)$, $(j_1, k_4, l_3, i_2)$.

This gives $4n - 4$ rounds. Finally, a further three rounds are:

   (v) all tables $(t_1, t_2, t_3, t_4)$,    $t \leqslant n$;
  (vi) all tables $(t_1, t_3, t_4, t_2)$,    $t \leqslant n$;
 (vii) all tables $(t_1, t_4, t_2, t_3)$,    $t \leqslant n$.

**Example 11.4.1** Since, by Example 5.5.2, a resolvable SAMDRR(38) exists, a Wh(152) exists.

**Corollary 11.4.2** *If $(n, 6) = 1$ or $n$ is an odd prime greater than 3, then a Wh(4n) exists.*

**Proof** This follows from Theorems 5.4.1, 5.4.3, and 11.4.1.

**Corollary 11.4.3** *There exists a Wh(4n) for all $n \leqslant 80$.*

**Proof** Of those values of $n$ listed at the end of the previous section, all odd values are dealt with by Corollary 11.4.2. The case $n = 38$ has been dealt with above, and $n = 44$ now yields to the product theorem since $176 = 4 \times 44$.

Thus, in view of Theorem 11.2.5, a Wh(4n) exists for all $n \geqslant 1$.

## 11.5  Whist tournaments for $4n + 1$ players

We now show that a Wh($4n + 1$) exists for all positive integers $n$. So far we have seen that such a design exists whenever $4n + 1$ is a prime power $\equiv 1$ (mod 4) and hence for all products of such prime powers, and also for $4n + 1 = 21, 33, 57,$ and 69.

Recall that Theorem 11.2.1 gave a particularly nice PBD result: if a PBD($v, K, 1$) exists where, for each $k \in K$, $k \equiv 1$ (mod 4) and a Wh($k$) exists, then a Wh($v$) also exists. We can use this to deal with some small values of $n$.

**Lemma 11.5.1** *A Wh($4n + 1$) exists for $4n + 1 = 45, 77, 129, 133$.*

**Proof** In each case we use Theorem 11.2.1. For 45 use a TD(5,9) to construct a PBD(45, {5, 9}, 1). For 77, use a PBD(77, {5, 13, 17}, 1) which can be obtained by choosing $g = 4$ in Exercises 3, number 15. For 129, use the PBD(129, {5, 9}, 1) constructed in Exercises 11, number 3. Finally, for 133, take a resolvable (100, 4, 1) design and, for each $i \leqslant 33$, adjoin $\infty_i$ to each block of the $i$th class to obtain a PBD(133, {5, 33}, 1).

Next, we obtain a result corresponding to Theorem 11.4.1.

**Theorem 11.5.2** *Suppose that a resolvable* SAMDRR($n$) *exists, where $n$ is odd. Then a* Wh($4n + 1$) *exists.*

**Proof** We construct a Wh($4n + 1$) on $X = \{\infty\} \cup \{m_i : 1 \leqslant m \leqslant n, 1 \leqslant i \leqslant 4\}$. Consider the $t$th round of the SAMDRR($n$), which omits $H_t$ and $W_t$, and which consists of games $H_i W_i$ v $H_j W_k$. Obtain from it the four rounds

(i) all tables $(i_1, l_3, j_1, k_3)$, $(i_2, l_4, j_2, k_4)$, and $(\infty, t_4, t_2, t_3)$;
(ii) all tables $(i_4, j_4, l_1, k_1)$, $(i_3, j_3, l_2, k_2)$, and $(\infty, t_1, t_3, t_4)$;
(iii) all tables $(i_1, k_2, l_2, j_1)$, $(i_3, k_4, l_4, j_3)$, and $(\infty, t_2, t_4, t_1)$;
(iv) all tables $(i_1, l_4, k_3, j_2)$, $(j_1, k_4, l_3, i_2)$, and $(\infty, t_3, t_1, t_2)$.

This gives $4n$ rounds. Finally take the further round consisting of all games $(t_1, t_4, t_3, t_2)$.

Note that this result ensures the existence of a Wh(93). In the general case where $(n, 6) = 1$, note that the choice $l = 2i - j$, $k = 2j - i$ corresponds to the SAMDRR of Theorem 5.4.1; the resulting Wh($4n + 1$) were first obtained by Watson (1954).

We can now deal with every $4n + 1 \leqslant 137$.

**Theorem 11.5.3** *A* Wh($4n + 1$) *exists for all $4n + 1 \leqslant 137$.*

**Proof** The remaining cases can all be dealt with by the product theorem:

$$65 = 5 \times 13, 85 = 5 \times 17, 105 = 5 \times 21, 117 = 9 \times 13.$$

Having dealt with small values of $4n + 1$, we next turn to the problem of dealing with $4n + 1 \geqslant 141$. We aim for a result like Theorem 11.2.5, and discover that we can obtain such a result by making use of a family of pairwise balanced designs already used in the construction of resolvable $(12t + 4, 4, 1)$ designs (Exercises 7, number 6).

**Theorem 11.5.4** *If $N(t) \geqslant 4$ and $0 \leqslant h \leqslant t$ then a* PBD($20t + 4h + 1$, $\{5, 4t + 1, 4h + 1\}, 1$) *exists.*

**Proof** Since $N(t) \geqslant 4$, a TD($6, t$) exists. Remove $t - h$ elements from one of its groups; the resulting design has five groups of size $t$ and one group of size $h$, and blocks of size 5 or 6; so a GDD($5t + h, \{5, 6\}, \{t, h\}$) is obtained. Take such a GDD on a set $X$ of $v = 5t + h$ elements, and let $X^*$ denote the set obtained from $X$ by replacing each $x \in X$ by four copies of $x, x_1, \ldots, x_4$. If in each group we replace each element by its four copies, we obtain groups of sizes $4t$ or $4h$. If we replace each element of each block by its four copies we do not obtain a GDD since each block will have 0 or four members of a given group in it. However, if, for each block $B$, we take the set of 20 or 24 elements arising from it and form a GDD $(4|B|, 5, 4)$ in which each group consists of the copies of one element of $B$, then we can replace each $B$ by the resulting blocks of size 5 and thus obtain a GDD$(4v, 5, \{4t, 4h\})$. Then, on adjoining a new element $\infty$ to each group, and taking all resulting groups as further blocks, we obtain a PBD$(4v + 1, \{5, 4t + 1, 4h + 1\}, 1)$ as required.

It remains therefore to check that a GDD$(4|B|, 5, 4)$ can be constructed from every block. Since $|B| = 5$ or 6 we need the existence of a GDD$(20, 5, 4)$ and a GDD$(24, 5, 4)$. By Theorem 4.7.2 we require the existence of $(21, 5, 1)$ and $(25, 5, 1)$ designs; and these certainly exist.

The first thing we do with this result is use it to deal with every $4n + 1$ between 140 and 620.

**Corollary 11.5.5**  *If* $141 \leqslant 4n + 1 \leqslant 617$, *a* Wh$(4n + 1)$ *exists.*

**Proof**  The idea is to express each $4n + 1$ in the form $20t + 4h + 1$ with $0 \leqslant h \leqslant t$ and $N(t) \geqslant 4$, and where a Wh$(4t + 1)$ and a Wh$(4h + 1)$ both exist. Suitable choices of $t$ and $h$ are given below.

$$
\begin{array}{llll}
\text{For } & 141 \leqslant 4n + 1 \leqslant 169 & \text{take } t = 7 & \text{and } 0 \leqslant h \leqslant 7 \\
& 173 \leqslant 4n + 1 \leqslant 193 & t = 8 & 3 \leqslant h \leqslant 8 \\
& 197 \leqslant 4n + 1 \leqslant 217 & t = 9 & 4 \leqslant h \leqslant 9 \\
& 221 \leqslant 4n + 1 \leqslant 265 & t = 11 & 0 \leqslant h \leqslant 11 \\
& 269 \leqslant 4n + 1 \leqslant 313 & t = 13 & 2 \leqslant h \leqslant 13 \\
& 317 \leqslant 4n + 1 \leqslant 361 & t = 15 & 4 \leqslant h \leqslant 15 \\
& 365 \leqslant 4n + 1 \leqslant 409 & t = 17 & 6 \leqslant h \leqslant 17 \\
& 413 \leqslant 4n + 1 \leqslant 457 & t = 19 & 8 \leqslant h \leqslant 19 \\
& 461 \leqslant 4n + 1 \leqslant 505 & t = 21 & 10 \leqslant h \leqslant 21 \\
& 509 \leqslant 4n + 1 \leqslant 601 & t = 25 & 2 \leqslant h \leqslant 25 \\
& 605 \leqslant 4n + 1 \leqslant 617 & t = 29 & 6 \leqslant h \leqslant 9.
\end{array}
$$

This completes the proof. Note that we used $N(15) \geqslant 4$; we could avoid this by using $t = 16$ instead of $t = 15$, but then 317 would have to be dealt with separately (easily, in fact, since 317 is prime). Note too that we used $N(21) \geqslant 4$, as shown in Example 4.3.3; otherwise $t$ is a prime power.

At last, we can now establish the general existence of Wh$(4n + 1)$.

**Theorem 11.5.6** (Baker)  *A* Wh$(4n + 1)$ *exists for all positive integers n.*

**Proof**  Suppose that the theorem has been established for all $4n + 1 < 120w + 20$, where $w \geqslant 5$. We show that the theorem is then true for all $4n + 1 < 120(w + 1) + 20$. Simply note that every $4n + 1$ between $120w + 20$ and $120(w + 1) + 20$ can be expressed in the form

$$4n + 1 = 20(6w + 1) + 4h + 1 \qquad (0 \leqslant h \leqslant 29)$$

where $h \leqslant 6w + 1$ since $w \geqslant 5$. In each case a $\text{PBD}(4n + 1, \{5,$ $4(6w + 1) + 1, 4h + 1\}, 1)$ exists by Theorem 11.5.4, and a $\text{Wh}(4n + 1)$ exists by Theorem 11.2.1. The result therefore follows by induction on $w$.

## 11.6 Directedwhist and triplewhist tournaments

In this closing section we introduce two specializations of whist tournaments which have been studied with renewed vigour in recent years. Visualize the players sitting round a table; the opponents of a player $X$ sit on either side of $X$, and altogether each player is an opponent of $X$ twice. Can we arrange that each player sits once on $X$'s left, and once on $X$'s right?

**Definition 11.6.1** A $\text{Wh}(v)$ is a *directedwhist* tournament $\text{DWh}(v)$ if each player has each other player as an opponent once on his left and once on his right.

Note that the opponent condition for the games $(a_i, b_i, c_i, d_i)$ to form the first round of a cyclic $\text{DWh}(4n + 1)$ in $G$ is:

the differences $a_i - b_i, b_i - c_i, c_i - d_i, d_i - a_i$ give every non-zero element of G exactly once.

These differences are known as the directed opponent differences.

**Theorem 11.6.1** *Let* $q = p^\alpha = 4n + 1$, *where* $p$ *is prime. Then a directedwhist tournament* $\text{DWh}(4n + 1)$ *exists.*

**Proof** Take the tournament of Theorem 2.3.3, with first round games given by (11.2). The directed opponent differences are $\theta^i - \theta^{n+i}$, $\theta^{n+i} - \theta^{2n+i}$, $\theta^{2n+i} - \theta^{3n+i}$, $\theta^{3n+i} - \theta^i$, $0 \leqslant i < n$, i.e. $(1 - \theta^n)$ times $\theta^i, \theta^{n+i}$, $\theta^{2n+i}, \theta^{3n+1}$, $0 \leqslant i < n$, i.e. every non-zero element of $\text{GF}(q) - \{0\}$ once.

It is now known that a $\text{DWh}(4n + 1)$ exists for all $n \geqslant 1$, and that a $\text{DWh}(4n)$ exists for all but a handful of small values of $n$.

**Example 11.6.1** (Finizio, 1991) Initial round of a cyclic $\text{DWh}(21)$.

$$(1, 8, 2, 14), (3, 7, 6, 11), (4, 10, 12, 15), (5, 19, 16, 17), (9, 20, 18, 13).$$

Another specialization, that of *triplewhist* tournaments, goes back to Moore (1896).

**Definition 11.6.2** In the game $(a, b, c, d)$ we say that $a, b$ ( and $c, d$) are *opponents of the first kind,* and $a, d$ (and $b, c$) are *opponents of the second kind.* A whist tournament Wh($v$) in which each player has every other player once as an opponent of the first kind, and once as an opponent of the second kind, is called a *triplewhist tournament,* and is denoted by TWh($v$).

It is known that a TWh($4n + 1$) exists for all sufficiently large $n$, and that a TWh($4n$) exists for all $n$ except possibly 14, 54, 62, and 70.

Note that the opponent condition for games to form the first round of a cyclic triplewhist tournament is:

the opponent pairs of the first kind form a starter

as do the opponent pairs of the second kind.

**Example 11.6.2**   The Wh(8) given by (11.1) is a TWh(8).

**Example 11.6.3** (Moore)   A cyclic TWh(16). Take initial round games:

$$(\infty, 5, 0, 10), (1, 4, 2, 8), (6, 13, 9, 7), (11, 12, 3, 14).$$

**Example 11.6.4** (Baker)   A cyclic TWh(29). Take initial round games:

$$(1, 2, 9, 27), (16, 3, 28, 26), (24, 19, 13, 10), (7, 14, 5, 15),$$
$$(25, 21, 22, 8), (23, 17, 4, 12), (20, 11, 6, 18).$$

Cyclic TWh($p$), $p$ prime, $p \equiv 1 \pmod 4$, have been studied recently, and it has been shown that, for example, a cyclic TWh($p$) exists for all primes $p \equiv 5 \pmod 8$, $p \geqslant 29$. (Anderson, Cohen, Finizio, 1995). The construction requires the existence of a primitive root $\theta$ such that $\theta^2 + \theta + 1$ and $\theta^2 - \theta + 1$ are both squares in GF($p$), but $\theta^2 - 1$ is not a square. See Exercises 11, number 9.

We now exhibit some nice connections between triplewhist tournaments and SAMDRRs.

**Theorem 11.6.2**   *If a resolvable SAMDRR($n$) exists then a TWh($4n$) exists.*

**Proof**   The whist tournaments constructed in Theorem 11.4.1 are in fact all triplewhist tournaments; check the opponent pairs of the first and second kinds separately.

Conversely, we can go from a triplewhist tournament to a SAMDRR.

**Theorem 11.6.3**   *If a* TWh($4n + 1$) *exists, then so does a resolvable* SAMDRR($4n + 1$), *and hence so does a* SOLS($4n + 1$) *with symmetric orthogonal mate.*

**Proof** Replace each game $(a, b, c, d)$ by the two games $H_aW_c$ v $H_bW_d$ and $H_cW_a$ v $H_dW_b$, and observe that opponent pairs of the first kind give same sex opponent pairs, while opponent pairs of the second kind give opposite sex opponent pairs. The partner pairs arise from the whist partner pairs.

It follows from the proofs of Theorems 5.1.5 and 5.4.2 that the resulting SOLS $A$ and its orthogonal mate $B$ are given by:

$a_{i,j} = i$'s partner in the triplewhist game in which $i$ and $j$ are first opponents; $a_{ii} = i$;

$b_{i,j} =$ round of the triplewhist tournament in which $i$ and $j$ are first opponents; $b_{ii} = i$.

Similarly, we obtain a SOLS $A$ and mate $B$ from a TWh($4n$); the only difference is that in this case, each $b_{ii}$ is defined to be the element not used in the labelling of the rounds.

**Example 11.6.5** Start with the cyclic TWh(8) with round 0 given by (11.1). We obtain the following SOLS $A$ and its orthogonal mate $B$:

$$
A = \begin{matrix}
0 & 2 & 6 & 5 & 3 & \infty & 4 & 1 \\
5 & 1 & 3 & 0 & 6 & 4 & \infty & 2 \\
\infty & 6 & 2 & 4 & 1 & 0 & 5 & 3 \\
6 & \infty & 0 & 3 & 5 & 2 & 1 & 4 \\
2 & 0 & \infty & 1 & 4 & 6 & 3 & 5 \\
4 & 3 & 1 & \infty & 2 & 5 & 0 & 6 \\
1 & 5 & 4 & 2 & \infty & 3 & 6 & 0 \\
3 & 4 & 5 & 6 & 0 & 1 & 2 & \infty
\end{matrix}
$$

$$
B = \begin{matrix}
\infty & 6 & 2 & 4 & 1 & 0 & 5 & 3 \\
6 & \infty & 0 & 3 & 5 & 2 & 1 & 4 \\
2 & 0 & \infty & 1 & 4 & 6 & 3 & 5 \\
4 & 3 & 1 & \infty & 2 & 5 & 0 & 6 \\
1 & 5 & 4 & 2 & \infty & 3 & 6 & 0 \\
0 & 2 & 6 & 5 & 3 & \infty & 4 & 1 \\
5 & 1 & 3 & 0 & 6 & 4 & \infty & 2 \\
3 & 4 & 5 & 6 & 0 & 1 & 2 & \infty
\end{matrix}
$$

Note that, because the TWh(8) was cyclic, both $A$ and $B$ are *pseudo-cyclic* in the sense that if we remove the last row and column the resulting matrix is cyclic, and the last row and column themselves have entries in cyclic order. Similarly, if we start with a cyclic TWh($4n + 1$), since there is no row or column corresponding to a player $\infty$, both $A$ and $B$ are themselves cyclic. (See Finizio, 1994)

As this example shows, the knowledge of a cyclic TWh($v$) makes it very easy to obtain a SOLS($v$) $A$ and an orthogonal mate $B$. But, even more, another Latin square orthogonal to all of $A, A'$, and $B$ can also be found.

**Theorem 11.6.4** (Baker) *If a* TWh($4n + 1$) *exists which is cyclic over* $Z_{4n+1}$, *then there exist Latin squares* $A, B, C$ *such that* $A, A', B, C$ *are mutually orthogonal.*

**Proof** Construct $A$ and $B$ as above, and define $C$ by $c_{i,j} \equiv i - j$ (mod $4n + 1$). Then $C$ is clearly a Latin square which has constant entries on each of its left-to-right diagonals. Since $A, A'$, and $B$ are all cyclic, it is clear that $C$ must be orthogonal to each.

Corresponding to Theorems 11.6.3 and 11.6.4 there are similar results for directedwhist tournaments.

**Theorem 11.6.5** *If a* DWh($4n + 1$) *exists, then so does a* SOLS($4n + 1$) *with symmetric orthogonal mate.*

**Proof** As for the triplewhist case, but this time define $A = (a_{ij})$ to be $i$'s left-hand opponent when $i$ and $j$ are partners.

**Example 11.6.6** The Wh(33) displayed in Example 11.1.4 is in fact a DWh(33). Thus $N(33) \geqslant 4$; for many years this was the best known lower bound for $N(33)$. The close connection between the DWh(33) and the SOLS(33) with orthogonal mate constructed by Baker in Example 5.5.1 can be seen by comparing the games of 11.1.4 with the columns of 5.5.1.

Finally, we would like to mention that recently Norman Finizio and I have discovered infinitely many examples of whist tournaments which are both directedwhist and triplewhist tournaments. Here is an example.

**Example 11.6.7** Initial round of a cyclic DWh(29) which is also a TWh(29).

$$(1, 19, 10, 9), (24, 21, 8, 13), (25, 11, 18, 22), (20, 3, 26, 6),$$
$$(16, 14, 15, 28), (7, 17, 12, 5), (23, 2, 27, 4).$$

We now bring our excursion into whist tournaments to an end. But, in closing, note the ingredients used in this chapter: finite fields, MOLS, PBD, GDD, SAMDRR, TWh. The interconnections between different areas of design theory have surely been displayed! It is tantalizing to note that Dudeney in his 1917 classic *Amusements in mathematics* poses a problem (266) relating to SAMDRR immediately after one on whist tournaments (265) which in turn follows one essentially on league schedules (264). One wonders how much he knew!

## 11.7  Exercises 11

1. Show that a resolvable $(v, 4, 3)$ design can exist only when $v$ is a multiple of 4. Use the results of this chapter to show that such designs exist for all $v \equiv 0 \pmod 4$.

2. Show that if a resolvable $TD(k, n)$ and a $Wh(k)$ exist, where $n \equiv 1 \pmod 4$ and $k \equiv 0 \pmod 4$, then a $Wh(kn)$ exists. (Use the ideas in the proof of the product theorem.) Use this to give alternative proofs of the existence of $Wh(20)$, $Wh(136)$, $Wh(204)$, $Wh(300)$. (Baker, 1975$a$)

3. Construction of a $PBD(129, \{5, 9\}, 1)$ with 822 blocks of size 5 and one block of size 9. Take set of elements $\{\infty\} \cup \{a_i : a \in Z_2, i \in I_4\} \cup \{b_i : b \in Z_{60}, i \in \{5, 6\}\}$. Take the 9-set $\{\infty, 0_1, 0_2, 0_3, 0_4, 1_1, 1_2, 1_3, 1_4\}$ and develop the following 5-sets modulo 60, with the convention that in $a_i, 1 \leq i \leq 4$, $a$ is always reduced mod 2.

   $\{\infty, 0_5, 0_6, 30_5, 30_6\}$   (30 translates),
   $\{0_5, 12_5, 24_5, 36_5, 48_5\}$   (12 translates),
   $\{0_5, 1_5, 6_5, 15_5, 16_6\}$, $\quad \{0_5, 2_5, 20_5, 52_5, 45_6\}$, $\quad \{0_5, 31_5, 35_5, 40_6, 55_6\}$,
   $\{0_5, 3_5, 16_5, 7_6, 35_6\}$, $\quad \{0_5, 21_5, 38_5, 27_6, 50_6\}$, $\quad \{0_5, 26_5, 2_6, 48_6, 52_6\}$,
   $\{0_5, 37_5, 11_6, 18_6, 31_6\}$, $\quad \{0_5, 21_6, 33_6, 38_6, 59_6\}$, $\quad \{0_6, 16_6, 49_6, 52_6, 58_6\}$,
   $\{0_1, 0_5, 41_5, 23_6, 58_6\}$, $\quad \{0_2, 0_5, 11_5, 8_6, 39_6\}$, $\quad \{0_3, 0_5, 7_5, 3_6, 44_6\}$,
   $\{0_4, 0_5, 27_5, 13_6, 14_6\}$. (Hamel *et al.*, 1993)

4. Suppose that every prime factor of $4m + 1$ is congruent to 1 (mod 4) so that there exists $k$ such that $k^2 \equiv -1 \pmod{4m + 1}$. Then a cyclic $DWh(4m + 1)$ can be constructed by taking initial round games $G(x) : (x, kx, -x, -kx)$. (Note that $G(x) = G(-x) = G(kx) = G(-kx)$.) (Watson, 1954)

5. Verify that the following form the initial round of a cyclic $TWh(41)$: $(1, 26, 31, 5)$, $(18, 17, 25, 8)$, $(37, 19, 40, 21)$, $(10, 14, 23, 9)$, $(16, 6, 4, 39)$, $(20, 11, 35, 7)$, $(32, 34, 15, 3)$, $(2, 38, 24, 13)$, $(36, 28, 22, 29)$, $(33, 12, 27, 30)$. (Liaw, 1996)

6. Verify that the following give the initial round of a cyclic $TWh(20)$: $(\infty, 13, 0, 17)$, $(8, 2, 10, 1)$, $(3, 18, 15, 4)$, $(7, 12, 16, 9)$, $(11, 14, 5, 6)$. (Finizio, 1991)

7. Verify that the following give the initial round of a cyclic $TWh(36)$: $(\infty, 16, 0, 32)$, $(28, 14, 3, 27)$, $(31, 13, 4, 2)$, $(20, 24, 34, 33)$, $(26, 1, 19, 6)$, $(7, 12, 8, 15)$, $(17, 23, 21, 29)$, $(18, 9, 30, 11)$, $(22, 10, 5, 25)$. (Finizio, 1994)

8. Show that the games $(\theta^{4i}, \theta^{4i+1}, \theta^{4i+2}, \theta^{4i+3})$, $0 \leq i \leq t - 1$, where $\theta$ is

a primitive root of the prime $p = 4t + 1$, $t$ odd, form the initial round of a cyclic Wh($p$) provided $\theta^2 + \theta + 1$ is not a square (mod $p$). Illustrate with the choices $(p, \theta) = (13, 2)$, $(29, 8)$.

9. Verify that if $\theta$ is a primitive root of the prime $p = 4t + 1$, $t$ odd, the games $(\theta^{4i}, \theta^{4i+1}, -\theta^{4i+1}, \theta^{4i-2})$, $0 \leqslant i \leqslant t - 1$, form the initial round of a cyclic TWh($p$), provided that $\theta^2 - 1$ is not a square but $\theta^2 \pm \theta + 1$ are both squares. Illustrate in the case $(p, \theta) = (29, 3)$. (Anderson, Cohen, Finizio, 1995)

10. The primitive roots of the previous exercise can be used to construct SSBS. Show that if $\theta^2 \pm \theta + 1$ are both squares, then $(\theta^{4i}, \theta^{4i+3})$, $(\theta^{4i+2}, \theta^{4i+1})$, $0 \leqslant i \leqslant t - 1$, form an SSBS.

11. Show that there is no cyclic DWh($4n$) in $\{\infty\} \cup Z_{4n-1}$.

# Bibliography

Alltop, W.O. (1972) An infinite class of 5-designs, *J. Combinatorial Theory, Ser. A*, **12**, 390–395.

Anderson, B.A., Schellenberg, P.J., and Stinson, D.R. (1984) The existence of Howell designs of even side, *J. Combin. Theory, Ser. A*, **36**, 23–55.

Anderson, I. (1974, second edition 1989) *A first course in combinatorial mathematics*, Oxford Univ. Press.

Anderson, I. (1990) *Combinatorial designs*, Ellis Horwood.

Anderson, I. (1991) Kirkman and $GK_{2n}$, *Bull. I.C.A.*, **3**, 111–112.

Anderson, I. (1995) Cyclic designs in the 1850s: the work of Rev. R.R. Anstice, *Bull. Inst. Combinatorics and its Applications*, **15**, 41–46.

Anderson, I., Cohen, S.D., and Finizio, N.J. (1995) An existence theorem for cyclic triplewhist tournaments, *Discrete Math.* **138**, 31–41.

Anstice, R.R. (1852–3) On a problem in combinations, *Cambridge and Dublin Math. J.*, **7**, 279–292 (1852) and **8**, 149–152 (1853).

Baker, R.D. (1975a) Whist tournaments, *Proc. 6th S.E. Conference on Combinatorics, Graph theory and Computing, Congressus Num.* **14**, 89–100.

Baker, R.D. (1975b) *Factorisation of graphs*, PhD thesis, Ohio State Univ.

Baumert, L.D., Golomb, S.W., and Hall, M. (1962) Discovery of a Hadamard matrix of order 92, *Bull. Amer. Math. Soc.*, **68**, 237–238.

Bays, S. and de Weck, E. (1935) Sur les systèmes de quadruples, *Comment. Math. Helvet.*, **7**, 222–241.

Beintema, M.B., Bonn, J.T., Fitzgerald, R.W., and Yucas, J.L., Orderings of finite fields and balanced tournaments, *Ars Combinatoria*, in press.

Bennett, F.E. (1987) Pairwise balanced designs with prime power block sizes exceeding 7, *Annals Discrete Math.*, **34**, 43–64.

Bhat, V.N. and Shrikhande, S.S. (1970) Nonisomorphic solutions of some balanced incomplete block designs, *J. Combin. Theory*, **9**, 174–191.

Bose, R.C. (1938) On the application of the properties of Galois fields to the construction of hyper-Graeco-Latin squares, *Sankhya*, **3**, 323–338.

Bose, R.C. (1939) On the construction of balanced incomplete block designs, *Ann. Eugenics*, **9**, 353–399.

Bose, R.C. (1942) On some new series of balanced incomplete block designs, *Bull. Calcutta Math. Soc.*, **34**, 17–31.

Bose, R.C. and Chowla, S. (1962) Theorems in the additive theory of numbers, *Comment. Math. Helvet.* **37**, 141–147.

Bose, R.C. and Shrikhande, S.S. (1959) On the falsity of Euler's conjecture about the nonexistence of two orthogonal Latin squares of order $4t + 2$, *Proc. Nat. Acad. Sci. U.S.A.*, **45**, 734–737.

Bose, R.C. and Shrikhande, S.S. (1960) On the construction of sets of mutually orthogonal Latin squares and the falsity of a conjecture of Euler, *Trans. Amer. Math. Soc.*, **95**, 191–209.

Bose, R.C., Shrikhande, S.S., and Parker, E.T. (1960) Further results on the construction of mutually orthogonal Latin squares and the falsity of Euler's conjecture, *Canad. J. Math.*, **12**, 189–203.

Brayton, R.K., Coppersmith, D., and Hoffman, A.J. (1976) Self-orthogonal Latin squares, *Atti del Convegni Lincei*, **17**, 509–517.

Bruck, R.H. and Ryser, H.J. (1949) The non-existence of certain finite projective planes, *Canad. J. Math.*, **1**, 88–93.

Bush, K.A. (1973) Construction of symmetric Hadamard matrices, *A survey of combinatorial theory* (ed. J.N. Srivastava *et al.*), North Holland, 81–83.

Byleen, K. (1970) On Stanton and Mullin's construction of Room squares, *Ann. Math. Stat.*, **41**, 1122–1125.

Cameron, P.J. (1973) Extending symmetric designs, *J. Combin. Theory, Ser. A*, **14**, 215–220

Cameron, P.J. and van Lint, J.H. (1991) *Designs, graphs, codes and their links*, Cambridge Univ. Press.

Chong, B.C. and Chan, K.M. (1974) On the existence of normalized Room squares, *Nanta Math.*, **7**, 8–17.

Chowla, S. and Ryser, H.J. (1950) Combinatorial problems, *Canad. J. Math.*, **2**, 93–99.

Colbourn, C.J. and Colbourn, M.J. (1980) On cyclic block designs, *Math. Report Canad. Acad. Sci.*, **2**, 21–26.

Colbourn, C.J. and Dinitz, J.H. (ed.) (1996) *The CRC Handbook of Combinatorial Designs*, CRC Press.

Constable, R.L. (1974) Positions in Room squares, in *Combinatorics* (ed. T.P. McDonough and V.C. Mavron), L.M.S. Lecture Note Series 13, C.U.P., 23–26.

Corradi, K.A. and Katai, I. (1969) A note on combinatorics, *Ann. Univ. Scient. Budapest de Rol. Eotvos Nomin.*, **12**, 100–106.

Davies, R.O. (1959) On Langford's problem, *Math. Gazette*, **43**, 253–255.

Denniston, R.H.F. (1974) Sylvester's problem of the fifteen schoolgirls, *Discrete Math.*, **9**, 229–233.

Denniston, R.H.F. (1976) Some new 5-designs, *Bull. London Math. Soc.*, **8**, 263–267.

de Palluel, C. (1788) Sur les avantages et l'économie que procurent les racines employées à l'engrais des moutons à l'étable, *Memoires d'Agric.*, 17–23.

de Werra, D. (1981) Scheduling in sports, *Studies on graphs and discrete programming* (ed. P. Hansen), North Holland, 381–395.

Du, Ding-Zhu and Hwang, F.K. (1988) Existence of symmetric skew balanced starters for odd prime powers, *Proc. Amer. Math. Soc.*, **104**, 660–667.

Dudeney, H.E. (1917) *Amusements in mathematics*, Nelson (reissued by Dover, 1958).

Dulmage, A.L., Johnson, D.M., and Mendelsohn, N.S. (1961) Orthomorphisms of groups and orthogonal Latin squares, *Canad. J. Math.*, **13**, 356–372.

Finizio, N.J. (1991) Orbits of cyclic Wh($\nu$) of $Z_N$-type, *Congressus Numer.*, **82**, 15–28.

Finizio, N.J. (1994) Some quick and easy SOLSSOMS, *Congressus Numer.*, **99**, 307–313.

Fisher, R.A. (1926) The arrangement of field experiments, *J. Ministry Agriculture.*, **33**, 503–513.

Fisher, R.A. (1940) An examination of the different possible solutions of a problem in incomplete blocks, *Ann. Eugenics*, **10**, 52–75.

Fisher, R.A. and Yates, F. (1938) *Statistical tables for biological, agricultural and medical research*, Oliver and Boyd.

Franklin, M.F. (1984) Cyclic generation of self-orthogonal Latin squares, *Utilitas Math.*, **25**, 135–146.

Frost, A. (1871) General solution and extension of the problem of the fifteen schoolgirls, *Quart. J. Pure Applied Math.*, **11**, 26–37.

Gilbert, E.N. (1961) Design of mixed doubles tournaments, *Amer. Math. Monthly*, **68**, 124–131.

Grannell, M.J., Griggs, T.S., and Mathon, R.A. (1993), Some Steiner 5-designs with 108 and 132 points, *J. Combinatorial Designs*, **1**, 213–238.

Hadamard, J. (1893) Résolutions d'une question relative aux déterminants, *Bull. Sci. Math.*, **17**, 240–246.

Hall, M. (1948) Distinct representatives of subsets, *Bull. Amer. Math. Soc.*, **54**, 922–926.

Hall, M. (1986) *Combinatorial theory* (second edition), Wiley.

Hall, M. and Ryser, H.J. (1951) Cyclic incidence matrices, *Canad. J. Math.*, **3**, 495–502.

Hall, P. (1935) On representatives of subsets, *J. London Math. Soc.*, **10**, 26–30.

Halmos, P.R. and Vaughan, H.E. (1950) The marriage problem, *Amer. J. Math.*, **72**, 214–215.

Hamel, A.M., Mills, W.H., Mullin, R.C., Rees, R., Stinson, D.R., and Jianxing Yin (1993) The spectrum of PBD($\{5, k^*\}, \nu$) for $k = 9, 13$, *Ars Combinatoria*, **36**, 7–26.

Hanani, H. (1960) On quadruple systems, *Canad. J. Math.*, **12**, 145–157.

Hanani, H. (1961) The existence and construction of balanced incomplete block designs, *Ann. Math. Stat.* **32**, 361–386.

Hanani, H., Ray-Chaudhuri, D.K., and Wilson, R.M. (1972) On resolvable designs, *Discrete Math.*, **3**, 343–357.

Hardy, G.H. and Wright, E.M. (1979) *An introduction to the theory of numbers* (fifth edition), Oxford Univ. Press.

Haselgrove, J. and Leech, J. (1977) A tournament design problem, *Amer. Math. Monthly*, **84**, 198–201.

Heffter, L. (1897) Über Tripelsysteme, *Math. Ann.*, **49**, 101–112.

Hill, R. (1986) *A first course in coding theory*, Oxford Univ. Press.

Hilton, A.J.W. (1969) On Steiner and similar triple systems, *Math. Scand.*, **24**, 208–216.

Hilton, A.J.W. (1972) A simplification of Moore's proof of the existence of Steiner triple systems, *J. Combinatorial Theory, Ser. A.*, **13**, 422–425.

Horton, J.D. (1971) Quintuplication of Room squares, *Aequationes Math.*, **7**, 243–245.

Horton, J.D. (1981) Room designs and one-factorisations, *Aequat. Math.*, **22**, 56–63.

Hughes, D.R. and Piper, F.C. (1985) *Design theory*, Cambridge Univ. Press.

Hung, S.H.Y. and Mendelsohn, N.S. (1974) On Howell designs, *J. Combinatorial Theory, Ser. A.*, **16**, 174–198.

Hwang, F.K. (1970) Some more contributions on constructing balanced Howell rotations, *Proc. Second Chapel Hill Conf. on comb. math. and applic.*, 307–323.

Hwang, F.K. (1989) How to design round-robin schedules, in *Combinatorics, computing and complexity, Math. Applic.* (*Chinese series*) **1** (ed. Ding Zhu Du and Hu Guoding), 142–160.

Kirkman, T.P. (1847) On a problem in combinations, *Cambridge and Dublin Math. J.*, **2**, 191–204.

Kirkman, T.P. (1850) Note on an unanswered prize question, *Cambridge and Dublin Math. J.*, **5**, 255–262.

Kirkman, T.P. (1853) Theorems on combinations, *Cambridge and Dublin Math. J.*, **8**, 39–45.

Kirkman, T.P. (1857) On the perfect $r$-partitions of $r^2 - r + 1$, *Trans. Hist. Soc. Lancs. and Cheshire*, **9**, 127–142.

Lam, C., Thiel, L., and Swiercz, S. (1989) The non-existence of a finite projective plane of order 10, *Canad. J. Math.*, **41**, 1117–1123.

Lamken, E.R. (1996) A few more partitioned balanced tournament designs, *Ars Combinatoria*, **43**, 121–134.

Lamken, E.R. and Vanstone, S.A. (1985) The existence of factored balanced tournament designs, *Ars Combinatoria*, **19**, 157–160.

Lander, E.S. (1980) Symmetric designs: an algebraic approach, *London Math. Soc. Lecture Notes Series*, **74**, Cambridge Univ. Press.

Langford, C.D. (1958) Problem, *Math. Gazette*, **42**, 228.

Liaw, Y.S. (1996) Construction of Z-cyclic triplewhist tournaments, *J. Combinatorial Designs*, **4**, 219–233.

Lu Jiaxi (1990) *Collected works of Lu Jiaxi on combinatorial designs*, Inner Mongolia People's Press, Huhhot, China.

MacNeish, H.F. (1922) Euler squares, *Ann. Math.*, **23**, 221–227.

Mann, H.B. (1965) Difference sets in elementary abelian groups, *Illinois Math. J.*, **9**, 212–219.

Mendelsohn, N.S. (1971) Latin squares orthogonal to their transposes, *J. Combinat. Theory, Ser. A*, **11**, 187–189.

Mills, W.H. (1975) Two new block designs, *Utilitas Math.*, **7**, 73–75.

Mills, W.H. (1978) A new 5-design, *Ars Combinatoria*, **6**, 193–195.

Moore, E.H. (1893) Concerning triple systems, *Math. Ann.*, **42**, 271–285.

Moore, E.H. (1896) Tactical Memoranda I–III, *Amer. J. Math.*, **18**, 264–303.

Mullin, R.C. and Nemeth, E. (1969) On furnishing Room squares, *J. Combinatorial Theory*, **7**, 266–272.

Murphy, I.S. (1975) A note on a matrix result of Ryser, *Canad. Math. Bull.*, **18**, 149–150.

Netto, E. (1893) Zur Theorie der Tripelsysteme, *Math. Ann.*, **42**, 143–152.

O'Keefe, E.S. (1961) Verification of a conjecture of T. Skolem, *Math. Scand.*, **9**, 80–82.

Paley, R.E.A.C. (1933) On orthogonal matrices, *J. Math. Phys.*, **12**, 311–320.

Parker, E.T. (1959*a*) Orthogonal Latin squares, *Proc. Nat. Acad. Sci. U.S.A.*, **45**, 859–862.

Parker, E.T. (1959*b*) Construction of some sets of mutually orthogonal Latin squares, *Proc. Amer. Math. Soc.* **10**, 946–949.

Parker, E.T. and Mood, A.M. (1955) Some balanced Howell rotations for duplicate bridge sessions, *Amer. Math. Monthly*, **62**, 714–716.

Peltesohn, R. (1939) Eine Lösung der beiden Heffterschen Differenzenprobleme, *Compositio Math.*, **6**, 251–257.

Rasch, D. and Herrendörfer, G. (1977) *Experimental design*, Reidel.

Ray-Chaudhuri, D.K. and Wilson, R.M. (1971) Solution of Kirkman's schoolgirl problem, *Proc. Symp. Math.*, **19**, 187–203.

Reiss, M. (1859) Über eine Steinersche combinatorische Aufgabe, *J. reine und angew. Math.*, **56**, 326–344.

Room, T.G. (1955) A new type of magic square, *Math. Gazette*, **39**, 307.

Rosa, A. (1966) Poznamka o cyklickych Steinerovych systemoch trojic, *Math. Fyz. Cas.*, **16**, 285–290.

Rouse-Ball, W.W. (1940) *Mathematical recreations and essays* (revised by H.S.M. Coxeter), Macmillan.

Russell, K.G. (1980) Balancing carry-over effects in round robin tournaments, *Biometrika*, **67**, 127–131.

Scarpis, V. (1898) Sui determinanti di valore massimo, *Rend. R. Ist. Lombardo Sci. Lett.*, **31**, 1441–1446.

Schellenberg, P.J. (1973) On balanced Room squares and complete balanced Howell rotations, *Aequationes Math.*, **9**, 75–90.

Schellenberg, P.J., van Rees, G.H.J., and Vanstone, S.A. (1977) The existence of balanced tournament designs, *Ars Combinatoria*, **3**, 303–318.

Schellenberg, P.J., van Rees, G.H.J., and Vanstone, S.A. (1978) Four pairwise orthogonal Latin squares of order 15, *Ars Combinatoria*, **6**, 141–150.

Schonheim, J. (1966) On maximal systems of *k*-tuples, *Stud. Sci. Math. Hungar.*, **1**, 363–368.

Schutzenberger, M.P. (1949) A non-existence theorem for an infinite family of symmetrical block designs, *Ann. Eugenics*, **14**, 286–287.

Shrikhande, S.S. (1950) The impossibility of certain symmetrical balanced incomplete block designs, *Ann. Math. Stat.*, **21**, 106–111.

Skolem, T. (1957) On certain distributions of integers in pairs with given differences, *Math. Scand.*, **5**, 57–68.

Skolem, T. (1958) Some remarks on the triple systems of Steiner, *Math. Scand.*, **6**, 273–280.

Spence, E. (1971) Some new symmetric block designs, *J. Combinatorial Theory, Ser. A*, **11**, 299–302.

Stanton, R.G. and Horton, J.D. (1972) A multiplication theorem for Room squares, *J. Combinatorial Theory, Ser. A*, **12**, 322–325.

Stanton, R.G. and Mullin, R.C. (1968) Construction of Room squares, *Ann. Math. Stat.*, **39**, 1540–1548.

Steiner, J. (1853) Combinatorische Aufgabe, *J. reine und angew. Math.*, **45**, 181–182.

Stevens, W.L. (1939) The completely orthogonalized Latin square, *Ann. Eugenics*, **9**, 82–93.

Stewart, I.N. (1973) *Galois theory*, Chapman and Hall.

Stinson, D.R. (1982) The existence of Howell designs of odd side, *J. Combinatorial Theory*, Ser. *A*, **32**, 53–65.

Stinson, D.R. (1984) A short proof of the non-existence of a pair of orthogonal Latin squares of order 6, *J. Combinatorial Theory*, Ser. *A*, **36**, 373–376.

Stinson, D.R. (1985) Room squares with maximum empty subarrays, *Ars Combinatoria*, **20**, 159–166.

Sylvester, J.J. (1867) Thoughts on inverse orthogonal matrices, *Philos. Mag.*, **34**, 461–475.

Szekeres, E. and Szekeres, G. (1965) On a problem of Schütte and Erdös, *Math. Gazette*, **49**, 290–293.

Tarry, G. (1900) Le problème des 36 officiers, *C.R. Acad. Franc. pour l'Avanc. de Sci. Naturel*, **1**, 122–123; **2**, 170–203.

Tierlinck, L. (1987) Nontrivial *t*-designs without repeated blocks exist for all *t*, *Discrete Math.*, **65**, 301–311.

Todorov, D.T. (1985) Three mutually orthogonal Latin squares of order 14, *Ars Combinatoria*, **20**, 45–48.

Turyn, R.J. (1972) An infinite class of Williamson matrices, *J. Combinatorial Theory*, Ser. *A*, **12**, 319–321.

Veblen, O. and Bussey, W.H. (1906) Finite projective geometries, *Trans. Amer. Math. Soc.*, **7**, 241–259.

Wallis, J.S. (1972) Hadamard matrices (in Wallis, Street, and Wallis).

Wallis, W.D. (1972) A doubling construction for Room squares, *Discrete Math.*, **3**, 379–399.

Wallis, W.D. (1973) A construction for Room squares, in *A survey of combinatorial theory* (ed. J. Srivastava), North Holland, 449–451.

Wallis, W.D. (1979) Spouse-avoiding mixed doubles tournaments, *Proc. New York Acad. Sci.*, **319**, 549–554.

Wallis, W.D. (1983) A tournament problem, *J. Austral. Math. Soc.*, Ser. *B*, **24**, 289–291.

Wallis, W.D., Street, A.P., and Wallis, J.S. (1972) Combinatorics: Room squares, sum-free sets, Hadamard matrices, *Lecture Notes in Math.* 292, Springer.

Wang, S.P. (1978) *On self-orthogonal Latin squares and partial transversals of Latin squares*, PhD thesis, Ohio State Univ.

Watson, G.L. (1954) Bridge problem, *Math. Gazette*, **38**, 129–130.

Williams, E.J. (1949) Experimental designs balanced for the estimation of residual effects of treatments, *Australian J. Scient. Research A* 2, 149–168.

Williamson, J. (1944) Hadamard's determinant theorem and the sum of four squares, *Duke Math. J.*, **11**, 65–81.

Wilson, R.M. (1974*a*) Concerning the number of mutually orthogonal Latin squares, *Discrete Math.*, **9**, 181–198.

Wilson, R.M. (1974*b*) A few more squares, *Congressus Numer.*, **10**, 675–680.

Witt, E. (1938) Über Steinersche Systeme, *Abh. Math. Sem. Univ. Hamburg*, **12**, 265–275.

Yates, F. (1936) Incomplete randomized blocks, *Ann. Eugenics*, **7**, 121–140.

# Index of $(v, k, \lambda)$ BIBDs explicitly constructed

(1.1.3 means Example 1.1.3; 1.11 means Exercises 1, number 11)

| $v$ | $k$ | $\lambda$ | |
|---|---|---|---|
| 6 | 3 | 2 | 1.1.1 |
| 7 | 3 | 1 | 1.1.3, 1.2.4 |
| 9 | 3 | 1 | 1.1.9, 2.4.2, 6.1.2, 7.3.1 |
| 10 | 3 | 2 | 1.11, 2.4.1 |
| 13 | 3 | 1 | 2.2.2, 6.1.3 |
| 15 | 3 | 1 | 1.2.1, 2.4.4, 6.1.1, 6.2.7, 6.3, 7.1.1, 7.3.2 |
| 16 | 3 | 2 | 2.4.3 |
| 19 | 3 | 1 | 6.2.6 |
| 21 | 3 | 1 | 2.4.5, 7.1.2 |
| 25 | 3 | 1 | 6.2.3 |
| 27 | 3 | 1 | 6.2.9 |
| 33 | 3 | 1 | 7.1.4 |
| 69 | 3 | 1 | 7.1, 7.2.2 |
| 8 | 4 | 3 | 2.2.6 |
| 13 | 4 | 1 | 1.1.6 |
| 16 | 4 | 1 | 1.3.7 |
| 19 | 4 | 2 | 2.2.1, 2.8 |
| 22 | 4 | 2 | 2.20 |
| 25 | 4 | 1 | 2.8, 2.16 |
| 25 | 4 | 2 | 2.12 |
| 28 | 4 | 1 | 2.19 |
| 37 | 4 | 1 | 2.8 |
| 100 | 4 | 1 | 7.6 |
| 11 | 5 | 2 | 2.1.4 |
| 21 | 5 | 1 | 2.1.3, 3.4.1 |
| 41 | 5 | 1 | 2.10 |
| 61 | 5 | 1 | 7.7 |
| 12 | 6 | 5 | 2.2.7 |
| 16 | 6 | 2 | 1.1.5, 1.26, 2.1.6 |
| 31 | 6 | 1 | 2.1.3 |
| 91 | 6 | 1 | 2.9 |
| 15 | 7 | 3 | 2.2, 4.2.2 |
| 22 | 7 | 4 | 2.8 |
| 17 | 8 | 7 | 2.2.4 |
| 57 | 8 | 1 | 2.1 |
| 37 | 9 | 2 | 2.5.3, 2.5 |
| 73 | 9 | 1 | 2.1 |
| 19 | 10 | 5 | 2.1.8 |
| 31 | 10 | 3 | 2.22 |
| 91 | 10 | 1 | 2.1 |
| 36 | 15 | 6 | 1.19, 3.3.7 |

# Index